Weeds of North America

Weeds
of North America

RICHARD DICKINSON

and **FRANCE ROYER**

THE UNIVERSITY OF CHICAGO PRESS

Chicago and London

Richard Dickinson lives in Toronto and has taught plant taxonomy for more than twenty-five years. France Royer is a photographer living in Edmonton, Alberta. Together they are the authors of *Wildflowers of Edmonton and Central Alberta, Wildflowers of Calgary and Southern Alberta,* and *Plants of Alberta.*

The University of Chicago Press, Chicago 60637

The University of Chicago Press, Ltd., London

© 2014 by France Royer and Richard Dickinson

All rights reserved. Published 2014.

Printed in China

23 22 21 20 19 18 17 16 15 14   1 2 3 4 5

ISBN-13: 978-0-226-07644-7 (paper)

ISBN-13: 978-0-226-07658-4 (e-book)

DOI: 10.7208/chicago/9780226076584.001.0001

Library of Congress Cataloging-in-Publication Data

Dickinson, Richard, 1960– author.

Weeds of North America / Richard Dickinson and France Royer.

pages : illustrations ; cm

Includes bibliographical references and index.

ISBN 978-0-226-07644-7 (pbk. : alk. paper)—

ISBN 978-0-226-07658-4 (e-book)

1. Weeds—North America—Identification.

2. Weeds—North America—Handbooks, manuals, etc.

I. Royer, France, 1951– author. II. Title.

SB613.N7D54 2014

632′.5097—dc23

2013038953

♾ This paper meets the requirements of ANSI/NISO Z39.48–1992 (Permanence of Paper).

CONTENTS

ACKNOWLEDGMENTS

The authors would like to thank the following agencies and individuals who provided assistance throughout the production of this book.

- Agriculture & Agri-Food Canada
- United States Department of Agriculture
- Dr. Erin Rosskopf, Agricultural Research Service, USDA
- Dr. Jim Broatch, Alberta Agriculture and Rural Development
- Larry Markle, Agricultural Research Service, USDA
- Dr. Nadia Talent, University of Toronto
- Bridget Simon, Washington State Noxious Weed Control Board
- Janet Nelson, Okanogan County Noxious Weed Control Board
- Sally Abella, METRO King County, Lake Stewardship Program, Washington
- Eric Lane, Colorado Department of Agriculture
- Roy Cranston and David Ralph, British Columbia Department of Agriculture
- Mike Cowbrough, Ontario Ministry of Agriculture, Food and Rural Affairs

We would also like to thank family and friends whose support throughout the production of this book is greatly appreciated.

INTRODUCTION

Since the rise of agriculture about 10,000 years ago, humankind has had to deal with unwanted plants competing with its cultivated crops, as well as affecting the health of families and livestock. It has only been in the last 400 years, however, that there has been an explosion in the spread of weeds. This marked increase coincides with colonial expansion and, more recently, an increase in world trade (Canadian Food Inspection Agency Summary Report 2008).

These troublesome plants introduced from other parts of the globe often thrive in their new environment, free of natural pests that keep them in check in their native range. Whether they affect crop yields, species diversity of the natural environment, or human health, weeds have a detrimental effect on the economy. Billions of dollars each year are spent on vegetation management, prevention, and education. Prior to implementing a management strategy, however, one must be able to identify the species, understand the biology, and institute the best control method. This book has been designed to provide assistance with the weed identification process.

This reference guide includes over 600 species from 69 plant families. Detailed descriptions of each species are included for easy recognition at any growth stage. Over 1,200 color photographs complement the informative text, aiding in identification. Closely related or similar species are described and compared using the most distinguishing characteristic between the two. Nontechnical terms have been used whenever possible, making the text easier to understand.

Scientific names used in this guide follow the Integrated Taxonomic Information System (www.itis.gov), a North American database of plant names that conforms with the International Code of Botanical Nomenclature.

Plant families are listed alphabetically by scientific name. The family description introduces general characteristics of the family, its worldwide distribution, and its well-known members. Within each family, species are listed alphabetically by scientific name. Synonyms appear in the text and index, allowing for easy cross-referencing. A list of regional common names is also included in the text and index.

Species selection for this guide was determined by federal, provincial, and state weed legislation, as well as those listed by various conservation and environmental associations dealing with invasive plants. Weed designation by jurisdiction is provided for each species. These designations are subject to frequent reviews and changes and should only be used as an indication of legislation at the time of printing.

Canada, Mexico, and the United States are members of the North American Plant Protection Agency (NAPPO). This organization ensures that phytosanitary measures are in place to deal with the introduction of invasive plants. In accordance with the Plant Protection Act (2000), the United States federal government designates certain plants as noxious weeds. In Canada, this designation takes place under the Weed Seeds Order (1986). In Mexico, several pieces of legislation deal with noxious plants. Within each country, states, provinces, territories, and municipalities have the power to designate weeds.

Identification keys have been constructed using easily identifiable characters and thumbprint photographs.

ABBREVIATIONS FOR PROVINCIAL AND STATE NAMES

Canada

Alberta	AB	Ontario	ON	Newfoundland	NL
British Columbia	BC	Quebec	PQ	Yukon	YK
Manitoba	MB	Prince Edward Island	PE	Northwest Territories	NT
Saskatchewan	SK	New Brunswick	NB	Nunavut	NU
		Nova Scotia	NS		

United States of America

Alabama	AL	Montana	MT
Alaska	AK	Nebraska	NB
Arizona	AZ	Nevada	NV
Arkansas	AR	New Hampshire	NH
California	CA	New Jersey	NJ
Colorado	CO	New Mexico	NM
Connecticut	CT	New York	NY
Delaware	DE	North Carolina	NC
Florida	FL	North Dakota	ND
Georgia	GA	Ohio	OH
Hawaii	HI	Oklahoma	OK
Idaho	ID	Oregon	OR
Illinois	IL	Pennsylvania	PA
Indiana	IN	Rhode Island	RI
Iowa	IA	South Carolina	SC
Kansas	KS	South Dakota	SD
Kentucky	KY	Tennessee	TN
Louisiana	LA	Texas	TX
Maine	ME	Utah	UT
Maryland	MD	Vermont	VT
Massachusetts	MA	Virginia	VA
Michigan	MI	Washington	WA
Minnesota	MN	West Virginia	WV
Mississippi	MS	Wisconsin	WI
Missouri	MO	Wyoming	WY

Mexico

Distrito Federal	DIF
Aguascalientes	AGS
Baja California	BCA
Baja California Sur	BCS
Campeche	CAM
Chiapas	CHI
Chihuahua	CHH
Coahuila de Zaragoza	COA
Colima	COL
Durango	DGO
Guanajuato	GTO
Guerrero	GRO
Hidalgo	HGO
Jalisco	JAL
México	MEX
Michoacan de Ocampo	MIC
Morelos	MOR
Nayarit	NAY
Nuevo Leon	NLE
Oaxaca	OAX
Puebla	PUE
Querétaro de Arteaga	QRO
Quintana Roo	ROO
San Luis Potos	SLP
Sinaloa	SIN
Sonora	SON
Tabasco	TAB
Tamaulipas	TMP
Tlaxcala	TLA
Veracruz de Ignacio de la Llave	VER
Yucatán	YUC
Zacatecas	ZAC

PLANT STRUCTURES

FLOWER STRUCTURE

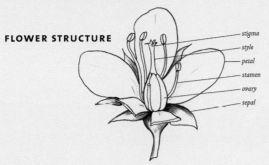

- stigma
- style
- petal
- stamen
- ovary
- sepal

SEEDLING STRUCTURE

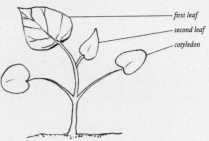

- first leaf
- second leaf
- cotyledon

LEAF TYPE

simple

pinnately compound

bipinnately compound

palmately compound

LEAF ARRANGEMENT

basal

alternate

opposite

whorled

IDENTIFICATION OF WEED SPECIES

Plants are identified by using a combination of characteristics, such as plant type, leaf arrangement, and flower color. Species with more than one type of leaf arrangement or flower color will appear in each of their respective groups.

KEY TO TREES AND SHRUBS

LEAVES ALTERNATE

FLOWERS GREEN

Morus alba,
page 488

Ailanthus
altissima, page 698

FLOWERS PINK OR PURPLE

Tamarix ramossissima,
page 724

FLOWERS WHITE

Schinus
terebinthifolius,
page 14

Toxicodendron
radicans,
page 16

Rosa multiflora,
page 688

FLOWERS YELLOW

Berberis vulgaris,
page 160

Elaeagnus
angustifolia,
page 330

Cytisus scoparius,
page 360

Ulex europaeus,
page 378

LEAVES OPPOSITE OR WHORLED

Rhamnus
cathartica,
page 658

Lonicera tatarica,
page 226

Lantana camara,
page 736

KEY TO VINES AND CLIMBING PLANTS

LEAVES ABSENT

Cuscuta ssp., page 284

LEAVES ALTERNATE

FLOWERS PINK OR PURPLE

Ipomoea hederacea, page 290   *Vicia cracca,* page 380   *Solanum dulcamara,* page 712

FLOWERS WHITE

Calystegia sepium, page 280   *Convolvulus arvensis,* page 282   *Ipomoea aquatica,* page 288   *Echinocystis lobata,* page 296   *Sicyos angulatus,* page 300

Abrus precatorius, page 358   *Cardiospermum halicacabum,* page 686

FLOWERS YELLOW

Momordica charantia, page 298

LEAVES OPPOSITE

Vincetoxicum rossicum, page 40

Dioscorea polystachya, page 316

KEY TO HERBACEOUS LAND PLANTS

LEAVES ABSENT OR REDUCED TO SCALES

Orobanche ramosa, page 502

LEAVES ALTERNATE

INFLORESCENCE A CYME
Flowers Blue, Pink, Purple, or Red

Anchusa officinalis, page 166

Echium vulgare, page 170

Lappula squarrosa, page 172

Commelina communis, page 272

Commelina diffusa, page 274

Urena lobata, page 474

Cynoglossum officinale, page 168

Flowers White

Malva rotundifolia, page 470

Polygonum aviculare, page 614

Solanum carolinense, page 710

Flowers Yellow

Oxalis corniculata, page 508

Potentilla norvegica, page 664

Potentilla recta, page 666

INFLORESCENCE A HEAD
Flowers Blue, Pink, or Purple

Cichorium intybus, page 92

Arctium minus, page 62

Carduus acanthoides, page 70

Carduus nutans, page 72

Centaurea stoebe, page 88

Rhaponticum repens, page 122

Cirsium arvense, page 94

Cirsium vulgare, page 96

Emilia sonchifolia, page 104

Onopordum acanthium, page 120

Silybum marianum, page 126

Flowers Green

Ambrosia trifida,
page 58

*Ambrosia
artemisiifolia,*
page 56

*Artemisia
absinthium,*
page 64

Artemisia vulgaris,
page 66

*Matricaria
discoidea,*
page 118

*Xanthium
strumarium,*
page 146

Flowers White

Anthemis cotula,
page 60

Centaurea diffusa,
page 84

*Conyza
canadensis,*
page 98

*Leucanthemum
vulgare,* page 116

*Tripleurospermum
inodorum,*
page 142

Flowers Yellow

*Centaurea
solstitialis,*
page 86

Chondrilla juncea,
page 90

Crepis tectorum,
page 100

Jacobaea vulgaris,
page 112

Lactuca serriola,
page 114

Senecio vulgaris,
page 124

Sonchus arvensis
ssp. *arvensis,*
page 128

Sonchus asper,
page 130

Sonchus oleraceus,
page 132

Tanacetum vulgare,
page 134

*Tragopogon
dubius,* page 138

INFLORESCENCE A PANICLE
Flowers Green

Amaranthus retroflexus, page 6

Amaranthus spinosus, page 8

Atriplex hortensis, page 258

Axyris amaranthoides, page 260

Chenopodium album, page 262

Ricinus communis, page 352

Flowers Red

Rumex acetosella, page 616

Rumex crispus, page 618

Flowers White

Fallopia japonica, page 610

INFLORESCENCE A RACEME OR SPIKE
Flowers Blue, Purple, or Pink

Campanula rapunculoides, page 220

Delphinium bicolor, page 648

Hesperis matronalis, page 196

Lepidium densiflorum, page 202

Galega officinalis, page 362

Trifolium repens, page 376

Flowers Green

Kochia scoparia,
page 264

Acalypha rhomboidea,
page 342

Flowers White

Alliaria petiolata,
page 178

Berteroa incana,
page 182

*Capsella bursa-
pastoris,* page 188

*Hesperis
matronalis (also
white),* page 196

*Lepidium
chalapense,*
page 200

Thlaspi arvense,
page 212

Melilotus albus,
page 368

*Oxytropis
monticola,*
page 370

*Asparagus
densiflorus,*
page 452

*Phytolacca
americana,*
page 524

*Fagopyrum
tataricum,*
page 606

*Persicaria
lapathifolia,*
page 612

Solanum nigrum,
page 714

Solanum viarum,
page 718

Flowers Yellow

Barbarea vulgaris,
page 180

Brassica juncea,
page 184

*Camelina
microcarpa,*
page 186

*Descurainia
sophia,* page 190

*Erucastrum
gallicum,*
page 192

*Erysimum
cheiranthoides,*
page 194

Isatis tinctoria,
page 198

Neslia paniculata,
page 204

*Raphanus
raphanistrum,*
page 206

Sinapis arvensis,
page 208

*Sisymbrium
altissimum,*
page 210

*Sesbania
herbacea,*
page 374

*Oenothera
biennis,* page 496

*Portulaca
oleracea,*
page 630

Linaria dalmatica,
page 532

Linaria vulgaris,
page 534

*Verbascum
thapsus,* page 692

*Solanum
rostratum,*
page 716

INFLORESCENCE A SOLITARY FLOWER
Flowers Blue

*Chaenorrhinum
minus,* page 530

Nicandra physalodes,
page 708

Flowers Green

*Amaranthus
blitoides,* page 4

Salsola kali,
page 266

Flowers White

Datura stramonium,
page 704

Flowers Yellow

Abutilon theophrasti, page 466

Hibiscus trionum, page 468

Sida rhombifolia, page 472

Argemone mexicana, page 514

Ranunculus acris, page 650

Ranunculus repens, page 652

Hyoscyamus niger, page 706

Viola arvensis, page 742

Tribulus terrestris, page 748

INFLORESCENCE AN UMBEL
Flowers Pink or Purple

Securigera varia, page 372

Allium vineale, page 450

Flowers White

Anthriscus sylvestris, page 22

Cicuta maculata, page 24

Conium maculatum, page 26

Daucus carota, page 28

Heracleum mantegazzianum, page 30

Flowers Yellow

Lotus corniculatus, page 364

Medicago lupulina, page 366

Chelidonium majus, page 516

INFLORESCENCE A CYATHIUM

Euphorbia cyparissias, page 348

Euphorbia esula, page 350

LEAVES BASAL

INFLORESCENCE A CYME

Erodium cicutarium, page 390

Iris pseudacorus, page 422

INFLORESCENCE A HEAD

Hieracium aurantiacum, page 108

Hypochaeris radicata, page 110

Taraxacum officinale, page 136

Tussilago farfara, page 144

INFLORESCENCE A RACEME OR SPIKE

Plantago lanceolata, page 536

Plantago major, page 538

Triglochin maritima, page 428

Zigadenus venenosus, page 454

INFLORESCENCE A SOLITARY FLOWER

Papaver rhoeas, page 518

LEAVES OPPOSITE

INFLORESCENCE A CYME
Flowers Pink

Impatiens glandulifera, page 154

Saponaria officinalis, page 238

Vaccaria hispanica, page 252

Erodium cicutarium, page 390

Flowers White

Apocynum androsaemifolium, page 36

Cerastium arvense, page 234

Gypsophila paniculata, page 236

Scleranthus annuus, page 240

Silene latifolia, page 242

Silene noctiflora, page 244

Silene vulgaris, page 246

Spergula arvensis, page 248

Stellaria media, page 250

Flowers Yellow

Hypericum perforatum, page 416

INFLORESCENCE A HEAD
Flowers Blue, Pink, or Purple

*Ageratum
conyzoides,*
page 54

Dipsacus fullonum
ssp. *sylvestris,*
page 322

Knautia arvensis,
page 324

Flowers Green

*Ambrosia
artemisiifolia,*
page 56

Ambrosia trifida,
page 58

*Xanthium
strumarium,*
page 146

Flowers White

Bidens pilosa,
page 68

Eclipta prostrata,
page 102

*Galinsoga
quadriradiata,*
page 106

Flowers Yellow

Tridax procumbens,
page 140

INFLORESCENCE A PANICLE

Atriplex hortensis,
page 258

Urtica dioica,
page 730

INFLORESCENCE A RACEME OR SPIKE
Flowers Blue, Pink, or Purple

Dracocephalum parviflorum,
page 434

Prunella vulgaris,
page 442

Galeopsis tetrahit,
page 436

Glechoma hederacea,
page 438

Lamium amplexicaule,
page 440

Stachys palustris,
page 444

Lythrum salicaria,
page 460

Epilobium hirsutum,
page 494

Flowers Green

Acalypha rhomboidea,
page 342

INFLORESCENCE A SOLITARY FLOWER

Anagallis arvensis,
page 642

Chaenorrhinum minus, page 530

Agrostemma githago,
page 232

INFLORESCENCE AN UMBEL

Asclepias syriaca,
page 38

INFLORESCENCE A CYATHIUM

Chamaesyce *Chamaesyce hirta,* *Euphorbia esula,*
glyptosperma, page 346 page 350
page 344

LEAVES WHORLED

INFLORESCENCE A CYME

Galium aparine, *Impatiens* *Spergula arvensis,*
page 674 *glandulifera,* page 248
page 154

INFLORESCENCE A RACEME OR SPIKE

Equisteum *Lythrum salicaria,*
arvense, page 336 page 460

INFLORESCENCE A SOLITARY FLOWER

Anagallis arvensis,
page 642

Anagallis arvensis,
page 642

INFLORESCENCE AN UMBEL

Mollugo verticillata,
page 482

KEY TO AQUATIC PLANTS

For plants with stems not emerging more than 30 cm above the water surface.

LEAVES ABSENT

Lemna trisulca,
page 46

LEAVES ALTERNATE

*Potamogeton
crispus,* page 636

LEAVES BASAL

Eichhornia crassipes, page 624

Pistia stratiotes, page 48

Hydrocharis morsus-ranae, page 410

LEAVES OPPOSITE

Elodea canadensis, page 406

LEAVES WHORLED

Salvinia molesta, page 680

Myriophyllum aquaticum, page 396

Myriophyllum spicatum, page 398

Egeria densa, page 404

Elodea canadensis, page 406

Hydrilla verticillata, page 408

KEY TO GRASSES AND GRASSLIKE PLANTS

Plants in this key have been arranged according to their inflorescence and presence or absence of auricles and ligules.

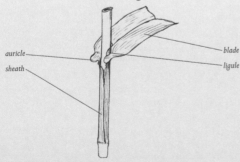

auricle — blade
sheath — ligule

Grass leaf structures

PANICLE COMPACT

AURICLES PRESENT

Aegilops cylindrica, page 546

Elymus repens, page 568

Hordeum jubatum, page 570

Taeniatherum cupul-medusae, page 594

AURICLES ABSENT

Cenchrus longispinus, page 554

Pennisetum glaucum, page 580

Phleum pratense, page 582

Setaria faberi, page 588

Setaria viridis, page 590

PANICLE OPEN

LIGULES PRESENT

Avena fatua, page 548

Bromus inermis, page 550

Bromus tectorum, page 552

Cynodon dactylon, page 556

Dactylis glomerata, page 558

Dactyloctenium aegyptium, page 560

Digitaria sanguinalis, page 562

Eleusine indica, page 566

Lolium persicum, page 572

Nassella trichotoma, page 574

Panicum capillare,
page 576

*Panicum
miliaceum,*
page 578

*Phragmites
australis,*
page 584

Poa annua,
page 586

*Sorghum
halepense,*
page 592

LIGULES ABSENT

*Echinochloa
crusgalli,*
page 564

*Cyperus
esculentus,*
page 306

Cyperus iria,
page 308

Cyperus rotundus,
page 310

Family and **Species Descriptions**

Amaranthaceae

AMARANTH FAMILY

Herbs or shrubs; leaves alternate or opposite, simple; stipules absent; inflorescence of solitary flowers, spikes, heads, or racemes, borne in terminal or axillary clusters; flowers perfect or imperfect, nonshowy, regular, often with three papery bracts at the base; four or five sepals, usually free; petals absent; four or five stamens, opposite the sepals; one pistil; ovary superior; fruit an achene, utricle, or capsule. Worldwide distribution: 60–65 genera/900 species; predominantly temperate and tropical regions.

Distinguishing characteristics: herbs; flowers nonshowy; floral bracts—three, papery.

Members of the amaranth family are of little economic importance. Several species are noxious weeds. Recently, herbicide resistance has been observed in several weedy members of this family.

redroot pigweed (*Amaranthus retroflexus*)

Amaranthus blitoides S. Wats.
PROSTRATE PIGWEED

Also known as: pigweed amaranth, tumbleweed pigweed, white pigweed, tumbleweed, pigweed, mat amaranth, spreading pigweed

Scientific Synonym(s): *A. graecizans* auct. non L., *A. pubescens* (Uline & Bray) Rydb.

QUICK ID: plant prostrate; leaf undersides with prominent white veins; flowers green and axillary

Origin: native to western North America

Life Cycle: Annual reproducing by seed; large plants producing up to 15,000 seeds.

Weed Designation: none

DESCRIPTION

Seed: Dark reddish-brown to black, disc-shaped, 1.5–1.7 mm across; surface smooth and dull. Viable in soil up to 10 years. Germination stimulated by light and alternating temperatures; can occur at 7°C; optimal temperature 25–35°C; optimal depth 0–10 cm.

Seedling: With narrow cotyledons, 4–10 mm long and 1–1.5 mm wide, pale green above and reddish-purple below; stem below cotyledons reddish-purple; first leaf oval with a shallow notch at tip.

Leaves: Alternate, oval to spoon-shaped, 1–5 cm long and 1–3 cm wide, stalks 5–25 mm long, dark green and shiny; tips with a slight indentation; underside with prominent white veins. Inflorescence a short axillary raceme; two-or-more-flowered.

Flower: Green, 1.5–3 mm across; male flowers with four or five sepals and three or four stamens; female flowers with four or five sepals and one pistil; bracts 1.5–5 mm long, soft with sharp points.

Plant: Plants often forming mats; stems prostrate to erect, 15–20 cm long, green or reddish-purple, somewhat fleshy.

Fruit: Lens-shaped capsule, 1.7–2.5 mm long; one-seeded.

REASONS FOR CONCERN

A common weed of gardens, roadsides, and waste areas, prostrate pigweed is a weed with serious consequences to row crops. It competes with the crop for moisture and nutrients. Prostrate pigweed is a host for beet curly top, beet yellows, and tobacco mosaic viruses. It is also capable of accumulating large amounts of nitrogen, which may cause nitrate poisoning in cattle.

Amaranthus retroflexus L.
REDROOT PIGWEED

Also known as: redroot, rough pigweed, green amaranth, redroot amaranth, Chinaman's greens, careless weed

Scientific Synonym(s):
A. delilei Loret,
A. retroflexus var. *salicifolius* L. I. M. Johnst.

QUICK ID: taproot reddish-pink; flower clusters terminal and dense; plants with a bristly rough texture

Origin: native to the southern United States and northern Mexico; observed in eastern Canada prior to 1900, and spreading to British Columbia by 1942

Life Cycle: Annual reproducing by seeds only; large plants capable of producing up to 150,000 seeds; mature fruit within 8 weeks of germination, allowing the possibility of two generations in a single growing season.

Weed Designation: CAN: MB, PQ

DESCRIPTION

Seed: Dark reddish-brown to black, disc-shaped, 1–1.3 mm wide; surface shiny.
Viable in soil up to 40 years.
Germination occurs very late in spring, often after cultivated crops have emerged; seeds below 2.5 cm of the soil surface do not emerge.

Seedling: With lance-shaped cotyledons; 3.5–10 mm long and 1–2 mm wide; underside reddish-purple; stem below cotyledons dark red near the soil surface; first leaves oval, prominently veined, and slightly notched at the tip; hairs on the leaf stalk and margin appear by the third leaf stage.

Leaves: Alternate, oval to diamond-shaped, 4–15 cm long and 1–7 cm wide; margins smooth to wavy; often appear drooping due to the long leaf stalks; underside with prominent white veins. Inflorescence a dense terminal panicle up to 20 cm long.

Flower: Green, 2–4 mm wide; male or female; male flowers with one to five greenish-purple spine-tipped sepals 4–8 mm long, and five stamens; female flowers with five sepals and one pistil; one to three bracts, 2.5–8 mm long, spine-tipped; giving the inflorescence a bristly appearance.

Plant: Plants up to 1 meter tall, indeterminate growth habit producing seeds until killed by frost; stems pale green or reddish, branched above, and rough-textured; taproot pinkish-red, short and fleshy.

Fruit: Capsule 1.5–2 mm long, single-seeded.

REASONS FOR CONCERN

Redroot pigweed is a large, fast-growing weed that can cause significant crop-yield reductions due to the prolific seed production and viability. Redroot pigweed accumulates high levels of nitrates, which may cause poisoning in some livestock, especially pigs. It is a host for the following pests: tarnished plant bug and European corn borer (pests of vegetables and ornamentals), *Orobanche ramosa* L. (p. 502; a parasitic plant of tomatoes), and green peach aphid. It is also a host for several vegetable viral diseases.

Amaranthus spinosus L.
SPINY AMARANTH

Also known as: thorny amaranth, thorny pigweed, careless weed, edelbur, needle burr, red amaranth, soldierweed

Scientific Synonym(s): none

QUICK ID: stems reddish, often striped; leaf bases with two sharp spines; flower cluster long and narrow

Origin: native to tropical North America

Life Cycle: Annual reproducing by seed only; large plants may produce up to 235,000 seeds.

Weed Designation: none

DESCRIPTION

Seed: Lens-shaped to circular, 0.9–1 mm long and 0.8–0.9 mm wide, dark black to reddish-brown; surface smooth and shiny, finely textured at high magnification.

Viable for 19 years in laboratory storage. Germination occurs in light or dark, often within a couple days after being shed; some seeds require 4–5 months to break dormancy.

Seedling: With lance-shaped cotyledons, 2–12 mm long and 0.4–2 mm wide, often purplish on underside, margins wavy; stem below cotyledons often purplish; first leaf oval with an indented tip; spines appear opposite the leaf by third leaf stage.

Leaves: Alternate, oval to lance-shaped, 3–7 cm long and 1.5–6 cm wide; base wedge-shaped, narrowing to a long petiole up to 7 cm long; margins smooth; underside conspicuously veined; spines 5–10 mm long and borne at leaf base.

Inflorescence a bristly spike, 5–15 cm long and 6–10 mm thick, upper section with male flowers, lower part predominantly female.

Flower: Green, male or female; five sepals, 1.2–2 mm long; five stamens; one pistil; three styles; bracts shorter than the sepals.

Plant: Plants 30–120 cm tall, preferring dry exposed sites; intolerant of shade, waterlogged soils, and cool temperatures; stems angular, somewhat woody and fibrous, green with reddish lines or grooves, much-branched; nodes with a pair of spines, each 1.5–2.5 cm long.

Fruit: Utricle, 1.5–2 mm long, bursting irregularly.

REASONS FOR CONCERN

Spiny pigweed is one of the most troublesome amaranth species in watermelon, sugarcane, and cotton production, as it hampers hand-harvesting. It is also a problem in feedlots and pastures as livestock tend to avoid infested areas. Plants also accumulate nitrates, which has been implicated in cattle poisonings. Forty-four countries report spiny amaranth as a serious weed.

OTHER SPECIES OF CONCERN IN THE AMARANTH FAMILY

ACHYRANTHES ASPERA L. — CHAFF FLOWER, DEVIL'S HORSEWHIP

Native to tropical America and Asia; perennial or annual herb; stems 40–200 cm, hairy; leaves elliptical to oval or rounded, 1–20 cm long and 2–6 cm wide, hairy; inflorescence a spike, 5–30 cm long, bracts papery and spine-tipped; flowers greenish; four or five sepals, 3–7 mm long; two to five stamens; one pistil; fruit a utricle, 2–4 mm long. Weed designation: none.

ALTERNANTHERA PHILOXEROIDES (MART.) GRISEB. — ALLIGATOR WEED

Native to South America; aquatic to semiaquatic perennial herb reproducing by stolons up to 500 cm; stems prostrate and forming mats; leaves narrowly elliptical to oblong or lance-shaped, 3.5–7.1 cm long and 0.5–2 cm wide, hairless, stalkless; inflorescence a globe-shaped head, 14–17 mm in diameter; bracts less than 3 mm long; flowers white; five sepals; five stamens; one pistil; fruit and seeds not produced in North American plants. Weed designation: USA: AL, AR, AZ, CA, FL, SC, TX.

Alternanthera philoxeroides

ALTERNANTHERA SESSILIS (L.) R.BR. EX DC. — SESSILE JOYWEED

Native to tropical America; annual or perennial herb 20–60 cm tall; stems ascending, hairy in lines; leaves elliptical to oblong or lance-shaped; 1.2–5 cm long and 0.5–2.2 cm wide; hairless; stalkless; inflorescence an oval to globe-shaped head, 5–11 mm across; bracts 1–2 mm long; flowers white; five sepals, 1–3.5 mm long; five stamens; one pistil; fruit a greenish utricle 1.3–1.7 mm long; seeds about 1 mm long. Weed designation: USA: federal, AL, AR, CA, FL, MA, MN, NC, OR, SC, VT.

AMARANTHUS PALMERI S. WATS. — PALMER'S AMARANTH

Native to southwestern United States and Mexico; annual; stems erect, 30–300 cm tall, branched, nearly hairless; leaves alternate, oblong to oval or diamond-shaped, 1.5–7 cm long and 1–3.5 cm wide, long-stalked, margins smooth; inflorescence a terminal panicle composed of linear spikes, often drooping; bracts of female flowers 4–6 mm long while those of the male flowers less than 4 mm long; male flowers with five tepals and five stamens; female flowers with five tepals and one pistil with two style branches; fruit a utricle, 1.5–2 mm long; seeds dark reddish-brown, 1–1.2 mm across, shiny. Weed designation: none.

Amaranthus palmeri

AMARANTHUS TUBERCULATUS (MOQUIN-TANDON) J. D. SAUER—ROUGH-FRUITED WATERHEMP, TALL WATERHEMP, ROUGH-FRUIT AMARANTH

Synonym: *A. rudis* J. D. Sauer
Native to the United States; annual; stems erect to ascending, 50–300 cm tall, branched; leaves alternate, oval to oblong or elliptical, 1.5–15 cm long and 0.5–3 cm wide, stalks 1–7 cm long, margins smooth; inflorescence a panicle of linear spikes; bracts 1–2 mm long, tips pointed; male flowers with five tepals, often spiny, and five stamens; female flowers with one or two small tepals (or absent) and one pistil with three style branches; fruit a utricle, 1.5–2 mm across; seeds dark reddish-brown, 0.7–1 mm across, shiny. Weed designation: none.

Amaranthus tuberculatus

AMARANTHUS VIRIDIS L.—GREEN AMARANTH, SLENDER AMARANTH

Native to South America; annual; stems erect to ascending, 20–100 cm tall, sparingly to densely branched, striped, hairless; leaves alternate, triangular to oval or diamond-shaped, 2–7 cm long and 0.5–5.5 cm wide, stalks 1–10 cm long, surface hairless except for a few hairs on the lower veins, margins smooth; inflorescence a slender spikelike panicle; male and female flowers scattered throughout the cluster; flowers green, three or four sepals, 1.3–1.8 mm long, five stamens, one pistil, bracts about 1 mm long, whitish-green, membranous with a short pale or reddish-green awn; fruit a globe-shaped capsule 1.3–1.5 mm long; seeds dark brown to black, shiny, 1–1.3 mm long. Weed designation: none.

Amaranthus viridis

Anacardiaceae

SUMAC FAMILY

Trees or shrubs, rarely woody vines; leaves alternate, simple or pinnately compound; stipules absent; inflorescence a panicle; flowers perfect or imperfect; regular; three to seven sepals (commonly five), fused; three to seven petals (commonly five); five to 10 stamens; petals and stamens inserted on a nectary disc; one pistil; ovary superior, rarely inferior; fruit a drupe. Worldwide distribution 79 genera/600 species; predominantly tropical with a few temperate species.

Distinguishing characteristics: trees and shrubs; leaves compound; flowers greenish-white.

The most renowned species in this family is poison ivy. Several other species in the family cause similar severe skin rashes and irritation, while other species—such as cashew, pistachio and mango—produce edible fruit and seeds. A few species are cultivated as ornamentals.

poison ivy (*Toxicodendron radicans*)

Schinus terebinthifolius Raddi
BRAZILIAN PEPPER

Also known as: Christmas
berry, Florida holly, pepper
tree
Scientific Synonym(s):
S. terebinthifolia Raddi

QUICK ID: tree up to 13 m
tall; leaves compound
with three to 11 leaflets;
berries red
Origin: native to Brazil,
Argentina, and Paraguay;
introduced as an
ornamental shrub into
Florida in the 1840s
Life Cycle: Perennial
(living about 35 years)
reproducing by seed
and suckers; research
has shown plants with
over 600 fruits per
inflorescence.
Weed Designation: USA: FL,
TX

DESCRIPTION
Seed: Kidney-shaped; dark brown and 0.3 mm in
diameter.
Viable in soil about 5 months.
Germination usually occurs within 20 days of
contact with the soil.
Seedling: With oblong cotyledons, margin curved
inward on one side; first two leaves simple with
a toothed margin; later leaves are compound;
tolerant to floods, may resprout after a fire.
Leaves: Alternate, compound with three to 11 leaflets
(seven to nine common), aromatic when crushed
(peppery or turpentine odor); leaflets elliptical to
oblong; 2.5–5 cm long and 1–3.5 cm wide; upper
surface dark green; lower surface pale; margins
somewhat toothed.
Inflorescence an axillary panicle 4–15 cm long;
borne on the current season's stem.
Flower: Flowers greenish-white, 1.2–2.5 mm long;
male or female; five sepals; five petals, white,
about 2 mm long; 10 stamens, borne in two
whorls; one pistil; male flowers lasting 1 day
while female flowers last 6 days; flowers are
often produced on plants 3 years old or more.
Plant: Evergreen shrub or tree to 13 m, often
multistemmed; barks grayish and scaly; under
favorable conditions trees reported to grow
30–50 cm per year; suckers produced and
increase the number of stems, forming an
impenetrable woodland.
Fruit: Red berry-like drupe, 4–7 mm across, glossy,
green, and juicy at first, becoming bright red on
ripening; red skin dries to become a papery shell
surrounding the seed; remaining on the tree for
up to 8 months.

REASONS FOR CONCERN

Brazilian-pepper quickly invades forests and roadsides displacing native vegetation. Over 700,000 acres have been infested in Florida. The attractive red fruit is eaten by birds and wildlife assisting in the spread of the plant. Research indicates that it may be allelopathic and suppress the growth of other plants. It is also suspected that chemicals found in the leaves, flowers, and fruit may irritate human skin and the respiratory system. Similar species: Peruvian pepper tree (*S. molle* L.), another introduced tree, 5–18 m tall, is distinguished by compound leaves 10–30 cm long with more than 15 leaflets. The leaflets, 1–6 cm long and 4–8 mm wide, are linear to lance-shaped, stalkless, and smooth-margined.

Toxicodendron radicans (L.) Kuntze
POISON IVY

Also known as: poison oak, poison creeper, three-leaved ivy, picry, mercury, markweed

Scientific Synonym(s): *Rhus radicans* L., *T. rydbergii* (Small ex Rydb.) Greene

QUICK ID: shrub or climbing vine; leaves alternate, compound with three shiny leaflets; flowers yellowish-green

Origin: native to North America

Life Cycle: Perennial reproducing by seed and rhizomes; rhizomes spreading at a rate of about 10 cm per year.

Weed Designation: USA: MN CAN: MB, ON, PQ

DESCRIPTION

Seed: White with gray stripes, round, 3–4 mm across. Viable for 6 years in soil.
Germination occurs at 20–30°C; flowers appear 3 years after germination.

Seedling: With oblong to lance-shaped cotyledons; first leaf compound with three leaflets; later leaves resembling mature leaves.

Leaves: Alternate, compound with three leaflets; leaflets 6.5–40 cm long, reddish-purple in spring and turning glossy green by midsummer, margins smooth to toothed; terminal leaflet stalked, lateral leaflets stalkless.
Inflorescence an axillary cluster 2.5–10 cm long.

Flower: Yellowish-green, 2–5 mm across; male, female or both; one pistil; five sepals; five petals; five stamens; one pistil.

Plant: Shrub 20–45 cm tall or climbing vine 30–120 cm tall, spreading by creeping rhizome and often producing large colonies; some plants never producing flowers or fruits.

Fruit: Globe-shaped drupe, creamy white, and waxy, 3–7 mm in diameter, often remaining on the stem throughout the winter.

REASONS FOR CONCERN

Some references indicate that just touching the plant causes poisoning, while others indicate that the plant tissue must be ruptured. Extreme care should be taken when traveling or working in areas where this plant is found. Burning of plants also releases the poison in the ash and smoke, so avoid inhalation. The poisonous compound is found in the sap of leaves, stems, and roots. The poison, urushiol, initially causes severe itching, then 24–48 hours later, the skin becomes red and blistered. Contaminated clothing should be hot laundered to neutralize the poisons. Poisoning can also occur when using contaminated tools.

OTHER SPECIES OF CONCERN IN THE SUMAC FAMILY

Toxicodendron pubescens

TOXICODENDRON PUBESCENS P. MILL.— ATLANTIC POISON OAK

Native to North America; shrub spreading by seed and rhizomes; bark light brown; stems 50–100 cm tall; twigs grayish-brown; buds covered in velvet hairs; leaves alternate, compound with three leaflets; leaflets up to 15 cm long, resembling oak leaves, three- to seven-lobed, shiny above, velvety below, turning yellow or orange in autumn; inflorescence a panicle, 2.5–7.6 cm long; flowers yellowish-green; five sepals; five petals; five stamens; one pistil; fruit a greenish-white drupe, about 6 mm across. Weed designation: USA: MI.

TOXICODENDRON VERNIX (L.) KUNTZE—
POISON SUMAC

Native to North America; shrub 1.5–5 m tall, exuding milky sap when cut or broken; branches grayish-brown; leaves alternate, stalks red, 15–30 cm long, compound with seven to 13 leaflets; leaflets oblong to oval or elliptical, 4–8 cm long and 2–3 cm wide, margins smooth; inflorescence a panicle 4–20 cm long; flowers white or greenish; five sepals; five petals; five stamens; one pistil; fruit a grayish-white drupe, 4–5 mm across; borne in arched clusters. Weed designation: USA: MI.

Toxicodendron vernix

Apiaceae
(Formerly Umbelliferae)

CARROT FAMILY

Herbs, often aromatic or strong-scented; stems hollow; leaves alternate, usually compound and sheathing at the base; stipules absent; inflorescence an umbel; flowers perfect, regular, or irregular; five sepals, very small or sometimes absent; five petals, usually white or yellow; five stamens, alternating with the petals; one pistil; two styles, the base bulbous (stylopodium); ovary inferior; fruit a schizocarp breaking into two one-to-two-seeded segments called mericarps. Worldwide distribution: 428 genera/3,000 species; predominantly found in the north temperate region.

Distinguishing characteristics: flowers borne in umbels; plants aromatic; fruit a schizocarp.

The carrot family is very important economically. It contains several food crops such as carrot, dill, parsley, celery, parsnips and caraway but also poisonous species such as fool's parsley, water hemlock, and poison hemlock. A number of species are noxious weeds.

cow parsnip (*Heracleum sphondylium* ssp. *montanum*)

Anthriscus sylvestris (L.) Hoffmann
WILD CHERVIL

Also known as: cow parsley, bur chervil, woodland beakchervil, keck, mother die, adder's meat, old rot, gypsy curtains, June flower

Scientific Synonym(s): *Chaerophyllum sylvestre* L.

QUICK ID: leaves fernlike; umbels with six to 15 rays; flowers white

Origin: native to Great Britain and Europe; introduced in wildflower seed mix; naturalized in Massachusetts by 1919

Life Cycle: Biennial or short-lived perennial reproducing by seed and lateral root buds; plants may produce up to 10,000 seeds.

Weed Designation: USA: MA, WA CAN: BC, NS, SK

DESCRIPTION
Seed: Dark brown, lance-shaped, 4–7 mm long. Viability for less than 2 years in soil. Germination occurs in the top 2 cm of soil; over 85% of seed germinates the first year.

Seedling: With narrow linear cotyledons (eight times longer than broad), tapered at base, hairless; first leaves hairy with three deeply lobed segments.

Leaves: Alternate, two to three times pinnately compound, fernlike, up to 30 cm long, hairy below, leaflets 10–50 mm long, oval and coarsely toothed.

Inflorescence a terminal umbel 2–6 cm across; six to 15 rays, each 15–30 mm long; bracts oval to lance-shaped, 3–6 mm long.

Flower: White, 3–4 mm across, often male flowers only in the middle of the umbel; five small sepals; five petals; five stamens; one pistil; two stigmas.

Plant: Plants 60–120 cm tall; stems erect, hollow, furrowed, hairy below, and hairless above; roots thick and tuberlike, up to 1.8 m deep.

Fruit: Schizocarp, 6–9 mm long, oblong, smooth, short-beaked; breaking into two mericarps at maturity.

REASONS FOR CONCERN

Wild chervil is an aggressive species that quickly displaces native vegetation. It is also suspected of causing photodermatitis in some people.

Cicuta maculata L.
WATER HEMLOCK

Also known as: musquash root, beaver poison, spotted cowbane, muskratweed, children's bane, poison parsnip, snakeroot, snakeweed, false parsley, poison hemlock

Scientific Synonym(s): *C. douglasii* (DC.) Coult. & Rose

QUICK ID: flowers white and in umbels; leaves compound with one to several leaflets; roots with horizontal chambers

Origin: native to North America

Life Cycle: Perennial reproducing by seed.

Weed Designation: USA: NV; CAN: MB

DESCRIPTION

Seed: Light brown, 2.7–3.3 mm long and 1.7–2 mm wide, elliptical with a flat side, surface with five yellowish- to orange-brown ribs; wrinkled between the ribs.
Viable in wet soil or water up to 4 years. Germination may occur under water; optimal germination occurs with alternating temperatures of 15°C and 21°C for 2 weeks.

Seedling: With needle-shaped cotyledons, 16–23 mm long and 1–2 mm wide, distinct parsnip odor when bruised; first leaves compound with three leaflets, margins toothed.

Leaves: Alternate, compound with several leaflets; leaflets 3–20 cm long and 0.5–3.5 cm wide, divided twice into narrow segments, margins coarsely toothed.
Inflorescence a terminal umbel, 3–10 cm across; 18–28 rays, each supporting smaller umbels with 12–25 rays; bracts few and narrow below each umbel.

Flower: White to greenish-white, about 2 mm across; five small, green sepals; five petals; five stamens; one pistil with two styles.

Plant: Plant of wet habitats; stem hollow, up to 2.2 m tall, often purplish spotted and streaked; rootstalk short and tuberlike, 3–10 cm long, with numerous horizontal chambers; yellow oily liquid in root chambers turns reddish-brown when exposed to air.

Fruit: Schizocarp, 2–4 mm long, breaking into two mericarps at maturity; floating on water until saturated.

REASONS FOR CONCERN

Water hemlock is the most toxic plant in North America. The roots and stems of water hemlock contain cicutoxin and are extremely poisonous to man and livestock. One root is sufficient to kill cattle and horses. Animals are usually poisoned in spring when the plants are easily uprooted and eaten. A violent death (respiratory paralysis) occurs within between 15 minutes and 6 hours after ingesting a lethal dose.

Similar species: water hemlock is easily confused with water parsnip (*Sium suave* Walt.), an edible native plant that grows in similar habitats. The leaves of water parsnip are divided once into narrow segments, while those of water hemlock are divided two or three times.

Conium maculatum L.
POISON HEMLOCK

Also known as: deadly hemlock, poison parsley, snakeweed, wode whistle, poison stinkweed, St. Bennet's herb, bad man's oatmeal, cashes, bunk, heck-how, poison root, beaver poison

Scientific Synonym(s): *Coriandrum maculatum* Roth, *Sium conium* Vest

QUICK ID: stem purple-spotted and hollow; leaves fernlike; flowers white, borne in umbels

Origin: native to Europe, northern Africa, and western Asia; introduced as an ornamental plant

Life Cycle: Biennial or perennial reproducing by seed only; large plants may produce up to 38,000 seeds.

Weed Designation: USA: CO, IA, ID, NM, NV, OH, OR, WA; CAN: federal, ON, SK

DESCRIPTION

Seed: Ovoid, grayish-brown, 2–3 mm long and 1.4–1.9 mm wide, surface dull, prominently five-ribbed, pale brown between ribs.
Viable in soil up to 5 years.
Germination occurs in light or dark; optimal temperature is 9°–34°C; about 15% of seed is dormant at maturity.

Seedling: With narrowly elliptical cotyledons, 4.5–21 mm long and 2.3–15 mm wide, light green, underside prominently net-veined; stem below cotyledons purple-tinged; parsniplike odor when crushed; first leaf two to three times deeply dissected, later leaves divided two to three times into narrow segments.

Leaves: Alternate and basal, 20–40 cm long, broadly triangular, composed of three to seven main divisions, then divided several times into narrow fernlike segments; leaflets oval to oblong, hairless, margins with teeth 4–10 mm long; stalks often splotched with purple, reduced in size upward, lower leaves with a sheath that surround the stem; ill-scented when bruised.
Inflorescence a terminal umbel 2–8 cm wide, appearing first, then overtopped by lateral umbels; 10–15 rays; bracts up to 10 mm long and 1–3 mm wide, margins smooth; secondary umbels with small bracts, but only on one side.

Flower: Flowers white, 2–4 mm across; sepals absent; five petals; five stamens; one pistil.

Plant: Plants up to 3 m tall; stems hollow and ridged, pale green, and often purple-spotted; stem and foliage with a strong mousy odor when crushed; taproot 20–25 cm long, white and branched; basal rosette of leaves produced in the first growing season.

Fruit: Sphere-shaped schizocarp 2–3 mm long; breaking into two mericarps at maturity.

REASONS FOR CONCERN

All parts of the plant are extremely poisonous. Livestock are highly susceptible to poisoning in the spring when the young plants are emerging. Symptoms of poisoning are depressed nervous system and paralysis. Death usually occurs within 2 hours of ingesting a lethal dose. Cattle must ingest 3.3 mg of plant material per kilogram of body weight to receive a lethal dose. For horses and sheep, lethal doses are 15.5 and 44 mg/kg, respectively.

Daucus carota L. var. carota
WILD CARROT

Also known as: Queen Anne's lace, bird's nest, devil's plague, lace flower

Scientific Synonym(s): none

QUICK ID: flowers borne in umbels, white with one black central flower; leaves compound with one to several leaflets

Origin: native to Europe, southwest Asia, and North Africa; first reported in United States in 1739; first Canadian reports from Ontario and Quebec in 1883

Life Cycle: Biennial (occasionally annual) reproducing by seed only; large plants producing up to 40,000 seeds.

Weed Designation: USA: IA, MI, OH, WA; CAN: federal, ON, PQ

DESCRIPTION

Seed: Elliptical with one flat side, yellowish- to grayish-brown, 3.2–4.2 mm long and 1.7–2.6 mm wide; surface with four rows of short curved spines and five lengthwise rows of white hairs. Viable in soil for at least 4 years. Germination occurs in dark when temperatures are below 20°C, and in light when above 20°C.

Seedling: With linear cotyledons about 20 mm long and 1 mm wide, hairless; first leaves compound with three leaflets, later leaves with numerous divisions; distinctive carrot odor when crushed.

Leaves: Alternate, divided two to four times into narrow segments, margins with short hairs, prominently veined below; basal leaves 5–40 cm long, long-stalked; upper leaves stalkless with white papery sheaths at the base. Inflorescence a flat-topped umbel; primary umbel 6–15 cm across with numerous secondary umbels; average plants often have as many as 100 umbels; bracts green and finely divided with three to five branches.

Flower: White, 1–3 mm across; five green sepals; five petals; five stamens; one pistil with two styles; a single purplish-black flower appears in the center of the umbel.

Plant: Plants up to 160 cm tall; stems erect, reddish-purple and bristly haired near the base, taproot white to brown, somewhat woody, less than 5 cm in diameter; stem and foliage with a characteristic carrot odor when crushed. The range of wild carrot is climatically controlled, as it requires at least 120 frost free days and 80–100 cm of precipitation.

Fruit: Schizocarp, 2–4 mm long; breaking into two mericarps at maturity; as fruits mature, the umbel becomes concave and resembles a bird's nest; at maturity, the "bird's nest" flattens and releases the seeds.

REASONS FOR CONCERN

Controlling wild carrot in cultivated carrot crops is very difficult as the two species are very closely related. It generally does not persist in cultivated crops because of its biennial life cycle. Wild carrot is common on roadsides, pastures and waste areas. It is a host for the following viruses: cucumber mosaic, tobacco mosaic, tomato big bud, vaccinium false bottom, and aster yellows. The carrot rust fly affects both wild and cultivated carrot.

Similar species: cultivated carrot (*D. carota* L. var. *sativa*) can be distinguished from wild carrot by its red, orange, or purple taproot.

Heracleum mantegazzianum Sommier & Levier
GIANT HOGWEED

Also known as: giant cow parsnip, cartwheel flower

Scientific Synonym(s): *H. giganteum* (Hornemann) hort. (non Fischer), *H. villosum* hort. (non Fischer)

QUICK ID: plants over 2 m tall; flowers in umbels over 30 cm across; base of stem with bristly hairs

Origin: native to the Caucasus Mountains of southwest Asia; introduced as an ornamental plant into Britain, then Europe and North America; first invasive reports for North America in 1917

Life Cycle: Short-lived perennial reproducing by seed only; large plants producing up to 100,000 seeds.

Weed Designation: USA: federal, AL, CA, CT, FL, MA, MN, NH, NC, OR, PA, SC, VT, WA; CAN: AB, BC, ON, SK; MEX: federal

DESCRIPTION

Seed: Elliptical to oblong or oval, flattened, 7–14 mm long and 5–11 mm wide, dark straw yellow, five-ribbed; four reddish-brown oil glands between the ribs; surface dull to slightly glossy. Viable in soil at least 2 years; some reports indicate 7–15 years. Germination occurs in the top 5 cm of soil; dormancy broken at 2°–4°C for 2 months.

Seedling: Seedling with lance-shaped cotyledons; first to fourth leaves shallowly three-lobed; later leaves deeply lobed and coarsely toothed.

Leaves: Alternate, lower stem leaves compound with three leaflets, upper deeply three-lobed margins toothed; basal rosette 1.5–2 m across; basal leaves up to 1 m wide, compound with three leaflets; leaflets deeply cut and toothed; stalks long, often covered in sharp-pointed bumps. Inflorescence a terminal umbel 30–80 cm across; 50–150 rays, 15–40 cm long; bracts linear to oval; lower umbels smaller, often composed of male flowers only; lateral umbels up to eight, often overtopping the terminal umbel.

Flower: White; five small sepals; five petals, 8–12 mm long; five stamens; one pistil; two styles; stalks 10–20 mm long, hairy.

Plant: Plants 4–6 m tall, requiring 2–5 years to flower, flowering once during lifetime, then dying; stems 3–10 cm across, hollow, often splotched with red or purple patches or bumps; bristly near the base; bristles sturdy and containing toxic sap; taproot 45–60 cm deep, pale yellow; crown often 15 cm in diameter.

Fruit: Schizocarp, 6–18 mm long and 4–10 mm wide; surface with three to five swollen brown resin canals.

REASONS FOR CONCERN

Seeds are often dispersed by water. The large plants shade out native plants, which may increase stream bank instability. Giant hogweed produces light-activated chemicals (furanocoumarins) that can cause minor to severe dermatitis when exposed to sunlight. Blisters appear within 48 hours and become purplish-black, producing scars that take up to 6 years to heal. This often leads to long-term sensitivity to sunlight. The highest concentration of toxins is in the leaves and roots, particularly early in the season.

Similar species: cow parsnip (*H. sphondylium* ssp. *montanum* [Schleich. ex. Gaudin] Briq.), a native perennial to North America, is distinguished by flowering stems less than 2 m tall, compound leaves with three leaflets, and umbels 15–30 cm across.

OTHER SPECIES OF CONCERN IN THE CARROT FAMILY

Aegopodium podagraria

Carum carvi

AEGOPODIUM PODAGRARIA L. — BISHOP'S GOUTWEED

Native to Europe and Asia; perennial with creeping rhizomes; stems erect, 40–90 cm tall; leaves alternate, compound with nine leaflets; leaflets oblong to oval, 3–8 cm long, often variegated with white, margins sharply toothed, basal leaves long-stalked, reduced in size and leaflet number upward; inflorescence an umbel, 6–12 cm across, long-stalked, 15–25 rays, bracts absent; flowers white; sepals absent; five petals; five stamens; one pistil; two styles; fruit a schizocarp, 3–4 mm long, breaking into two mericarps at maturity. Weed designation: USA: CT, MA, VT.

CARUM CARVI L. — CARAWAY

Native to Europe and Asia; biennial with stout taproot; stems hairless, 20–100 cm tall; leaves alternate, divided three or four times into narrow leaflets; leaflets linear, 5–15 mm long; inflorescence a compound umbel; stalks 5–13 cm long; seven to 14 rays, 2–4 cm long; bracts, when present, threadlike; flowers white; sepals absent; five petals; five stamens; one pistil; two styles; stalks 1–12 mm long; fruit a schizocarp, 3–4 mm long, breaking into 2 mericarps at maturity. Weed designation: USA: CO.

PASTINACA SATIVA L. — WILD PARSNIP

Native to Europe and Asia; biennial; stems stout, 30–150 cm tall; leaves alternate, compound with five to 15 leaflets; leaflets oblong to oval, 5–10 cm long, margins sharply toothed to lobed; lower leaves long-stalked, upper leaves sheathing; inflorescence a compound umbel; primary umbel 10–20 cm across; 15–25 rays; bracts absent; flowers yellow; sepals minute or absent; five petals; five stamens; one pistil; two styles; fruit a schizocarp, 5–7 mm long, breaking into two mericarps at maturity. Weed designation: USA: OH.

TORILIS ARVENSIS (HUDS.) LINK — SPREADING HEDGEPARSLEY

Native to Mediterranean region of Europe, Asia, and Africa; annual covered in short bristly hairs; stems 30–80 cm tall; leaves alternate, triangular in outline, compound and divided twice into numerous leaflets; leaflets oblong to lance-shaped, 1–6 cm long, margins toothed; inflorescence an umbel, 10–40 mm across; three to five rays; bracts present; flowers white or pinkish, about 2 mm across; sepals absent; five petals; five stamens; one pistil; two styles; fruit a schizocarp, about 3 mm long; breaking into two mericarps at maturity. Weed designation: USA: WA.

Pastinaca sativa

Torilis arvensis

Apocynaceae
DOGBANE FAMILY

Trees, shrubs, herbs, or cactuslike succulents, often with milky juice; leaves opposite or whorled (occasionally alternate), simple; stipules absent; inflorescence a raceme, cyme, or umbel; flowers perfect, regular; five sepals, united; five petals, united into a bell-shaped corolla; five stamens, inserted on the petals (in milkweeds, the stamen filaments fused to the stigma, and bearing petaloid appendages [corona]; corona often with hornlike appendages); one pistil; ovary superior; fruit a follicle; seeds with silky hairs. Worldwide distribution 415 genera/4,555 species; primarily tropical.

Distinguishing characteristics: herbs with milky juice; flowers bell-shaped; fruit a follicle.

Recent taxonomic research includes the milkweed family (Asclepiadaceae) in the dogbane family.

A few genera (*Vinca* L., *Plumeria* L., and *Catharanthus* G. Don) are grown for their ornamental value. Native North American species such as *Asclepias tuberosa*, *A. incarnata*, and *A. curassavica* provide food and habitat for the monarch butterfly.

spreading dogbane (*Apocynum androsaemifolium*)

flower

fruit

common milkweed
(*Asclepias syriaca*)

Apocynum androsaemifolium L.
SPREADING DOGBANE

Also known as: wandering milkweed, honeybloom, milk ipecac, rheumatism weed, western wallflower, flytrap dogbane

Scientific Synonym(s):
A. ambigens Greene,
A. pumilum (Gray) Greene,
A. scopulorum Greene ex Rydb.

QUICK ID: plants with milky juice; leaves opposite; flowers white, bell-shaped

Origin: native to North America

Life Cycle: Perennial reproducing by seed and rhizomes.

Weed Designation: CAN: MB

DESCRIPTION

Seed: Oval, light brown, 2.5–3 mm long, white silky hairs 1–2 cm long attached at one end.
Viable less than 3 years in soil.
Germination requires cold stratification; optimal temperatures 18–21°C.

Seedling: Seedling with lance-shaped cotyledons, 3–8 mm long and 1–2 mm wide; stem below cotyledon reddish to purplish; milky juice present at fourth leaf stage.

Leaves: Opposite, oval to oblong, 1–12 cm long and 0.5–6 cm wide; dark green; short-stalked; often drooping; midvein on the lower surface slightly hairy; leaves turning bright yellow or red in autumn.
Inflorescence a cyme of two to five flowers borne at the end of stems.

Flower: White-to-pink striped, fragrant, bell-shaped, 6–10 mm long; five sepals, united; five petals, united, the lobes recurved; five stamens; one pistil.

Plant: Plants up to 150 cm tall, preferring dry, sandy soils; stems smooth and reddish-green; exuding milky juice when broken; rhizomes creeping, up to 25 cm deep.

Fruit: Slender follicle, 7–20 cm long; splitting lengthwise and releasing numerous seeds.

REASONS FOR CONCERN

Spreading dogbane is a common weed of pastures, roadsides, orchards, and waste areas. It is now recognized as a problem in minimal- to zero-till cultivated fields. Some reports indicate that spreading dogbane may be poisonous to livestock. It is also host for several *Prunus* viruses.

Asclepias syriaca L.
COMMON MILKWEED

Also known as: silkweed, cottonweed, wild cotton, Virginia silk, broadleaf milkweed

Scientific Synonym(s):
A. cornuti Dcne.,
A. intermedia Vail,
A. kansana Vail

QUICK ID: plants with milky juice; leaves opposite; flowers pale pink, borne in globe-shaped clusters

Origin: native to eastern North America

Life Cycle: Perennial reproducing by seed and long creeping rhizomes; average plants producing up to 3,000 seeds.

Weed Designation: CAN: MB, NS, PQ; MEX: federal

DESCRIPTION

Seed: Seeds 7–8.5 mm long, 4–5 mm wide and less than 1 mm thick; pear-shaped, flattened, orangish-brown; fine silky white hairs, 3–3.5 cm long connected to the narrow end of the seed; surface dull with fine wrinkles.
Viability in soil about 8% after 7 years. Germination occurs between 15° and 40°C (optimal temperature 20°–35°C); seeds must be dormant for 1 year prior to germination.

Seedling: With elliptical cotyledons, 8–18 mm long and 3.5–7 mm wide, dull green above and shiny beneath; stem below cotyledons often tinged with purple; first leaves opposite, dull bluish-green, margin with short hairs; root buds form 18–21 days after emergence, and 28% of seedling resprout if the tops have been removed.

Leaves: Opposite, broadly oval, 7–26 cm long and 5–18 cm wide, short-stalked, margins smooth, underside prominently veined and covered in velvety hairs, base rounded.
Inflorescence a globe-shaped umbel borne in leaf axils or at the end of stems; one or two follicles produced per umbel.

Flower: Dull pinkish-purple, fragrant, about 1 cm across; five sepals; five petals; five stamens; one pistils, with two short styles; sepals and petals (6–9 mm long) bent backward along the stalk of the flower; stamens highly modified and united to form a tube around the styles, often referred to as a corona; less than 5% of flowers produce mature fruits.

Plant: Plants up to 2 m tall; stems stout, unbranched and covered in short silky hairs, exuding white milky juice when broken; rhizomes white and creeping, up to 120 cm deep, producing numerous buds giving rise to new plants.

Fruit: Grayish-green follicle, 7.5–10 cm long and 2.5 cm wide, each containing 150–425 seeds; surface covered in soft woolly hairs and soft spiny projections.

REASONS FOR CONCERN

Common milkweed is host for cucumber mosaic, strawberry mottle, tobacco streak, and Prunus A & G viruses. Common milkweed, an invader of rangeland, is suspected of being poisonous to livestock. It is also found in cultivated crops. In large infestations, the silky hairs on the fruit are reported to plug air intakes on farm machinery.

Similar species: showy milkweed (*A. speciosa* Torr.) is distinguished by a heart-shaped leaf base. Weed designation: CAN: MB.

Vincetoxicum rossicum (Kleopov) Borhidi
DOG-STRANGLING VINE

Also known as: pale swallowwort, European swallowwort, swallowwort

Scientific Synonym(s): *Cynanchum rossicum* Kleopow

QUICK ID: climbing vine; leaves opposite; flowers pink to maroon

Origin: Native to Ukraine and Russia; introduced as a garden plant; first report from British Columbia in 1885 and Ontario in 1889; first observed in New York State in 1897.

Life Cycle: Perennial vine reproducing by seeds and creeping rhizomes; infestations producing up to 2,100 seeds per square meter in open fields, and 1,330 seeds per square meter in shade.

Weed Designation: USA: CT, MA, NH, VT

DESCRIPTION

Seed: Light to dark brown, 4–6.5 mm long, 2.4–3.1 mm wide, flattened, oblong to oval, winged; tuft of white hairs 2–3 cm long.
Viability for 3–4 years under field conditions. Germination enhanced with alternating temperature and light intensities; germination in top 1 cm of soil; germination rates between 45% and 95%.

Seedling: With oval to elliptical cotyledons; stems about 5 cm long before the first leaves appear; over 75% of seeds producing two or more seedlings.

Leaves: Opposite, oval to elliptical, 6–12 cm long, 2.5–7 cm wide, margins smooth, hairy on the leaf underside margin and veins; stalks 5–20 mm long.
Inflorescence an umbel-like cyme, five- to 20-flowered; stalk 1.5–5 cm long, hairy.

Flower: Pink, red, maroon to brown, 5–7 mm across; five sepals, 1–1.5 mm long; five petals, 2.5–5 mm long, scarcely fleshy, twisted in bud; corona fleshy and lobed, pink to orange or yellow; gynostegium pale yellow to greenish-yellow; five stamens; two pistils.

Plant: Perennial vine, surviving minimum January temperatures of −36°C (Ottawa, Ontario); rhizomes short; crown woody; stems 60–250 cm long, erect to twining or scrambling.

Fruit: Follicle 2.8–7 cm long, smooth; often producing two fruits per flower.

REASONS FOR CONCERN

Dog-strangling vine is suspected of being allelopathic, thereby affecting the biodiversity of native habitats. It is also suspected of causing a decline in monarch butterfly (*Danaus plexippus* Linnaeus) populations. Being a member of the milkweed family, monarch butterflies lay their eggs on dog-strangling vine. The larvae die soon after feeding on the plant.

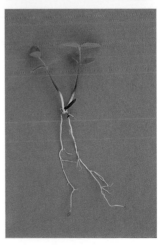

OTHER SPECIES OF CONCERN IN THE DOGBANE FAMILY

Apocynum cannabinum

APOCYNUM CANNABINUM L. — HEMP DOGBANE

Native to North America; perennial; stems erect, 30–100 cm tall, purplish, branched above; leaves opposite, oblong to lance-shaped, 3–10 cm long, 0.5–4 cm wide, short-stalked, margins smooth; inflorescence a terminal or axillary cyme, flower stalks less than 3 mm long, bract lance-shaped, about 2 mm long; flowers white to greenish, tube-shaped, 3–5 mm long; five sepals; five petals; five stamens, anthers orange; one pistil; fruit a follicle, 5–20 cm long, borne in pairs; seeds with a tuft of white hairs. Weed designation: none.

Asclepias verticillata

ASCLEPIAS VERTICILLATA L. — WHORLED MILKWEED

Native to North America; perennial with creeping rhizomes and fleshy roots; stems erect, 20–60 cm tall; leaves in whorls of three to six, linear, up to 7 cm long and 2–3 mm wide, margins curled under, short-stalked to stalkless; inflorescence an umbel, 10- to 20-flowered; stalks 0.5–3 cm long; zero to two bracts, linear; flowers greenish-white; five sepals; five petals, 4–5 mm long; hoods white, about 1.5 mm long; horns white, about 1.7 mm long; anther column greenish-white, about 2 mm long; two pistils, greenish, about 2 mm long; stalks purplish, 8–9 mm long; fruit a follicle up to 10 cm long and 1 cm wide. Poisonous to livestock. Weed designation: none.

Vinca minor

VINCA MINOR L. — COMMON PERIWINKLE

Native to southern Europe; evergreen perennial; stems trailing, 30–100 cm long, often forming mats; leaves opposite, leathery, and dark green, elliptical to lance-shaped, 3–5 cm long, short-stalked, margins smooth; inflorescence a solitary axillary flower borne in one axil of the leaf pairs; flowers blue, 2–3 cm across, 8–12 mm long; five sepals; five petals; five stamens; one pistil; fruit a narrow follicle; seeds without hairs. Weed designation: none.

VINCETOXICUM NIGRUM (L.) MOENCH—
BLACK DOG-STRANGLING VINE,
LOUISE'S SWALLOWWORT

Native to southwestern Europe; perennial vine with short rhizomes; roots pale, thick, and fleshy; stems 40–200 cm long, erect to twining or scrambling; leaves opposite, oblong to oval, 5–12 cm long and 2–6.5 cm long, margins smooth, underside not hairy; stalks 1–2 cm long; inflorescence a cyme, four- to 10-flowered, stalk 5–15 mm long, pubescent; flowers dark purple to black, 5–9 mm across, fleshy, stalks hairy; five sepals, 1–1.5 mm long; five petals, 1.5–3 mm long; corona same color as petals, not lobed; gynostegium pale green to yellowish; five stamens; two pistils; fruit a follicle, 4–8 cm long, rarely producing two fruits per flower; seeds dark brown, oblong to oval, 6–8 mm long and 3–4.7 mm wide, wing membranous, apical hairs 2–3 cm long. Weed designation: USA: CT, MA, NH, VT.

Vincetoxicum nigrum

Araceae

ARUM FAMILY

Shrubs, herbs, woody vines, and aquatic herbs; rhizome or tuber-bearing; leaves absent or alternate, simple to pinnately or palmately lobed, often with a sheathing base; palmately or parallel-veined; inflorescence a spadix (fleshy spike), often subtended by a petal-like spathe; flowers perfect or imperfect, regular or irregular; zero, four, or six tepals (sepals and petals indistinguishable from one another); six stamens (occasionally one, two, four, or eight); one pistil; ovary superior or inferior, often sunken into the fleshy stem of the spadix; fruit a berry or utricle. Worldwide distribution: 117 genera/4,100–5,422 species; greatest diversity in the tropical and subtropical regions.

Distinguishing characteristics: inflorescence a spathe and spadix; spathe often showy; fruit a berry or utricle.

Several members—anthurium, caladium, dieffenbachia, philodendron, pothos, and calla lily—are often grown as ornamentals and houseplants. A couple of well-known temperate-region wildflowers—jack-in-the-pulpit and skunk cabbage—are also members of this family.

common duckweed
(*Lemna minor*)

wild calla (*Calla palustris*)

Lemna trisulca L.
STAR DUCKWEED

Also known as: ivy
   duckweed, ivy-leaved
   duckweed
Scientific Synonym(s): none

QUICK ID: floating
   aquatic plant; leaves
   absent; flowers small and
   inconspicuous
Origin: native to North
   America and throughout
   temperate regions of the
   world
Life Cycle: perennial
   reproducing by budding
   and seeds
Weed Designation: none

DESCRIPTION
Seed: Oval, black, less than 0.5 mm long, distinctly
   12–18-ribbed; rarely produced in nature.
   Viability unknown.
   Germination unknown.
Seedling: Rarely produced in nature.
Leaves: Absent.
   Inflorescence a solitary flower borne on the edge
   of the thallus.
Flower: Rarely seen because of their small size;
   male or female; male flowers with two stamens;
   female flowers with a single pistil.
Plant: Submersed aquatic, not differentiated into
   stem and leaves; thallus (plant body) 6–12 mm
   long, faintly three-nerved, joined together to
   form T-shaped colonies; one root, 1–25 mm long;
   New plants formed on the edge of the parent
   plant, buds increase in numbers and form large
   colonies; in autumn buds sink to the bottom,
   overwinter, and rise to the water surface in
   spring to form new colonies.
Fruit: Capsule, 0.6–0.9 mm long, one- to seven-
   seeded, not opening at maturity.

REASONS FOR CONCERN

Duckweeds affect water quality for recreational purposes. Large populations reduce the amount of light available to submerged aquatics.

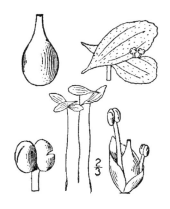

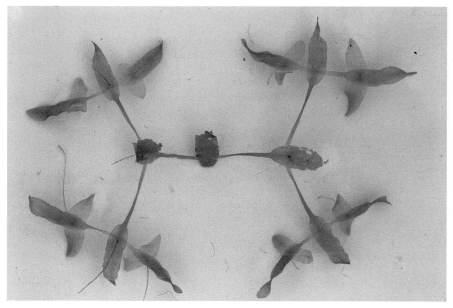

Pistia stratiotes L.
WATER LETTUCE

Also known as: Nile cabbage, tropical duckweed, water cabbage, Greek pistra, watertrough

Scientific Synonym(s):
P. spathulata Michx.

QUICK ID: floating aquatic plant resembling a head of lettuce; leaves grayish-green; young plants attached to parent by stolons

Origin: Uncertain, believed to have been brought to North America from tropical South America, Africa (Egypt), or southeast Asia (Sri Lanka) in ship ballasts; today spread is caused by shipping and sale from aquaria and nursery trade.

Life Cycle: Perennial reproducing by seed and brittle stolons.

Weed Designation: USA: AL, CA, CT, FL, SC, TX

DESCRIPTION

Seed: Light green to brown, pubescent, oval to oblong, about 2 mm long.
Viability unknown; seeds can withstand temperatures of −5°C for several weeks prior to being destroyed.
Germination occurs with optimal temperatures between 20 and 25°C; seeds float for 2 days after being released, then sink to bottom of the water body.

Seedling: Seedling with a thin hairless cotyledon, floating to surface on germination, about 5 days after sinking to the bottom of the water body; first leaf stout and hairy, taking over the floating function of the plant; first roots appear with the first leaf.

Leaves: Alternate, but appearing basal, 2.5–15 cm long and wide, grayish-green due to the surface having soft white hairs; spongy near base, which assists in keeping the plant afloat—absent on stranded plants; five to 15, prominently veined. Inflorescence a short spadix partly fused to spathe; spathe white to pale green, hairy on outside, hairless inside, 7–12 mm long and about 5 mm wide; six to eight male flowers, borne in a whorl around a central stalk; female flowers solitary.

Flower: White, less than 5 mm across; male or female; male flowers with two fused stamens; female flowers with a single pistil; sepals and petals absent.

Plant: Free-floating aquatic herb, surviving short periods on mud banks, optimal growth at 22–30°C, intolerant of frost; stolons brittle, up to 60 cm long; roots feathery, up to 100 cm long and 2–7 mm in diameter, short-branched.

Fruit: Green berry, 5–8 mm diameter, 4–15-seeded; opening irregularly, and shedding seeds.

REASONS FOR CONCERN

Large populations of water lettuce clog waterways, irrigation canals, and hydroelectric dams. It also reduces the air to water zone and light penetration, which affects the growth of native species. Plants also harbor several species of mosquito which are a principal vector of malaria. It is a serious weed of rice paddies throughout the world.

OTHER SPECIES OF CONCERN IN THE ARUM FAMILY

Colocasia esculenta

***COLOCASIA ESCULENTA* (L.) SCHOTT.—TARO, COCOYAM, EDDO, DASHEEN, MALANGA**
Origin uncertain; corms starchy; stolons produced at surface and spreading horizontally; leaves green to dark green above, pale below; shieldlike, 17–70 cm long and 10–40 cm wide; often reddish-purple at the stalk attachment; stalks 30–180 cm long, spongy; primary veins parallel, secondary veins netted; inflorescence a spathe 20–35 cm long, tube green, blade orange; spadix 9–15 cm long; female flowers green, borne near the base of the spadix; one pistil; sterile flowers white to pale yellow; male flowers pale orange, borne at the tip of the spadix; three to six stamens; fruit an orange berry. Escaped food crop found in wetlands throughout the tropics; probably brought to the Caribbean and North America from Africa during the slave trade. Weed designation: none.

***LANDOLTIA PUNCTATA* (G. MEYER) LES. & D. J. CRAWFORD (SYNONYM: *SPIRODELA PUNCTATA* [G. MEYER] C. H. THOMPSON)—GIANT DUCKWEED, DOTTED DUCKMEAT**
Native to Australia and southeast Asia; plant body (thallus) floating on surface of water, 1.5–8 mm long, 1–4 mm wide, three to seven veins, upper surface without a red spot; two to seven roots, up to 7 cm long; turions absent; flowers male or female; male flowers with two stamens; female flowers with a single pistil; fruit a utricle, 0.8–1 mm long; seeds with 10–15 ribs. Weed designation: USA: TX.

Lemna minor

***LEMNA MINOR* L.—COMMON DUCKWEED**
Native to North America; plant body (thallus) floating on surface of water, 1–8 mm long, 0.4–4 mm wide, three to five veins, margins smooth; one root; flowers male or female, green, one to two per thallus; male flowers with two stamens; female flowers with one pistil; fruit a utricle, 0.8–1 mm long; seeds with eight to 15 ribs. Weed designation: none.

SPIRODELA POLYRRHIZA SCHLEIDEN — GREATER DUCKWEED, DUCK-MEAL

Native to North America; plant body (thallus) floating on surface of water, 3–10 mm long, 3–8 mm wide, seven to 21 veins, upper surface often with a red spot in the center; seven to 21 roots, up to 3 cm long; turions round, 1–2 mm across, brownish-green; flowers male or female; flowers green, one per thallus; male flowers with two stamens; female flowers with a single pistil; fruit a utricle, 1–1.5 mm long; seeds with 12–20 ribs. Weed designation: none.

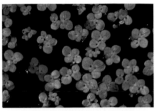

Spirodela polyrrhiza

Asteraceae

ASTER FAMILY

Herbs, occasionally shrubs or woody vines; leaves alternate (including basal rosettes), opposite or whorled, simple or compound; stipules absent; inflorescence a capitulum or head, often resembling a single flower, three types of heads produced; involucral bracts borne in one or more rows; flowers perfect or imperfect; two types of flowers, ray and disc florets; ray florets often straplike, heads composed of ray florets only referred to as ligulate; disc florets tubular, heads with disc florets only called discoid; heads containing both ray and disc florets referred to as radiate; sepals absent; five petals, united; five stamens, fused at the top and forming a tube around the style; one pistil; one style; ovary inferior; fruit an achene; pappus present or absent, when present represented by scales, hairs, or bristles. Worldwide distribution: 920 genera/22,000 species; found on every continent.

Distinguishing characteristics: flowers borne in heads; fruit an achene, often with pappus.

Some of the economically important plants of this family are sunflowers, artichokes, lettuce, and calendulas. Hundreds of other species are grown for ornamental value. Several species are noxious weeds.

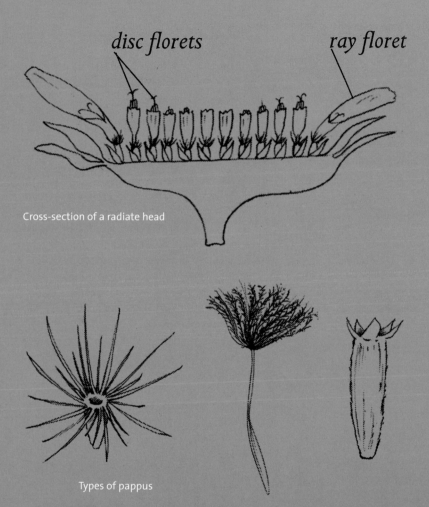

disc florets *ray floret*

Cross-section of a radiate head

Types of pappus

Ageratum conyzoides L.
FLOSSWEED

Also known as: tropical whiteweed, tropical ageratum, billy goat weed, chickweed, ageratum

Scientific Synonym(s):
A. album Stend;
A. caeruleum Hort. ex. Poir.; *A. coeruleum* Desf.; *A. cordifolium* Roxb.; *A. hirsutum* Lam.; *A. humile* Salisb.; *A. latifolium* Cav.; *A. maritimum* H. B. K.; *A. mexicanum* Sims.; *A. obtusifolium* Lam.; *A. odoratum* Vilm., *Cacalia mentrasto* Vell.

QUICK ID: plants densely haired; leaves opposite; flower heads light blue to violet or white

Origin: native of Central and South America

Life Cycle: Perennial or annual reproducing by seeds only; large plants producing up to 40,000 seeds; life cycle may be completed in less than 2 months.

Weed Designation: none

DESCRIPTION

Seed: Black, ribbed, 1.5–2 mm long, white-capped at one end; pappus of five scales, white to cream-colored, 0.5–3 mm long.
Viable for less than 1 year.
Germination occurs at 20°–25°C.

Seedling: With oval to fan-shaped cotyledons; first leaves opposite, oval to elliptical.

Leaves: Opposite, oval to triangular, 2–10 cm long, 5–50 mm wide; stalks 5–50 mm long, hairy; surfaces with long hairs; lower surface sticky-haired; margins toothed; prominently veined; crushed foliage ill-scented.

Flower: In terminal or axillary heads, flat-topped cluster with 30–50 heads; stalks hairy; heads 5 mm across and 4–6 mm high; 60–75 disc florets, light blue, white, or violet; 30–40 bracts, borne in two or three rows, linear, sparsely haired, about 3 mm long and 0.8–1.2 mm wide, green with reddish-purple tips.

Plant: Plant 5–120 cm tall; stems erect; lower branches often forming roots; nodes enlarged; young parts covered in long crinkly hairs; roots fibrous; life cycle may be completed within 2 months.

Fruit: Achene.

REASONS FOR CONCERN

Flossweed is a serious pest of pastures and crops such as corn, peanuts, rice, sugarcane, bananas, citrus, wheat, and sorghum. The plant is reported to have allelopathic affects on adjacent plants and the establishment of seedlings of other species. It is reported as a weed in 46 countries.

Similar species: another species, garden floss flower (*A. houstonianum* Mill.), commonly grown as an ornamental, is distinguished by blue or violet flowers, flower heads with 75–100 disc florets, hairy involucral bracts, and densely haired triangular leaves. This species is capable of producing more seed and may be weedier than *A. conyzoides*, depending on growing conditions.

Ambrosia artemisiifolia L.
COMMON RAGWEED

Also known as: ragweed, wild tansy, hogweed, bitterweed, mayweed, hay-fever weed, blackweed, roman wormwood, carrotweed, small ragweed, short ragweed

Scientific Synonym(s): *A. elatior* L., *A diversifolia* (Piper) Rydb., *A. glandulosa* Scheele, *A. monophylla* (Walt.) Rydb., *A. paniculata* Michx.

QUICK ID: terminal clusters of green flowers; fruit burlike and spiny; leaves deeply divided into narrow segments

Origin: native to North America; weedy nature observed in Michigan in 1838; Ontario in 1860

Life Cycle: Annual reproducing by seed only; large plants producing up to 62,000 seeds.

Weed Designation: USA: IL, MI, OR; CAN: federal, MB, ON, PQ

DESCRIPTION

Seed: Oval, light brown, 4–5 mm long and 1.6–2.5 mm wide; enclosed by a woody structure with a ring of short spines at the wide end of the fruit; pappus absent.
Viable in soil for at least 39 years.
Germination occurs on the soil surface or as deep as 15.5 cm; optimal depth is 2.5–10 cm.

Seedling: With club-shaped cotyledons, 7–13 mm long and 2–4 mm wide, purple dotted on the margin; stem below cotyledons spotted with purple or brownish-purple; first leaves opposite, five-lobed, covered in stiff hairs; later leaves alternate, deeply lobed, bristly haired.

Leaves: Opposite below and alternate above; oval in outline, divided two or three times into narrow segments; blades 25–80 mm long and 20–30 mm wide; stalks 25–35 mm long; young plants often yellowish-green and hairless, older plants grayish-green due to the presence of numerous stiff hairs.

Flower: Heads male or female; male flowers green, borne in terminal clusters, nodding, up to 100 flowers per head, releasing large amounts of pollen; bracts 12–30, borne in one to eight rows; female heads inconspicuous, 2–5 mm across, 10–20-flowered, stalkless, borne in axils of upper leaves, or near the base of the male inflorescence; five to 16 bracts and borne in a single row.

Plant: Plants 10–250 cm tall; stems branched above, rough-haired throughout, shallow-rooted.

Fruit: Achene; oval, woody and burlike, 3.5–4 mm long, grayish-brown, beak or spine about 1 mm long found at one end; terminal spine surrounded by a ring of short spines.

REASONS FOR CONCERN

Common ragweed produces large amounts of very light pollen, which may remain on air currents for up to 200 km. The pollen from this species is one of the most common causes for hay fever. Common ragweed is host for aster yellows, beet leaf curl, and cucumber mosaic.

Ambrosia trifida L.
GREAT RAGWEED

Also known as: kinghead,
giant ragweed,
crownweed, wild hemp,
horseweed, bitterweed,
tall ambrosia, buffalo
weed
Scientific Synonym(s):
A. aptera DC.

QUICK ID: terminal
clusters of green flowers;
fruit burlike and spiny;
leaves divided into three
to five lobes
Origin: annual native to
North America
Life Cycle: Annual
reproducing by seed;
average plants producing
over 275 seeds; population
has increased over the last
200 years, coinciding with
agricultural expansion in
North America.
Weed Designation: USA: CA,
DE, IL; CAN: federal, MB

DESCRIPTION
Seed: Egg-shaped, grayish-brown, 6–8 mm long and
2–3.8 mm wide; surface smooth.
Viability of seed buried at 20 cm soil depth has
been determined to be at least 9 years.
Germination occurs in the top 16 cm of soil
(optimal depth 2 cm, surface germination rare);
temperature range is 8°–41°C (optimal 10°–24°C).
Seedling: With spoon-shaped cotyledons, 2–4 cm
long and 1–1.5 cm wide; stem below cotyledons
shiny green with purple blotches; first pair of
leaves lance-shaped with toothed margins;
second pair deeply three-lobed and rough-haired.
Leaves: Opposite below, alternate above; 5–25 cm
across, three- to five-lobed, prominently three-
veined; stalks 10–70 mm long; margins coarsely
toothed; surfaces with a rough sandpapery
texture.
Flower: Heads male or female, greenish; male heads
borne in terminal spikes, each head surrounded
by five to 12 bracts, each with three prominent
black ribs; female heads 6–13 mm long, one-
flowered, borne in clusters of one to four at
the base of male heads, or in leaf axils; five to
16 bracts in a single series; wind pollinated,
produces large amounts of pollen.
Plant: Plants 50–400 cm tall (6 m tall on fertile soil);
stems branched, rough-haired.
Fruit: Achene; egg-shaped to five-sided, woody and
burlike, 5.6–6.3 mm long, greenish-tan with
purple streaks on the upper portion, beak or
spine found at the widest end, encircled by five
spines; one to three smaller spines may be found
between the outer ring and the central beak.

REASONS FOR CONCERN

Great ragweed is a highly competitive species in moist cropland and pastures. It is the major cause of hay fever in late summer and early fall. Great ragweed is also host for aster yellows, chrysanthemum stunt, tobacco mosaic, tobacco ring spot, and tobacco streak viruses.

Anthemis cotula L.
STINKING MAYWEED

Also known as: dog's chamomile, dog fennel, mayweed, stinking chamomile, stinkweed, dillweed, stinking daisy, hog's fennel, fetid chamomile

Scientific Synonym(s): *Maruta cotula* (L.) DC., *A. foetida* Lam., *Chamaemelum cotula* (L.) All.

QUICK ID: flowers white, daisy-like; leaves fernlike; plants ill-scented

Origin: native to the Mediterranean region of southern Europe

Life Cycle: Annual or winter annual reproducing by seed only; large plants producing up to 960,000 seeds.

Weed Designation: USA: CO, NV; CAN: federal, PQ

DESCRIPTION

Seed: Seeds 1–2 mm long, brown, oblong, ten-ribbed, furrows glandular-dotted; pappus absent. Viable in soil for 6 years.

Germination occurs on the soil surface; optimal temperature 20°C; greatest number of seeds germinating in the second and third year after being shed.

Seedling: With elliptical to lance-shaped cotyledons, 3–8 mm long, 1–2.5 mm wide, thickened and united at base to form a fleshy cup; stem below cotyledons green or dull purple; first two leaves appearing opposite but alternate, clasping at base, surface pale dull green and very thick, bearing a few divisions; later leaves becoming more dissected.

Leaves: Alternate, 2–6 cm long and 5–30 mm wide, pale green to yellowish-green, finely divided into narrow segments, stalkless; strong disagreeable odor when crushed.

Flower: Heads resembling daisies, 1.2–2.6 cm across, borne singly at tips of branches, stalks 4–6 cm long; 10–20 ray florets, white, sterile, three-toothed, 5–11 mm long; disc 5–10 mm across, with 60–300 disc florets; disc florets yellow, 2–2.5 mm long; 21–35 bracts, borne in three to five rows, tapered to tip.

Plant: Plants 10–60 cm tall; stems erect, profusely branched throughout, hairless, often reddish-green; taproot short and thick; stem and foliage ill-scented when crushed.

Fruit: Achene.

REASONS FOR CONCERN

Stinking mayweed is a common weed of gardens, nurseries, roadsides, and agricultural crops. Reports indicate that the sap can cause skin irritation in humans. It is suspected to be allelopathic to adjacent plants.

Similar species: corn chamomile (*A. arvensis* L.) resembles stinking mayweed. It is distinguished by the lack of a strong odor, branches only near the base, and leaves 1.5–3.5 cm long and 8–16 mm wide. Heads have five to 20 fertile white ray florets, 5–15 mm long. Weed designation: USA: CO.

Arctium minus (Hill) Bernh.
COMMON BURDOCK

Also known as: lesser
burdock, wild rhubarb,
clothbur, beggar's
buttons, smaller burdock,
cuckoo button, cockle
button, hardock, hurr-
burr, cuckold dock

Scientific Synonym(s): *Lappa
minor* Hill.

QUICK ID: floral bracts
with hooked bristles;
stems hollow; fruiting
heads remain closed at
maturity

Origin: native to northern
Europe; first reported
from New England in
1638; common in Ontario
by 1860

Life Cycle: Biennial
reproducing by seed only;
plants may produce up to
18,000 seeds.

Weed Designation: USA: CO,
WY; CAN: AB, MB, PQ, SK

DESCRIPTION

Seed: Club-shaped, mottled brown, 5.7–6.1 mm
long and 2.2–2.4 mm wide; pappus consisting of
barbed yellow bristles 1–3.5 mm long, deciduous.
Viable in soil for up to 3 years.
Germination increased by alternating
temperatures and exposure to light; seeds have
no dormancy.

Seedling: With elliptical cotyledons, 15–29 mm
long and 6–7 mm wide, dull green in color; stem
below the cotyledons often purplish-green; first
leaf oval with prominent veins on the upper
surface; leaf underside covered in downy hairs.

Leaves: Alternate and reduced in size upward; basal
leaves oval to oblong-shaped, 10–50 cm long and
5–40 cm wide, white woolly below, stalks hollow
near the base; basal rosette up to 1 m wide
produced in the first year of growth.

Flower: Heads borne in spikelike clusters in axils
of upper leaves; stalks 0.5–9.5 cm long; heads
1–3 cm across; 20–40 disc florets, purple, 7.5–12
mm long; bracts numerous, borne in nine to 12
overlapping rows, hooked, reduced in size inward;
not spreading at maturity.

Plant: With flowering stems 50–300 cm tall; stem
thick, hollow, and grooved; taproot large and
fleshy.

Fruit: Achene; fruiting heads about 1.4 cm across,
attach themselves to clothing and fur with the
aid of hooked floral bracts.

REASONS FOR CONCERN

Common burdock is not a problem in cultivated land or pastures. It is found growing along fence lines, river banks, and waste areas. The hooked bristles of the floral bracts assist in dispersal of the seeds to new locations. Common burdock is host for cucumber mosaic and tobacco streak viruses.

Similar species: greater burdock (*A. lappa* L.) is distinguished from common burdock by flower heads that are 3–4.5 cm across and outer bracts that spread at maturity. Weed designation: CAN: AB, BC, MB, PQ. Another species, woolly burdock (*A. tomentosum* Mill.), is distinguished from common burdock by floral bracts covered in soft woolly hairs. Flower heads are 2–3 cm across and produce numerous seeds, each 5–6.5 mm long. Both species may produce up to 10,500 seeds. Weed designation: CAN: AB, BC, MB, PQ.

woolly burdock
(*Arctium tomemtosum*)

Artemisia absinthium L.

ABSINTHE

Also known as: wormwood, absinthium, madderwort, warmot, vermouth, absinthe sagewort, common sagewort

Scientific Synonym(s): none

QUICK ID: plants sage scented; flower heads nodding; leaves grayish-green and divided two to three times into narrow segments

Origin: native to Europe; introduced as a medicinal garden plant by the Spanish in early 1800s; seed was available for sale in the United States in 1832; spread to Manitoba by 1860; not recognized as a weed until the mid-1950s; also grown as a flavoring and psychoactive ingredient for absinthe liquor, which is illegal in some jurisdictions, and associated neurotoxin is addictive and deadly if consumed in large amounts

Life Cycle: Perennial reproducing by seed and creeping roots; plants may produce up to 50,000 seeds.

Weed Designation: USA: CO, ND, WA; CAN: MB, SK

DESCRIPTION

Seed: Club-shaped, light brown, 0.8–1.1 mm long and 0.4–0.6 mm wide; surface smooth.
Viable for 3–4 years in soil.
Germination occurs in the top 2 cm of soil; emergence continues throughout the growing season.

Seedling: With oval cotyledons, 2 mm long and 1 mm wide, surface with a powdery appearance; first two or three leaves oval with a few toothlike lobes on the margin; later leaves deeply lobed, covered in dense soft hairs.

Leaves: Alternate, 3–10 cm long and 1–4 cm wide, grayish-green on both sides, reduced in size upward; basal and lower stem leaves long-stalked, divided two to three times into narrow segments, each segment 1.5–4 mm wide, margins smooth, tips rounded; uppermost leaves lance-shaped, simple, 1–2 cm long, stalkless.

Flower: Heads grayish-green with yellowish-brown centers, nodding, 2–3 mm high and 3–5 mm wide, borne in axils of upper leaves, forming a large panicle with up to 1,500 heads; heads averaging 36 disc florets, with an outer ring of nine to 20 female florets surrounding the inner 30–50 perfect florets; two to 20 bracts borne in four to seven rows, grayish.

Plant: Shrubby and up to 2 m tall, silvery gray throughout, hairy when young, becoming hairless with age, strong sagelike odor present; stems branched; root woody, up to 1.25 cm in diameter and spreading 1.8 m in all directions.

Fruit: Achene.

REASONS FOR CONCERN

Absinthe is not a problem in cultivated crops. It is a serious weed in pastures and increases as livestock graze on more palatable species. Cattle that consume absinthe produce off-flavored milk and dairy products. Small amounts of absinthe seed in grain have caused shipments to be rejected for export.

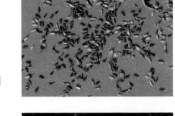

Artemisia vulgaris L.
MUGWORT

Also known as: common wormwood, green ginger, motherwort, mugweed, sailor's tobacco, chrysanthemum weed, felon herb

Scientific Synonym(s):
A. indica Willd.,
A. lavandulaefolia Miq.

QUICK ID: plants ill-scented; flower heads borne in dense spikes; flowers yellow to reddish-brown

Origin: native to Europe and Asia; introduced into North America as a medicinal plant and vermifuge

Life Cycle: Perennial reproducing by rhizome and rarely by seed; plants may produce up to 450,000 seeds per year.

Weed Designation: CAN: MB, PQ

DESCRIPTION

Seed: Brown to yellowish-brown, 1–5 mm long and less than 1 mm wide, ridges with minute bristles at one end; pappus absent.
Viable seeds are rarely produced in nature; research has shown that seed obtained from a 200-year-old excavation site germinated. Germination of young seeds require light, while older seeds germinate in total darkness; optimal temperature is 25°C, range is 7°–35°C.

Seedling: Seedlings rarely produced in nature; cotyledons egg-shaped, about 5 mm long, stalkless; first leaves opposite, bristly haired beneath, egg-shaped to round, long-stalked, margins with conspicuous teeth; rhizome production starts about 4 weeks after germination; research shows that, at 7 weeks, 12 rhizomes with a total length of 260 cm may be present; at 12 weeks secondary plants start to emerge from the soil; flowering begins at 4 months of age.

Leaves: Alternate, green, simple to divided and coarsely toothed; upper surface hairless, white woolly below; lower leaves stalked, the upper sessile; 5–10 cm long and 3–7 cm wide with large divided lobes; aromatic when crushed.

Flower: Heads erect, 2–3 mm across and 3–4 mm long, 15–30-flowered; borne in dense spikes 20–30 cm long and 7–15 cm wide; large plants with up to 7,400 heads; disc florets 1.5–3 mm long, greenish-yellow to reddish-brown; outer seven to 10 florets female, the inner five to 20, perfect; two to 20 lance-shaped bracts margins membranous, borne in four to seven rows.

Plant: plants ill-scented; extremely variable; 50–190 cm tall; stems red, brown, or purplish, ridged and

angular, much-branched and leafy, somewhat woody at base, clump-forming; rhizome coarse.
Fruit: Achene.

REASONS FOR CONCERN

Mugwort is a weed of nursery fields, vineyards, and fencerows. It can tolerate mowing to 3.5 cm height, often below the mower blade. The creeping rhizome of mugwort makes control in perennial crops difficult.

Bidens pilosa L.
BEGGAR-TICKS

Also known as: cobbler's pegs, hairy beggar-ticks, Spanish needles, broomstick, devil's-needles

Scientific Synonym(s): *B. odoranta* Cav., *B. pilosa* var. *bimucronata* (Turcz.) Shultz-Bip., *B. pilosa* var. *minor* (Blume) Sherff

QUICK ID: plants of moist sites; leaves opposite; flowers daisy-like

Origin: native to Europe

Life Cycle: Annual reproducing by seed only; large plants often produce between 3,000 and 6,000 seeds.

Weed Designation: none

DESCRIPTION

Seed: Seeds of two types; outer florets producing reddish-brown seeds 3–5 mm long; inner florets producing dark brown or blackish seeds 8–16 mm long; pappus absent or consisting of two to four barbed awns, 1–4 mm long.
Viability in soil about 80% after 5 years. Germination requires light; no dormant period is required, allowing three to four generations per season in warmer climates.

Seedling: With strap-shaped cotyledons; stem below cotyledons purple-tinged; first leaves opposite, oval.

Leaves: Opposite, 3–7 cm long and 1.2–1.8 cm wide, stalks 1–3 cm long, pinnately compound with three to five leaflets; leaflets oval, terminal leaflet largest and up to 9 cm long and 3 cm wide, margins toothed.

Flower: Twenty-one to 42 heads, 7–10 mm across, borne on stalks 1–9 cm long; ray florets white or yellowish, absent or four to seven per head, each 2–15 mm long; 20–80 yellow disc florets; eight to 21 bracts, 2–5 mm long, borne in two rows.

Plant: Plants of wet or moist sites; stems erect, 30–180 cm tall, branches ridged, green or with brown stripes.

Fruit: Achene; pappus of two to four awns.

REASONS FOR CONCERN

Beggar-ticks is a common weed of field and plantation crops. It is a serious weed for sugarcane, corn, coffee, potatoes, and citrus in Mexico. It is reported to be a serious weed of 31 crops in more than 40 countries.

Carduus acanthoides L.
PLUMELESS THISTLE

Also known as: welted thistle, curled thistle, acanthus thistle, giant plumeless thistle, spiny plumeless thistle

Scientific Synonym(s):
C. axillaris Gaudin,
C. polyanthus Schreber

QUICK ID: plants spiny; flowers pink to purplish; taproot hollow near the ground surface

Origin: native to Europe and North Africa; first reported from New Jersey in 1879; first Canadian report from Ontario in 1907

Life Cycle: Annual or biennial reproducing by seed only; large plants may produce up to 10,000 seeds.

Weed Designation: USA: AR, AZ, CA, CO, IA, MD, MN, NE, NC, OR, SD, WA, WV, WY; CAN: AB, BC

DESCRIPTION

Seed: Gray to golden brown, 2.5–3 mm long, 1.2–1.3 mm wide, surface with eight longitudinal nerves; pappus of white barbed bristles 11–13 mm long. Viability in soil about 10 years.
Germination occurs when temperatures reach 20°C.

Seedling: Seedling with round to oval cotyledons, broad pale green to white veins present; first leaves oval to lance-shaped, margins prickly.

Leaves: Alternate, 10–30 cm long and less than 8 cm wide, lower leaves with winged stalks, the upper stalkless, margins deeply lobed and spiny; spines to 5 mm long; lower surface covered in woolly hairs.

Flower: Heads 15–25 mm wide and 18–25 mm high, borne single or in corymbs of two to five heads; stalks 4–10 cm long, spiny-winged; 50–80 disc florets, pink to purple, 13–20 mm long; bracts numerous, linear and less than 2 mm wide, borne in seven to 10 rows; outer bracts spiny; inner bracts soft, rarely spiny.

Plant: Plants 30–400 cm tall; stems branched, densely haired, winged with spines 2–8 mm long; taproot large, hollow at or near the ground surface.

Fruit: Achene.

REASONS FOR CONCERN

Plumeless thistle populations tend to increase in pastures and rangeland because livestock refuse to graze near the spiny plants. Dense stands are often impenetrable.

Similar species: another plumeless thistle, *C. crispus* L., an introduced biennial from Europe and Asia, has brittle stems that are weakly spined, broader leaves usually covered in cottony woolly hairs, and heads 15–18 mm high. Weed designation: USA: AR, IA, WV.

plumeless thistle
(*Carduus crispus*)

NODDING THISTLE

Also known as: musk thistle, Italian thistle, plumeless thistle, buck thistle, Queen Anne's thistle, musk bristle thistle

Scientific Synonym(s):
C. macrocephalus Desf., *C. macrolepis* Petermann

QUICK ID: plants spiny; flower heads purple; leaf bases extend down the stems producing a spiny wing

Origin: native to Europe, Asia, and North Africa; introduced into North America in the early 1900s in ship ballasts; first reported from Pennsylvania in 1853

Life Cycle: Biennial (occasionally an annual) reproducing by seed; large plants may produce up to 20,000 seeds; large terminal head may produce 1,200–1,500 seeds (average is 165–256).

Weed Designation: USA: AR, CA, CO, IA, ID, IL, KS, KY, MD, MN, MO, NE, NV, NM, NC, ND, OH, OK, OR, PA, SD, UT, WA, WV, WY; CAN: federal, AB, MB, ON, SK

DESCRIPTION

Seed: Elliptical, pale yellow to orange-brown, 3.4–4.5 mm long and 1.3–1.9 mm wide; surface smooth and glossy, several ridges running its length; pappus of straight white hairs 13–25 mm long, deciduous.
Viable at least 10 years in soil.
Germination occurs in dark at 20°–30°C, and in light at 15°–20°C; no after-ripening dormancy.

Seedling: With lance-shaped cotyledons, 7.5–15 mm long and 2.5–6 mm wide, dull green, broad white veins present; first leaf lance-shaped, margin with numerous prickles; later leaves spatula-shaped, margins irregularly toothed and prickly.

Leaves: Alternate, deeply lobed, 10–40 cm long and 15 cm wide, dark green with a white midrib, lobes with three to five spiny points, each prominently yellow or white-spined at the tip; leaf bases extend down the stem giving it a spiny winged appearance; basal rosette of six to eight leaves produced in the first season of growth, up to 30 cm long and 15 cm wide, both surfaces covered in fine woolly hairs.

Flower: Heads nodding, 2–6 cm long and 1.5–8 cm across, borne at the end of stems in groups of one to three; stalks unwinged, 2–30 cm long, covered in soft white hairs; disc florets 15–18 mm long, red to purple; bracts numerous and borne in seven to 10 overlapping rows, tips purplish-green, spines 1–4 mm long; outer row 2–8 mm wide and reflexed.

Plant: Plants 40–250 cm tall; stems unbranched, winged and densely spiny, spines 2–10 mm long; taproot large and fleshy, often hollow near the soil surface.

Fruit: Achene.

REASONS FOR CONCERN

Nodding thistle is capable of reducing yield on pastures and rangeland up to 100%. Once established, nodding thistle forms large colonies that crowd out forage plants. Plants are so spiny that livestock often avoid grazing near it.

Similar species: Italian thistle (*C. pycnocephalus* L.) may be confused with nodding thistle. It is distinguished by erect heads borne in groups of one to five at ends of branches, winged peduncles, and seeds 4–6 mm long. Weed designation: USA: CA, OR, WA.

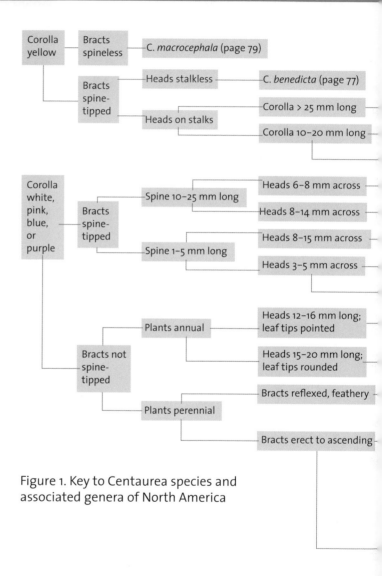

Figure 1. Key to Centaurea species and associated genera of North America

The genus *Centaurea* has 350–600 species of thistle-like plants. Native to Europe, western Asia, and North Africa, they have been introduced throughout the world. There are 19 introduced species in North America. The distinguishing characteristics of this genus are flower heads with six to many rows of involucral bracts that have variously fringed to spine-tipped margins and tips. Due to their invasive nature, an identification key has been included (fig. 1). Descriptions for all introduced species are included. Full treatments have been provided for the most common species. One species, Russian knapweed (*Rhaponticum repens*, p. 122), has been included in the key due to its similarity to the genus *Centaurea*.

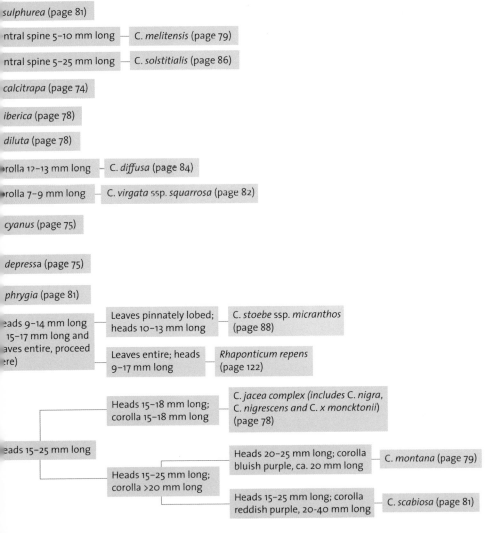

sulphurea (page 81)

ntral spine 5–10 mm long — C. melitensis (page 79)

ntral spine 5–25 mm long — C. solstitialis (page 86)

calcitrapa (page 74)

iberica (page 78)

diluta (page 78)

rolla 12–13 mm long — C. diffusa (page 84)

rolla 7–9 mm long — C. virgata ssp. squarrosa (page 82)

cyanus (page 75)

depressa (page 75)

phrygia (page 81)

eads 9–14 mm long
15–17 mm long and
aves entire, proceed
re)

Leaves pinnately lobed; heads 10–13 mm long — C. stoebe ssp. micranthos (page 88)

Leaves entire; heads 9–17 mm long — Rhaponticum repens (page 122)

Heads 15–18 mm long; corolla 15–18 mm long — C. jacea complex (includes C. nigra, C. nigrescens and C. x moncktonii) (page 78)

eads 15–25 mm long

Heads 15–25 mm long; corolla >20 mm long

Heads 20–25 mm long; corolla bluish purple, ca. 20 mm long — C. montana (page 79)

Heads 15–25 mm long; corolla reddish purple, 20–40 mm long — C. scabiosa (page 81)

Centaurea benedicta
Page 77

Centaurea melitensis
Page 80

Centaurea diffusa
Page 84

Centaurea calcitrapa
Page 77

Centaurea nigra
Page 81

Centaurea stoebe ssp. *micranthos*
Page 88

Centaurea cyanus
Page 77

Centaurea nigrescens
Page 81

Centaurea solstitialis
Page 86

Centaurea jacea
Page 79

Centaurea scabiosa
Page 82

Rhaponticum repens
Page 122

Centaurea benedicta (L.) L.—blessed thistle

Native to Europe, western Asia and North Africa; annual; stems erect, 60–180 cm tall, often reddish, covered in soft woolly hairs; leaves alternate, oblong, toothed, or pinnatifid, 7–25 cm long, 1–5 cm wide, pale green and prominently white-veined, lower leaves stalked, the upper stalkless; margins with spiny teeth; flower heads large; disc florets 19–24 mm long, yellow; bracts numerous and leaflike, borne in six to 12 overlapping rows, spine-tipped, inner spines branched, more than 5 mm long; fruit an achene, brown, cylindrical, 8–11 mm long, shiny and strongly 20-ribbed; pappus of two rows of awns, outer awns 10, 9–10 mm long, inner awns 10, 2–5 mm long. Weed designation: USA: SC.

Centaurea benedicta

Centaurea calcitrapa L.—purple star thistle, caltrops

Native to southern Europe and North Africa; annual, biennial, or short-lived perennial; stems 20–100 cm, branched and often forming mounds; leaves alternate, 5–20 cm long, pinnately divided one to three times, hairy and dotted with glands; upper leaves linear to oblong, smooth to shallowly lobed; heads borne in leafy corymbs, 15–20 mm long and 6–8 mm wide; 25–40 disc florets, purple, 15–24 mm long; bracts borne in overlapping series; outer bracts greenish-brown, spiny-fringed at base, tipped by a spine 10–25 mm long; inner bracts spineless; fruit an achene, 2.5–3.4 mm long, pappus absent. Unpalatable to livestock and increasing on rangeland; forming dense stands due to spiny nature of mature heads. Weed designation: USA: AZ, CA, NV, NM, OR, WA.

Centaurea calcitrapa

Centaurea cyanus L.—bachelor's button, cornflower

Native to southern Europe; annual; stems 20–100 cm tall, loosely woolly haired; leaves alternate, linear to lance-shaped, 3–10 cm long, reduced in size upward, margins smooth; heads borne in open flat-topped clusters, bell-shaped, 10–16 mm long; 25–35 disc florets, blue, occasionally white to purple, 20–25 mm long; bracts green with white to dark brown or black margins, fringed with small teeth about 1 mm

Centaurea cyanus

long; fruit an achene, straw-colored to blue, 4–5 mm long, pappus of numerous stiff bristles. Weed designation: none.

Centaurea depressa Bieb.—low cornflower
Native to southwestern and central Asia; annual; stems 20–60 cm tall, branched at the base, often covered in gray woolly hairs; basal and lower leaves oblong, 5–10 cm long, stalked, margins smooth to pinnately lobed; mid- to upper leaves linear to lance-shaped or oblong, stalked, margins smooth; heads oval to bell-shaped, 15–20 mm long, borne singly on long stalks; 25–35 disc florets, dark blue to purple, 15–30 mm long; bracts with silvery white to brown margins fringed with slender teeth, 1.5–2 mm long; fruit an achene, brown, 4.5–6 mm long, pappus of bristles 2–8 mm long and scales about 1.5 mm long. Weed designation: none.

Centaurea diluta Ait.—North African knapweed
Native to southwestern Europe and North Africa; annual or perennial; stems 10–200 cm tall, branched near the top; basal and lower leaves 10–15 cm long, stalked, margins pinnately lobed; midleaves oval to oblong, 2–8 cm long, short-stalked to stalkless, margins smooth to pinnately lobed; upper leaves oblong, margins smooth to lobed; heads borne in open cymes; heads oval, 8–15 mm across; disc florets numerous, pink to purple, 20–30 mm long; outer bracts with brown papery margins fringed by spine-tipped teeth, the spines 1–5 mm long; inner bracts spineless; fruit an achene, brown, 3–3.5 mm long, pappus of numerous white bristles, 3–5 mm long. Weed designation: none.

Centaurea iberica Trev. ex Sprengel—Iberian star thistle, knapweed
Native to southeastern Europe and central Asia; annual, biennial, or short-lived perennial; stems 20–200 cm tall, profusely branched; basal and lower leaves 10–20 cm long, pinnately lobed or dissected one to two times, stalked, covered in woolly hairs and glandular dots; midleaves stalkless, lance-

shaped; upper leaves linear to oblong, margins smooth to toothed or shallowly lobed; heads borne singly or few in leafy cymes; heads, oval 10–18 mm long; disc florets numerous, white, pink or pale purple, 15–20 mm long; outer bracts spiny-fringed at base and tipped by a spine 0.5–3 cm long; inner bracts spineless; fruit an achene, white or brown, 3–4 mm long, pappus of white bristles, 1–3 mm long. Weed designation: USA: WA.

Centaurea jacea L.—brown knapweed
Native to Europe and Asia; perennial; stems 30–150 cm tall, soft to bristly haired; leaves alternate, oblong to lance-shaped or elliptical, 5–25 cm long; margins smooth to toothed or irregularly lobed; heads borne in leafy corymbs, 15–18 mm long; 40–100 disc florets, purple, 15–18 mm long; bracts borne in several overlapping series; outer bracts light brown, margins pale and membranous, smooth to coarsely toothed; inner bracts irregularly toothed or lobed; fruit an achene, 2.5–3 mm long, pappus absent. Weed designation: USA: WA; CAN: AB, ON.

Centaurea jacea

Centaurea macrocephala Pusch. ex Willd.—globe centaurea, bighead knapweed, yellow bachelor's button, cornflower
Native to eastern Europe and western Asia; perennial; stems 50–170 cm tall, often soft woolly haired; basal and lower leaves oblong to lance-shaped, 10–30 cm long, covered in soft hairs and glandular dots, margins smooth to toothed; stem leaves lance-shaped to oval, 5–10 cm long, stalkless, margins smooth to wavy; heads borne single on short stalks, often close to the upper leaves; heads oval, 25–35 mm long; disc florets numerous, yellow, 3.5–4 mm long; bracts pale green to brownish with brownish papery margins, abruptly enlarged, 1–2 cm wide, the margins fringed and tipped by a spine 1–2 mm long; fruit an achene, 7–8 mm long, pappus of numerous flat bristles, 5–8 mm long. Weed designation: USA: WA.

Centaurea melitensis

Centaurea melitensis L.—tocalote, Maltese star
thistle, Maltese centaury, Napa thistle
Native to Mediterranean region of Europe, Asia, and
North Africa; annual; stems 10–100 cm tall, often
covered in soft gray woolly hairs and glandular dots;
basal and lower leaves absent at flowering; leaves
alternate, oblong to lance-shaped, 2–15 cm long,
margins smooth to toothed or pinnately lobed;
upper stem leaves linear to oblong, 1–5 cm long,
margins smooth to toothed; heads borne in leafy
corymbs, 10–15 mm long; disc florets numerous,
yellow, 10–12 mm long; outer bracts purplish- to
greenish-brown, spiny-fringed at base, and tipped
by a spine 5–10 mm long; inner bracts smooth
margined to spine-tipped; fruit an achene, dull
white to brown, 2–3 mm long, pappus of numerous
white stiff bristles. Weed designation: USA: NM, NV.

Centaurea montana L.—mountain cornflower, bluet
Native to Europe; perennial spreading by rhizomes
and stolons; stems 25–80 cm tall, often branched,
covered in thin cobwebby or woolly hairs; basal
and lower leaves 10–30 mm long, stalks winged,
margins smooth to toothed or pinnately lobed; mid-
to upper leaves oval to oblong, stalkless, margins
smooth to slightly toothed; heads borne singly or
a few in corymbs on stalks up to 7 cm long; heads
oval to bell-shaped, 20–25 mm long; 35–60 disc
florets, blue, white, pink, or purple, 2–4.5 cm long;
bracts brown to black with fringed margins; fruit an
achene, brown, 5–6 mm long, pappus of bristles less
than 1.5 mm long. Weed designation: none.

Centaurea nigra L.—black knapweed, lesser knapweed

Native to Europe; perennial; stems 30–150 cm tall, branched near tips, hairy to bristly; leaves alternate, oblong to lance-shaped or elliptical, 5–25 cm long, margins smooth to toothed or pinnately lobed; upper leaves linear to lance-shaped with entire to toothed margins; heads borne in leafy corymbs, oval to bell-shaped, 15–18 mm long and wide; 40–100 disc florets, purple, 15–18 mm long; outer bracts with dark brown to black comblike fringed margins; inner bract tips truncated, irregularly toothed or lobed; fruit an achene, tan, 2.5–3 mm long; pappus of numerous black bristles, often deciduous. Weed designation: USA: WA CAN: AB, ON.

Centaurea nigra

Centaurea nigrescens Willd.—Tyrol knapweed, short-fringed knapweed, Vochin knapweed

Native to Europe; perennial; stems 30–150 cm tall, hairless to woolly haired; basal and lower leaves oblong to lance-shaped, 5–25 cm long, stalked, margins smooth to toothed or pinnately lobed; upper leaves linear to lance-shaped, stalkless, margins smooth to toothed; heads borne in few-headed corymbs on leafy stems; heads oval to bell-shaped, 15–18 mm long; 40–100 disc florets, purple (rarely white), 15–18 mm long; outer bracts dark brown to black, the margins comblike with six to eight pairs of wiry lobes; inner bract tips truncated, irregularly toothed or lobed; fruit an achene, brown, 2.5–3 mm long, pappus absent, or if present, easily detached. Weed designation: USA: CO, ID, OR, WA CAN: AB.

Centaurea nigrescens

Centaurea phrygia L.—wig knapweed

Native to eastern Europe; perennial; stems 15–80 cm tall, simple to branched; basal and lower leaves oval to lance-shaped, 3–15 cm long, covered in cobwebby to woolly hairs, stalked, margins smooth to toothed; upper leaves stalkless, sometimes clasping; heads borne singly on stalks; heads, oval to spherical, 15–20 mm long; disc florets numerous, pink to purple, 20–25 mm long; outer bracts with brown or black margins and featherlike lobed tips; inner bracts with irregularly toothed or lobed tips; fruit an achene, brown, 3–4 mm long, pappus absent or composed of numerous bristles, less than 2 mm long. Weed designation: none.

Centaurea scabiosa L.— greater knapweed, hardheads

Native to eastern Europe; perennial; stems 30–150 cm tall, hairless to bristly haired; basal and lower leaves 10–25 cm long, bristly and glandular-dotted, margins divided one to two times into linear or oblong segments; heads borne singly or few in open cymelike clusters; heads oval to bell-shaped, 15–25 mm long; disc florets numerous, reddish-purple to white, 20–40 mm long; outer bracts covered in cobwebby hairs, margins black with slender teeth at tip; inner bracts with brown papery margins; fruit an achene, brown, 4.5–5 mm long, pappus of stiff white bristles, 4–5 mm long. Weed designation: none.

Centaurea scabiosa

Centaurea sulphurea Willd.—Sicilian star thistle, sulfur-colored Sicilian thistle, sulfur knapweed

Native to southwestern Europe; annual; stems 10–100 cm tall, often branched, soft to stiff-haired; basal and lower leaves oblong to lance-shaped, 10–15 cm long, soft to stiff-haired and glandular-dotted, margins pinnately lobed; stem leaves alternate linear to lance-shaped, 1–6 cm long, stalkless, margins smooth to spine-tipped toothed; heads borne singly or few in open corymb; heads oval, 12–30 mm long; disc florets numerous, yellow, 25–35 mm long; outer bracts with brown to purplish-black appendages, each with a palmately radiating cluster of spines, central spine 1–2.5 cm long, dark brown to black; inner bracts spineless to spine-tipped; fruit an achene, dark brown, 5–8 mm long, pappus of brown to black bristles, 6–7 mm long. Weed designation: USA: AZ, CA.

Centaurea virgata Lam. ssp. *squarrosa* (Boissier) Gugler—squarrose knapweed

Native to western Asia; perennial; stems 20–50 cm tall with a sandpapery texture, branched; basal and lower leaves 10–15 cm long, covered in woolly hairs and glandular dots, stalked, margins pinnately divided one to two times into narrow segments, often absent at flowering; mid- to upper leaves stalkless, simple to pinnately divided; heads borne in panicle-like clusters; heads oval to cylindrical, 7–8 mm long, 3–5 mm across, falling off at maturity; 10–16 disc florets, pink to pale purple, 7–9 mm long; outer bracts green, often purple-tinged, margins papery and fringed with spines, terminal spine 1–3 mm long; inner bracts spineless; fruit an achene, brown, 2.5–3.5 mm long, pappus of white bristles, 1–2.5 mm long. Weed designation: USA: AZ, CA, NV, UT; CAN: AB.

Centaurea diffusa Lam.
DIFFUSE KNAPWEED

Also known as: spreading knapweed, white knapweed

Scientific Synonym(s): *Acosta diffusa* (Lam.) Soják

QUICK ID: floral bracts spine-tipped; flowers white to pinkish-white; leaves divided once into narrow segments

Origin: native to eastern Europe and western Asia; first North American reports were from Washington in 1907 and British Columbia in 1936

Life Cycle: Annual, biennial, or short-lived perennial reproducing by seed only; plants produce up to 18,000 seeds.

Weed Designation: USA: AZ, CA, CO, ID, MT, NE, NM, ND, NV, OR, SD, UT, WA, WY; CAN: federal, AB, BC, MB, ON, SK; MEX: federal

DESCRIPTION

Seed: Oblong, brown to black, 2–3 mm long; surface with light-colored lines; pappus less than 1 mm long or absent.
Viability about 20 months in dry storage. Germination is enhance by the presence of light; temperature range 7°–34°C (optimal 10°–28°C; emergence from the top 3 cm of soil.

Seedling: With oval cotyledons; first leaves lance-shaped, margins smooth, surface hairless; later leaves deeply lobed and covered in short stiff hairs.

Leaves: Alternate, 10–20 cm long, 1–5 cm wide; lower leaves divided one to two times into narrow segments; upper leaves undivided; reduced in size upward; surface with a few short scattered hairs; basal rosette leaves may be present in young flowering plants but often absent.

Flower: Heads numerous, 3.5–5 mm across, 10–15 mm high, borne singly at the ends of branches; 25–35 disc florets, white to pinkish or red, 12–13 mm long; bracts numerous, borne in six to eight series; outer bracts 10–13 mm long, greenish-yellow, spine-tipped with hairy margins; spine about 1.5–4 mm long.

Plant: Preferring semiarid climates, rarely producing more than one stem; mature plants forming a tumbleweed; stems 20–150 cm tall, dull grayish-green and rough-haired.

Fruit: Achene.

REASONS FOR CONCERN

The spine-tipped floral bracts may cause damage to noses and mouths of livestock who graze on these plants. Diffuse knapweed invades grassland and pastures. Research has shown that the roots produce a chemical that prevents other species from growing in the area. This trait has enabled diffuse knapweed to infest over 75,000 acres in British Columbia and over 3.3 million acres in Idaho, Oregon, and Washington.

Similar species: squarrose knapweed (*C. triumfetti* All.), a native of Europe, may be mistaken for diffuse knapweed. It is distinguished by heads with 10–14 pink to pale purple disc florets, each 7–9 mm long, and bract margins with straw-colored spines, the terminal spine 1–3 mm long. Weed designation: USA: CO, OR.

Centaurea solstitialis L.
YELLOW STAR THISTLE

Also known as: Barnaby's thistle, yellow cockspur

Scientific Synonym(s): *Leucantha solstitialis* (L.) A. & D. Love

QUICK ID: floral bracts with stiff yellow spines; flowers yellow; leaves deeply lobed

Origin: native to southern Europe and western Asia; probably introduced as a contaminant in alfalfa seed; reported from California in 1852

Life Cycle: Annual, biennial, or perennial reproducing by seed only; large plants producing up to 150,000 seeds.

Weed Designation: USA: AZ, CA, CO, ID, MT, ND, NM, NV, OR, SD, UT, WA; CAN: federal, AB, BC, ON, SK

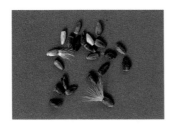

DESCRIPTION

Seed: Mottled light brown to dark brown; pappus bristly, 3–5 mm long, persistent or deciduous. Viable for 10 years in soil.
Germination occurs on the soil surface and initiated by autumn rains; light enhances germination; temperature range is 6°–30°C (optimal 10°–20°C); some seeds viable after 8 days after the flowers appear.

Seedling: With oblong to spatula-shaped cotyledons, 6–9 mm long and 3–5 mm wide; first leaves oblong to lance-shaped with smooth margins, later leaves pinnately lobed; rosette with six to 15 deeply lobed leaves, 4–20 cm long; surface covered with fine cottony hairs; sensitive to competition for light, nutrients, water, and space.

Leaves: Alternate, covered in gray woolly hairs; upper leaves linear to lance-shaped, 10–25 mm long, sharp pointed; lower leaves deeply lobed, 4–20 cm long and 1–5 cm wide; leaf bases extend down the stem; basal leaves deeply lobed 5–20 cm long.

Flower: Heads terminal and solitary, 2–3 cm across; disc florets numerous, yellow, 13–20 mm long; bracts 1–2.5 cm long, numerous, borne in six to many rows; outer and middle bracts spine-tipped; spines stiff, yellow, 11–22 mm long; plants producing up to 1,000 heads, with an average of 30–80 seeds per head.

Plant: Plants 10–100 cm tall; stems grayish-green, branched from the base, ridged or winged, covered in cottony gray hairs; roots to 1 m deep.

Fruit: Achene; outer disc florets producing dark-colored achenes without pappus; inner florets producing light-colored achenes with pappus 1.5–3 mm long.

REASONS FOR CONCERN

Yellow star thistle is toxic to horses and causes a disorder called chewing disease. Horses must consume 50–150% of an animal's weight in dry-weight plant material over a period of 1–3 months to produce symptoms. Toxic effects are cumulative and symptoms occur rapidly once the threshold has been reached. Symptoms include fatigue, lowered head, uncontrolled twitching of the lower lip, tongue flicking, involuntary chewing movements, and an unnatural open position of the mouth. Without intervention, affected horses are unable to eat or drink and eventually die from starvation or dehydration. The disease can occur when horses are allowed to graze infested pastures, where there are inadequate amounts of green forage, or fed contaminated hay over a period of time.

Similar species: Maltese star thistle (*C. meliten-sis* L., p. 80), an introduced species from Malta, has purplish floral bracts with spines 5–10 mm long and yellow florets 10–12 mm long. Another Mediterranean species, caltrops (*C. calcitrapa* L., p. 77) is distinguished by heads with 25–40 purple disc florets 15–24 mm long and bracts with yellowish spines, 1–2.5 cm long.

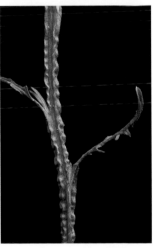

Centaurea stoebe L. ssp. micranthos (Gugler) Hayek
SPOTTED KNAPWEED

Also known as: knapweed
Scientific Synonym(s):
  C. maculosa Lam.,
  C. biebersteinii DC., *Acosta maculosa* auct. non Holub

QUICK ID: floral bracts black-tipped, not spiny; flowers purple, rarely white; leaves covered in translucent dots
Origin: native to Europe; introduced into Victoria, British Columbia in 1893
Life Cycle: Biennial or perennial (up to 9 years) reproducing by seed only; plants may produce up to 25,000 seeds.
Weed Designation: USA: AZ, CA, CO, CT, ID, MA, MT, ND, NE, NM, NV, OR, SD, UT, WA, WY; CAN: federal, AB, BC, MB, ON, SK; MEX: federal

DESCRIPTION

Seed: Oblong, grayish-brown, 2.8–3.3 mm long and 1.2–1.6 mm wide; surface with 10–15 faint ridges and several white hairs about 0.2 mm long; pappus of numerous white feathery or comblike bristles, each up to 2 mm long.
Viable at least 7 years in soil.
Germination occurs between 7°C and 34°C (optimal 10°–28°C); emergence occurs from top 5 cm of soil, best germination occurs on the soil surface; moisture is the limiting factor in germination.
Seedling: With oval cotyledons; first leaves lance-shaped, undivided, and hairless; later leaves deeply lobed.
Leaves: Alternate, 5–15 cm long, somewhat hairy, and covered in translucent dots; lower leaves divided into narrow segments, the upper undivided; basal leaves, when present, long-stalked; basal rosette leaves compound with several irregularly lobed segments.
Flower: Heads 1.5–2.5 cm across, borne at the ends of branches and in leaf axils; disc florets purplish, outer ring of florets sterile, 15–25 mm long, inner florets fertile, 12–15 mm long; bracts 1–1.4 cm long, margins comblike with five to seven pairs of black hairs, each 1–2 mm long.
Plant: Up to 180 cm tall, may remain in rosette stage up to 4 years; biennial plants produce one to six stems, while perennials may produce 15 or more stems; stems green, often purple-striped, branched, surface with a slight sandpapery texture.
Fruit: Achene.

REASONS FOR CONCERN

Spotted knapweed is not a serious concern in cropland as cultivation provides good control. In pastures, spotted knapweed's early spring growth allows it to outcompete other plants for moisture and nutrients. It also produces a chemical that does not allow other species to grow in the immediate area. This has enabled spotted knapweed to infest over 50,000 acres of rangeland in British Columbia and over 7.5 million acres in Idaho, Montana, Oregon, and Washington.

Similar species: black or lesser knapweed (*C. nigra* L., p. 81), a native of Europe, is distinguished by dark brown to black bracts with margins divided into numerous wiry lobes. Heads have 40–100 purple disc florets, each 15–18 mm long. Another species, meadow knapweed (*C. x moncktonii* E. C. Britton), believed to be a hybrid between black knapweed and brown knapweed (*C. jacea* L., p. 79), has tan to dark brown bracts with comblike to fringed margins.

Chondrilla juncea L.
RUSH SKELETONWEED

Also known as:
    skeletonweed, gum
    succory, hogbite, naked
    weed, devil's grass
Scientific Synonym(s): none

QUICK ID: plants
    appearing leafless;
    flowers yellow; basal
    leaves resembling those
    of dandelion

Origin: Native to
    southwestern Russia;
    introduced into eastern
    United States in 1870 and
    Spokane, Washington, in
    1938. Today over 2 million
    hectares are infested in
    western United States.

Life Cycle: Biennial or
    perennial (3–4 years
    in native range, up to
    20 years in Australia)
    reproducing by spreading
    roots and seed; large
    plants producing up
    to 20,000 seeds; may
    produce 250–300 seeds in
    its first growing season.

Weed Designation: USA: AZ,
    CA, CO, ID, MT, NV, OR, SD,
    WA; CAN: AB, BC

DESCRIPTION

Seed: Pale brown to black, cylindrical, 3–5 mm long,
prominently five-ribbed, beaked; pappus white,
3–5 mm long, deciduous.
Viable in soil for less than 5 years; 90% seeds
germinate the first year, 2% by the third year;
seeds lack dormancy and can germinate within
24 hours of being shed.
Germination on soil surface is rare; optimal
depth at 1 cm depth, decreasing to 3% at 3.5
cm, with no emergence from 5 cm; emergence
reduced in water-saturated or heavy clay
soils and during drought conditions; optimal
germination temperature between 10°C and
40°C; germination in fall in mild winter areas or
early spring in colder climates.

Seedling: With oval to spatula-shaped cotyledons;
remain at cotyledon stage for several weeks
while roots develop and grow at 1–2 cm per day;
roots up to 18 cm long after 1 month; first leaves
elliptical with backward pointing teeth.

Leaves: Alternate, lower leaves 2–10 cm long and
1–8 mm wide, reduced to narrow bracts near
the top, may appear leafless when in flower
(skeleton-like); basal leaves 5–13 cm long and
1.5–3.5 cm wide, resembling dandelion with
irregular backward-pointing teeth; margins often
purple-tinged, irregularly lobed; withering as the
flowering stem emerges.

Flower: Heads borne in terminal groups of two to
five, each 1.5–2.5 cm across; ray florets yellow,
seven to 15 per head (usually 11), 12–18 mm
long; five to nine bracts, borne in two unequal
rows; outer bracts much smaller than the inner,
9–12 mm long, covered in white woolly hairs;
temperature of at least 15°C necessary to initiate
flowering; flowers open 1 day only; flowering

continues until plant killed by frost; seeds produced 9–15 days after flower opening.

Plant: Plant 30–130 cm tall, rushlike, preferring well-drained soils, cannot withstand temperatures colder than −20°C, requires 25–100 cm of annual precipitation; stems stiff and wiry, nearly leafless, branched, bristly reddish-brown downward-pointing hair at base, hairless above, exuding milky juice when broken; taproot thick, penetrating to 3 m depth, lateral branches less than 8 cm long, short-lived; new plants rising from lateral buds in top 60 cm of soil; new plants produced from root fragments as small as 1–2 cm long.

Fruit: Achene; maturing 10–20 days after flowering.

REASONS FOR CONCERN

Rush skeletonweed has an extensive, deep root system that makes it difficult to control. It is the most serious weed in the Australian wheat belt, where it has been reported to reduce wheat yields by 80%. The wiry stems also interfere with harvesting by becoming entangled in machinery.

Cichorium intybus L.
CHICORY

Also known as: succory, blue sailors, blue daisy, coffeeweed, bunk

Scientific Synonym(s): none

QUICK ID: flowers blue, borne in headlike clusters; plants with milky juice; 13–15 floral bracts

Origin: native to the Mediterranean region of Europe; introduced as a food plant (leaves for salad greens, and roots roasted as a coffee additive)

Life Cycle: Perennial reproducing by seed only; often blooming in the first season of growth.

Weed Designation: USA: CO; CAN: federal

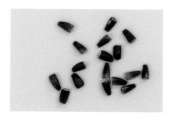

DESCRIPTION

Seed: Triangular, grayish-brown, 1.9–2.8 mm long, 0.7–1.4 mm thick; surface slightly shiny and smooth with about 10 lengthwise ridges; pappus with two rows of white bristle-like scales, each 0.2 mm long.
Viability at least 1 year in laboratory storage. Germination occurs in the top 2 cm of soil; optimal soil temperature range of 17–23°C.

Seedling: With oval to heart-shaped cotyledons, 6.5–12 mm long and 3–5 mm wide, upper surface granular; first leaf lance-shaped, bitter milky juice present; later leaves thin and pale green, clasping the stem.

Leaves: Alternate, reduced in size upward, rough-haired, upper leaves 3–7 cm long, stalkless and clasping the stem; basal leaves, 10–20 cm long and 2–12 cm wide; margins irregularly toothed and lobed; rosette resembling dandelion, except that leaves are rough and hairy.

Flower: Heads showy, 3–4 cm across, borne in dense axillary clusters of one to four, stalks 0–2 mm long, opening in morning and closing before noon; eight to 25 ray florets, blue or white; 10–15 bracts, whitish-green, borne in two overlapping rows; five outer bracts, 4–7 mm long, yellowish at base; eight to 10 inner bracts, 6–12 mm long, margins spiny.

Plant: Plants 20–150 cm tall; stems hollow, branched, often reddish-green, somewhat woody; taproot large and deep-penetrating, often grown as a coffee substitute.

Fruit: Achene.

REASONS FOR CONCERN

Chicory is readily eaten by livestock but gives a bitter taste to milk and butter. It is host for aster yellows, lettuce mosaic, and tomato-spotted wilt viruses.

Cirsium arvense (L.) Scop.
CANADA THISTLE

Also known as: creeping thistle, field thistle, cursed thistle, corn thistle, small-flowered thistle, green thistle

Scientific Synonym(s): *Serratula avensis* L., *C. setosum* (Willd.) Bess. ex Bieb., *C. incanum* (Gmel.) Fisch., *Carduus arvensis* (L.) Robson

QUICK ID: plants with prickly stems and leaves; rhizome extensive and creeping; flower heads purplish-pink, less than 2.5 cm across

Origin: native to Europe; introduced from France in the late 1600s; Vermont designated weed status in 1795

Life Cycle: Perennial reproducing by rhizomes and seed; large plants may produce over 40,000 seeds.

Weed Designation: USA: AK, AR, AZ, CA, CO, CT, DE, HI, IA, ID, IL, IN, KS, KY, MD, MI, MN, MO, MT, NC, ND, NE, NM, NV, OH, OK, OR, PA, SD, UT, WA, WI, WY; CAN: federal, AB, BC, MB, ON, PQ, SK

DESCRIPTION

Seed: Elliptical, light brown, 2–4 mm long, 1–1.2 mm wide; surface shiny and lined; pappus of numerous feathery bristles 13–32 mm long, often shed before seed reaches maturity.
Viable in soil up to 21 years.
Germination takes place in the top 1 cm of the soil; optimal temperatures 25°–30°C; combination of high light intensity and temperatures of 15°–20°C increases germination.

Seedling: With oval cotyledons, 6–14 mm long, 3–5.5 mm wide, stalkless, hairless; first leaves slightly hairy, margins irregularly lobed and spiny; by second leaf stage, roots may be up to 15 cm long; 34 days after germination (about the 3rd leaf stage) an extensive root system with numerous buds has developed; at 4 months the seedlings rhizome is over 100 cm long with an average of 19 buds, these buds produce new shoots in future growing seasons.

Leaves: Alternate, oblong, 5–30 cm long, 1–6 cm wide, stalkless, surface curled and wavy, margins prickly toothed and irregularly lobed; spines 1–7 mm long; underside often covered in soft woolly hairs.

Flower: Heads terminal or axillary; stalks 2–70 mm long; disc florets numerous, pinkish-purple or white; male heads oblong, about 12 mm across, disc florets 12–18 mm long; female heads flask-shaped, 10–25 mm across, producing about 45 seeds, disc florets 14–20 mm long; bracts borne in six to eight overlapping rows, spines about 1 mm long; seeds mature and capable of germination 8–11 days after flowering.

Plant: Male or female, 30–120 cm tall; stems leafy, hollow, becoming hairy with age; rhizome deep and creeping, approximately eight buds per meter, 1-year-old plants may have as many as 200 buds; single plant capable of producing over 6 m of rhizome per year; northern limits of range coincides with areas where the average January temperature is −18°C.

Fruit: Achene; male flowers do not produce fruit.

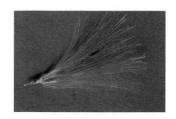

REASONS FOR CONCERN

Canada thistle is an aggressive weed and has the potential to reduce crop yields by 100%. The rhizomes are unaffected by cultivation as they grow below normal tillage depths. Canada thistle may harbor bean aphids and stalk borers of corn and tomatoes. It also is host for raspberry Scottish leaf curl and tobacco mosaic virus. Canada thistle is a weed of 27 crops in over 35 countries.

Cirsium vulgare (Savi) Ten.
BULL THISTLE

Also known as: spear thistle, plume thistle, bur thistle, roadside thistle, bank thistle, bird thistle, blue thistle, black thistle, button thistle, common thistle, Fuller's thistle

Scientific Synonym(s): *C. lanceolatum* (L.) Scop., non Hill, *Carduus lanceolatus* L., *Carduus vulgaris* Savi

QUICK ID: flowers purple; heads 3–5 cm across; stems spiny-winged

Origin: native to Europe, Asia, and North Africa

Life Cycle: Biennial reproducing by seed only; large plants producing up to 120,000 seeds; up to 400 heads per plant, and 100–300 seeds per head.

Weed Designation: USA: AR, CO, IA, MD, MN, NM, OR, PA, WA; CAN: MB, ON, PQ, SK

DESCRIPTION

Seed: Oblong to elliptical, orange-brown to gray-brown, 3.2–4 mm long, 1.2–1.3 mm wide; surface shiny with numerous lengthwise lines; pappus of white feathery bristles 1.4–3 cm long, deciduous and falling off as a unit.
Viable up to 5 years in soil.
Germination occurs in the fall or early spring; emergence from soil depths of 5 cm; occurs under a wide temperature range (5°–40°C), with or without light; can germinate under low moisture.

Seedling: With oval to oblong cotyledons, 5.5–20 mm long, 3–7.5 mm wide, upper surface granular, underside pale; first leaf lance-shaped, about two to four times longer than the cotyledons, margins irregularly toothed and ending in a weak prickle, upper surface covered in short white hairs; can tolerate temperatures of –2°C.

Leaves: Alternate, 7–15 cm long, densely spiny, underside covered in white woolly hairs, margins coarsely lobed, the lobes triangular with three or four long sharp spines 2–10 mm long and several short spines between; leaf bases extending down the stem, spiny; basal leaves elliptical to oblong or lance-shaped, 10–40 cm long, 6–15 cm wide, margins coarsely lobed and toothed; upper surface covered in woolly hairs, underside covered in cobweblike hairs; rosettes up to 65 cm across in the first growing season, susceptible to hard frosts but regrowing from the roots.

Flower: Heads 2.5–7.5 cm across, borne solitary or in groups of two or three; stalks 1–6 cm long; disc florets purple, 25–35 mm long; bracts linear to lance-shaped, 2.5–4 cm high, borne in 10–12 overlapping rows, spine-tipped with outward-pointing yellow spines, spines 2–5 mm long; woolly hairs scattered throughout.

Plant: Plants 30–300 cm tall, prickly; stems branched, surface with loose white cobweblike hairs; taproot up to 70 cm deep, often branched, fleshy; plants require a cold period to initiate flowering, those in nitrogen-poor soil may require more than 2 years to flower; plants in grazed pastures often produce more seed than plants in adjacent ungrazed areas.

Fruit: Achene.

REASONS FOR CONCERN

Bull thistle is a serious weed of waste places, fencerows, and cereal crops. It is reported as a weed of pastures in 19 countries and recognized as one of the most important weeds of pastures.

Similar species: European swamp thistle (*C. palustre* [L.] Scopoli), is a threat to northern forests and wetlands by forming impenetrable colonies that displace native vegetation. It is distinguished by spiny leaves 15–30 cm long and 3–10 cm wide, heads 1–1.5 cm across borne on stalks less than 1 cm long, purple disc florets 11–13 mm long, and bracts borne in five to seven rows. Weed designation: USA: AR, IA; CAN: AB, BC.

European swamp thistle
(*Cirsium palustre*)

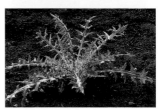

Conyza canadensis (L.) Cronq.
CANADA FLEABANE

Also known as: horseweed, fleabane, bitterweed, hogweed, mares tail, blood stanch, colt's tail, fireweed

Scientific Synonym(s): *Leptilon canadense* (L.) Britt., *Erigeron canadensis* L.

QUICK ID: flowers white to yellow, 3–5 mm across; stems with numerous branches; crushed leaves and stems carrot scented

Origin: native to North America; widespread throughout the world

Life Cycle: Annual, winter annual or biennial reproducing by seeds; plants producing up to 10,000 seeds.

Weed Designation: none

DESCRIPTION

Seed: Oblong, flattened, yellowish-brown, 1–1.2 mm long and about 0.3 mm wide; pappus bristles 15–25, grayish-white to tan, 2–3 mm long. Viability unknown.
Germination best at 20°–24°C; emergence from the top 5 mm of soil, no seedlings emerging from depth below.

Seedling: With oval cotyledons 2–3.5 mm long, 1–2 mm wide, hairless; first leaves spatula-shaped, upper surface and margins hairy.

Leaves: Alternate, oblong to lance-shaped, 2–10 cm long and 4–10 mm wide, reduced in size upward, lower leaves short-stalked, often bristly haired, middle and upper leaves 2–8 cm long and 1 cm wide, stalkless, margins smooth.

Flower: Heads numerous, 3–5 mm across, borne in crowded branched terminal clusters; ray florets white, short, and often hidden by the bracts; disc florets yellow, three to 30 per head; 20–40 bracts, borne in two to four overlapping rows, 2–4 mm long, tips pale green, hairless.

Plant: Plants 7.5–180 cm tall; stems profusely branched and leafy, covered in short bristly hairs; crushed leaves and stems reported to smell like carrots; roots fibrous.

Fruit: Achene.

REASONS FOR CONCERN

Canada fleabane is a serious weed in cultivated crops, pastures, and meadows. The leaves and flowers contain herpene, a chemical that irritates the nostrils of horses. The leaves of Canada fleabane may cause skin irritations to some people. It is also host for aster yellows, tobacco mosaic, and tobacco ring spot viruses.

Similar species: many-flowered horseweed (*C. floribunda* Kunth.), a native of South America, may be confused with Canada fleabane. It is distinguished by reddish-brown outer bracts with long hairs, achenes with red nerves, heads with 10–20 florets, and leaves 5–10 cm long and 5–15 mm wide.

Crepis tectorum L.
NARROW-LEAVED HAWK'S BEARD

Also known as: annual hawk's beard, yellow hawk's beard

Scientific Synonym(s): *Hieracium tectorum* Karsch

QUICK ID: milky juice; flowers yellow, 10–15 mm across; leaves with backward-pointing lobes

Origin: introduced from Siberia into North America prior to 1890; first Canadian report from New Brunswick in 1877

Life Cycle: Annual or winter annual reproducing by seed; large plants producing up to 49,000 seeds.

Weed Designation: CAN: MB, SK

DESCRIPTION

Seed: Dark purplish-brown, spindle-shaped, 3–4 mm long, prominently ten-ribbed; pappus of numerous white bristles 4–5 mm long. Viability less than 2.5 years in soil. Germination occurs between 10°C and 35°C (optimal 20°–25°C)

Seedling: With oval cotyledons 10–12 mm long and 2–3 mm wide, hairless; first leaves spatula-shaped, margins with a few downward-pointing teeth.

Leaves: Alternate, 1–4 cm long, 5–15 mm wide, stalkless and clasping the stem; margins often rolled under; basal leaves lance-shaped, 10–15 cm long and 4 cm wide, stalked, margins vary from numerous backward-pointing teeth to deeply lobed segments; rosette often absent at flowering time.

Flower: Heads 10–15 mm across, borne on branches often in groups of five, with five to 100 heads per stem; 30–70 ray florets, yellow, 10–13 mm long; five to 18 bracts, 6–9 mm long, borne on one or two rows; 12–15 inner bracts, 5–9 mm long, hairs bristly and black.

Plant: Plants 20–100 cm tall; stems hairless, leafy, exuding milky sap when broken; taproot branched, extending deep into the soil.

Fruit: Achene.

REASONS FOR CONCERN

Narrow-leaved hawk's beard is primarily a weed of forage crops, pastures, roadsides, and waste areas. Occasionally it is a serious weed in fall-sown crops where the winter annual phase is not controlled by spring cultivation.

Similar species: smooth hawkweed (*C. capillaris* [L.] Wallr.), native to Europe, may be confused with narrow-leaved hawk's beard. It is distinguished by rosette leaves 5–30 cm long and 1–4.5 cm wide, 10–30 flower heads per stem, and eight to 16 bracts, each 6–7 mm long.

Eclipta prostrata (L.) L.
ECLIPTA

Also known as: false daisy, swamp daisy, white eclipta

Scientific Synonym(s): *E. alba* (L.) Hassk., *E. erecta* L., *E. punctata* Linnaeus, *Verbesina prostrata* L., *V. alba* L.

QUICK ID: leaves opposite; ray and disc florets white and borne in heads; 10–12 involucral bracts

Origin: native to tropical areas worldwide

Life Cycle: Annual reproducing by seeds only; large plants producing up to 17,000 seeds.

Weed Designation: none

DESCRIPTION

Seed: Brown, oblong to oval, 1.8–3 mm long, three- or four-sided; pappus absent or consisting of a minute crown.
Viability at least 5 months in laboratory storage. Germination requires light; greatest emergence in fields with high moisture content; occurs with temperatures between 10°C and 35°C; no dormant period required.

Seedling: With elliptical to spatula-shaped cotyledons, less than 1 cm long and less than 0.5 mm wide, slightly thickened, hairless, prominently veined on the lower surface; stem below cotyledon light green to light purple; first leaves opposite, oval to oblong, margin with short teeth.

Leaves: Opposite, elliptical to linear or lance-shaped, 2–13 cm long and 0.5–3 cm wide, tip and base tapered, stalks short or absent, margins with or without a few small teeth, both surfaces with short appressed hairs.

Flower: Heads 4–6 mm across, borne singly or in groups of two or three at the end of branches, stalks 5–40 mm long; 20–40 ray florets, white, less than 2 mm long, female only, producing three-sided achenes; 15–30 disc florets, greenish-white, perfect, producing four-sided achenes; eight to 12 bracts, green, narrow, hairy, borne in two or three overlapping rows; outer bracts longer than the inner.

Plant: Prostrate to erect or spreading, capable of surviving saline conditions; stems 20–100 cm long, thick and succulent, branched, green to reddish-brown or purple, often rooting at lower nodes; taproot shallow with fibrous roots.

Fruit: Achene; borne in green buttonlike heads.

REASONS FOR CONCERN

Eclipta is a weed in low-lying fields, riverbanks, irrigation canals, and ditches. It is a common pest of rice paddies, sugarcane, corn, and taro crops. Research has shown that it can reduce peanut crop yields by 75%. It is reported to be a serious weed of 17 crops in 35 countries.

Also known as: lilac tasselflower; red tasselflower, red groundsel; purple sow thistle

Scientific Synonym(s): *E. javanica* (Burm. f.) C. B. Robins., *Cacalia sonchifolia* L.

QUICK ID: plant with sow thistle–like leaves; flowering stems branched; flowers not yellow

Origin: native to Europe, western Asia, and North Africa

Life Cycle: Annual reproducing by seed.

Weed Designation: none

DESCRIPTION

Seed prism-shaped, reddish-brown to whitish-tan, 2.4–3 mm long, distinctly 10-ribbed (five hairy alternating with five smooth); pappus feathery, white, 6–8 mm long.
Viability unknown.
Germination optimal at 25°–30°C; surface germination best, emergence from 5–10 mm possible; no emergence from 4 cm depth.

Seedling: Seedling with oval to lance-shaped cotyledons; first leaf oval to round, margin with blunt teeth; surface, margin, and stalk densely haired.

Leaves: Alternate, oval to triangular, resembling those of sow thistle (*Sonchus* ssp., p. 128); upper leaves not lobed but coarsely toothed, 5–12 cm long, 1.5–4.5 cm wide, stalkless and somewhat clasping; lower leaves 1.5–16 cm long and 1–8 cm wide, pinnately lobed, stalks with narrow wings.

Flower: Heads urn-shaped, 12–14 mm long and 4–5 mm wide, long-stalked, borne on branches with three to six heads; 15–40 disc florets, purple, scarlet, red, pink, white, orange, or lilac, 30–60 per head; outer disc florets female, inner florets perfect; seven to 10 bracts (commonly eight), green, 8–12 mm long, borne in one or two rows.

Plant: Plants 20–90 cm tall, leafiest near the base; stems erect, branched at base, hairless; taproot branched.

Fruit: Achene; outer florets producing reddish-brown achenes while inner florets produce whitish-tan.

REASONS FOR CONCERN

Emilia is a serious weed of asparagus, nurseries, bananas, and horticultural crops. It is also a host for yellow-spot virus of pineapple, which is transmitted by onion thrips.

Similar species: Florida tasselflower (*E. fosbergii* Nicolson), native to the tropics of Asia, may be mistaken for emilia. It is distinguished by leaves distributed equally along the stem, upper leaves with smooth margins, and flower heads with 50–60 pink, purple, or red florets.

Galinsoga quadriradiata Cav.
HAIRY GALINSOGA

Also known as: Frenchweed, quickweed, ciliate galinsoga, shaggy soldier, fringed quickweed

Scientific Synonym(s):
G. ciliata (Raf.) Blake, *Adventina ciliata* Raf., *G. aristulata* Bickn., *G. caracasana* (DC.) Schultz-Bip., *Vargasia caracassana* A. P. Candolle, *G. bicolorata* St. John & White

QUICK ID: leaves opposite; flowers pale pink to white, 5–8 mm across; heads with three to eight white ray florets

Origin: native to Peru or Mexico; first reported from Philadelphia in 1836 and Montreal in 1893

Life Cycle: Annual reproducing by seed; plants capable of producing up to 30,000 seeds.

Weed Designation: CAN: PQ

DESCRIPTION
Seed: Oblong (somewhat four-sided), dark brown to black, 1.5–2.5 mm long; pappus of 6–20 grayish-white scales, 0.5–1.7 mm long.
Viability at least 20 years in cultivated soil. Germination occurs in the top 2 cm of soil; germination increased when seeds are exposed to light; no dormant period is required for germination.

Seedling: With round to square cotyledons, 2.5–7 mm long and 2–4 mm wide, margins with short gland-tipped hairs; young root extensively branched; first pair of leaves thin, covered in short soft hairs, three-nerved, margins shallowly toothed; internodes often tinged with purple.

Leaves: Opposite, oval to triangular, 1–6 cm long and 15–45 mm wide, reduced in size upward, base wedge-shaped, stalks 2–15 mm long, surface with scattered hairs, prominently three-veined, margins toothed; upper leaves stalkless.

Flower: Heads 5–8 mm across, stalks 5–20 mm long; three to eight (commonly five) ray florets, white to pale pink, three-lobed, sterile; eight to 50 disc florets yellow, each 1.1–1.5 mm long, five-lobed; six to 16 bracts, borne in two overlapping rows, covered in sticky hairs; flowering stalks covered in red sticky hairs; flowers appear within 4 weeks of germination.

Plant: Plants 8–90 cm tall, highly susceptible to frost; stems erect, covered in gray woolly hairs, often rooting at nodes when in contact with soil; can complete life cycle within 50 days.

Fruit: Achene; ray seeds with six to 15 pappus scales; disc seeds with 14–20 pappus scales.

REASONS FOR CONCERN

Hairy galinsoga is host for cucumber mosaic, tobacco mosaic, tomato aspermy, tomato spotted wilt, curly top virus, and aster yellows virus. It is a common weed of vegetable crops in eastern North America.

Similar species: yellow weed or small-flowered galinsoga (*G. parviflora* Cav.) is distinguished by its hairless stem and leaves. The seeds of yellow weed have no pappus. Weed designation: USA: AK.

yellow weed
(*Galinsoga parviflora*)

Hieracium aurantiacum L.
ORANGE HAWKWEED

Also known as: devil's paintbrush, orange paintbrush, red daisy, missionary weed, king devil, fox and cubs

Scientific Synonym(s): *Pilosella aurantiaca* (L.) F. W. Schultz & Schultz-Bipontinus

QUICK ID: stems with milky juice; leaves basal; flowers orange

Origin: native to Europe; first observed in Vermont in 1875

Life Cycle: Perennial reproducing by stolons, rhizomes, and seed; large plants may produce up to 1,500 seeds.

Weed Designation: USA: CO, ID, MT, OR, WA; CAN: AB, BC

DESCRIPTION

Seed: Oblong, purplish-black, 1.2–2 mm long, surface with ridges running lengthwise; pappus of one row of 25–30 dirty white capillary bristles 3.5–4 mm long.
Viable in soil for up to 7 years.
Germination occurs in the spring and rosettes formed by autumn; seeds germinating in the autumn have a lower winter survival rate than those that germinate in spring.

Seedling: With oval cotyledons, 2–3.5 mm long and 1–2 mm wide, hairless, short-stalked; stem below cotyledons often reddish-purple; first three leaves spatula-shaped, upper surface with a few long hairs, lower surface hairless.

Leaves: Basal, three to eight (occasionally more), club-shaped, 5–15 cm long and 1–3.5 cm wide; surfaces covered with stiff hairs, margins smooth; one or two small leaves may be present on the flowering stem.

Flower: Heads 2–2.5 cm wide, borne in a compact terminal clusters of three to 30; 25–120 ray florets, burnt orange, 10–14 mm long; 13–30 bracts, borne in two or three overlapping rows, covered in black bristly hairs.

Plant: Plants 20–70 cm tall, covered in stiff black hairs; stems exuding milky juice when broken.

Fruit: Achene; each head producing 12–50 seeds.

REASONS FOR CONCERN

Orange hawkweed is a serious weed of lawns, pastures, and roadsides. The presence of stolons and rhizomes enables the plant to colonize large areas quickly, producing large mats of rosettes.

Similar species: king devil (*H. caespitosum* Dumort.) has stems up to 80 cm tall, are leafless, and are covered in black hairs. The two species are distinguished by the king devil's yellow flowers. The heads are about 1 cm across. Weed designation: USA: ID, MT, OR, WA; CAN: PQ.

king devil
(*Hieracium caespitosum*)

Hypochaeris radicata L.
SPOTTED CATSEAR

Also known as: long-rooted catsear, flatweed, coast dandelion, gosmore, hairy catsear, false dandelion

Scientific Synonym(s): *Hypochoeris radicata* L.

QUICK ID: leaves basal; stems with one to seven flower heads; milky juice present

Origin: native to Europe, Asia, and North Africa; first observed in British Columbia in 1884

Life Cycle: Perennial reproducing by seed; plants may produce up to 2,300 seeds.

Weed Designation: USA: WA

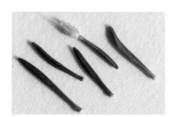

DESCRIPTION

Seed: Golden brown, 6–10 mm long, cylindrical, prominently 10–12-ribbed; pappus of 40–60 white bristles borne in two rows; outer bristles stiff while inner featherlike and 10–12 mm long. Viable in soil less than 3 years. Germination requires light; seeds have no dormancy.

Seedling: Seedling with oblong to lance-shaped cotyledons, 8–22 mm long and less than 1 cm wide; first leaves oblong, hairy.

Leaves: Basal, oblong to lance-shaped, 3–35 cm long and 5–70 mm wide, upper surface hairy; margins toothed or pinnatifid.

Flower: Heads 20–30 mm wide, borne in clusters of one to seven; 20–100 disc florets, yellow or grayish-green, 10–15 mm long; 20–30 bracts, borne in three or four rows, 3–20 mm long, margins papery; flowers may appear within 2 months after germination.

Plant: Plant 15–60 cm tall; one to 15 stems from the crown, branched above, base thickened and woody; roots fibrous and enlarged appearing taprooted.

Fruit: Achene.

REASONS FOR CONCERN

Spotted catsear is a troublesome weed of lawns. It is capable of a low growth habit, which escapes mower blades. It is also believed to have possible allelopathic effects on other plants.

Similar species: fall hawkbit (*Leontodon autumnalis* L., p. 150) native to Europe and Asia, is often confused with spotted catsear. It is distinguished by small scalelike leaves on the flowering stems, shallow to deeply lobed leaves, 20–30 bright yellow disc florets 13–16 mm long, and 18–20 bracts borne in two rows.

Jacobaea vulgaris Gaertn.
TANSY RAGWORT

Also known as: stinking willie, staggerwort, common ragwort

Scientific Synonym(s): *Senecio jacobaea* L.

QUICK ID: plants ill-scented; flower heads yellow, borne in flat-topped clusters

Origin: native to Europe; introduced into Oregon in 1922; first Canadian report from Nova Scotia in 1850

Life Cycle: Winter annual, biennial or short-lived perennial reproducing by fleshy roots and seed; large plants capable of producing over 150,000 seeds.

Weed Designation: USA: AZ, CA, CO, CT, ID, MA, MT, OR, WA; CAN: federal, AB, NS

DESCRIPTION

Seed: Oblong to cylindrical, light brown, 1.5–3 mm long; prominently ribbed; pappus of numerous soft white bristles, 3–6 mm long, deciduous; seeds produced by ray florets hairless, those produced by disc florets hairy, pappus persistent. Viable in soil for up to 20 years.
Germination occurs shortly after seeds are shed; optimal emergence occurs from the 1–2 cm soil depth; temperature range is 5°–30°C; seeds produced by ray florets disperse later and germinate slower than disc seeds; seeds subjected to frost, drought, and burial become dormant.

Seedling: With oval cotyledons, 2–4 mm long, surface granular; first leaves opposite, oval, 6–8 mm long, margins wavy, prominently veined; later leaves with smooth to deeply lobed margins.

Leaves: Alternate, 4–23 cm long, 2–11 cm wide, divided two or three times into narrow irregular lobes; margins wavy-toothed; underside with cobweblike hairs; lower leaves long-stalked, the upper stalkless; reduced in size upward.

Flower: Heads about 2.5 cm across, borne in flat-topped clusters of 20–60; 10–15 (usually 13) ray florets, yellow, 8–12 mm long and 2 mm wide, female; five to 80 disc florets, yellow, 7–10 mm long; 13 bracts, 3–15 mm long and 1 mm wide, black-tipped, borne in a single row; large plants may produce up to 2,500 heads, each with an average of 55 seeds.

Plant: Plant 30–120 cm tall, ill-scented; stems branched, often purple tinged, hairless at flowering time; taproot and crown producing spreading fleshy roots about 15 cm long, fragments producing new shoots; survives droughtlike conditions, and winter temperatures of −20°C with good snow cover.

Fruit: Achene.

REASONS FOR CONCERN

Tansy ragwort contains a toxic compound implicated in numerous livestock fatalities. Ingestion of the plant causes livestock to stagger, hence the common name staggerwort. The toxin, pyrrolizidine alkaloids, accumulate in the liver. A lethal dose is 3–7% of the animal's body weight. Leaves and flowers are the most toxic. Toxins are not destroyed in hay curing or silage production. Cattle and horses are affected, while sheep show no symptoms.

Lactuca serriola L.
PRICKLY LETTUCE

Also known as: common wild lettuce, compass plant, milk thistle, horse thistle, wild opium, China lettuce

Scientific Synonym(s):
L. scariola L.

QUICK ID: stems with milky juice; leaf underside with a prickly midrib; flowers yellow

Origin: introduced from Europe; introduced into North America about 1860

Life Cycle: Annual, winter annual, or biennial reproducing by seed only; large plants producing up to 87,000 seeds.

Weed Designation: CAN: MB, PQ, SK

DESCRIPTION

Seed: Lance-shaped, olive brown, 2.7–3 mm long and 0.7–0.9 mm wide, prominently ribbed on both sides; pappus of numerous white bristles 3–5 mm long.
Viable less than 3 years in soil.
Germination occurs over a wide temperature range, 10°–30°C; emergence from 2 cm possible, optimal is 1 mm depth.

Seedling: With oval cotyledons, 5–9 mm long and 2–3 mm wide, upper surface and margin hairy; first leaf oval, prominently veined, hairy.

Leaves: Alternate, oblong, 5–30 cm long and 2.5–10 cm wide, bluish-green, deeply lobed with backward pointing lobes, margins sharply toothed to prickly; stem leaves clasping; midvein on underside with a row of sharp yellowish spines; leaves often point north and south giving rise to the common name, compass plant.

Flower: Heads 3–10 mm across and 8–10 mm high, borne in a large pyramid-shaped panicle; five to 20 ray florets, yellow; five to 13 bracts, borne in three or four overlapping rows, 9–16 mm long, hairless, reflexed in fruit.

Plant: Plants 30–180 cm tall; stems pale green, hollow, woody and prickly at base; taproot white, large, up to 2 m deep.

Fruit: Achene.

REASONS FOR CONCERN

A serious weed of cropland, orchards, and gardens, prickly lettuce has been reported to reduce crop yields drastically. Reports indicate that cattle feeding on large amounts of prickly lettuce will develop pulmonary emphysema, a chronic respiratory condition. Prickly lettuce is also host for cucumber mosaic, tomato spotted wilt, aster yellows, lettuce mosaic, and tobacco ring spot viruses.

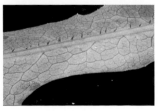

Leucanthemum vulgare Lam.

OXEYE DAISY

Also known as: white daisy, white weed, field daisy, midsummer daisy, marguerite, poorland flower, moonpenny, poverty weed, dog daisy

Scientific Synonym(s): *Chrysanthemum leucanthemum* L., *L. leucanthemum* (L.) Rydb.

QUICK ID: daisy-like flower with white ray florets and yellow disc florets; leaves with wavy to lobed margins; leaves with clasping bases

Origin: native to Europe; introduced into North America as a garden plant

Life Cycle: Perennial reproducing by seed and creeping rhizomes; large plants producing up to 26,000 seeds.

Weed Designation: USA: CO, MT, OH, WA, WY; CAN: federal, AB, SK

DESCRIPTION

Seed: Oval, dark brown to black, 1–2 mm long; surface prominently 10-ribbed; pappus absent. Viable in soil for 39 years.
Germination is inhibited by continuous darkness; no dormant period required.

Seedling: With oval cotyledons, 3–5.5 mm long and 1–2.5 mm wide, hairless; first leaves appearing opposite, spoon-shaped, margins wavy; later leaves with toothed margins.

Leaves: Alternate, glossy, 3–8 cm long, 2–15 mm wide, reduced in size upward, stalks short and clasping the stem, margins toothed to shallowly lobed, often lacerated at base; basal leaves spatula-shaped, 1.2–15 cm long and 8–20 mm wide, stalks 10–30 mm long, margins toothed to deeply lobed.

Flower: Heads borne singly at ends of stems, 2–6 cm across; 15–35 ray florets, white, 10–20 mm long, female; 120–200 disc florets, yellow, about 3 mm long, perfect; 35–60 bracts, green with brown margins, borne in three or four overlapping rows; both ray and disc florets producing seed.

Plant: Plants 10–100 cm tall; stems numerous from base, hairless; rhizomes creeping, shallow-rooted.

Fruit: Achene.

REASONS FOR CONCERN

Once established, oxeye daisy can quickly replace up to 50% of grass in a pasture. It is not a problem in cultivated crops as it has a shallow root system that is easily damaged by cultivation. Oxeye daisy is host for chrysanthemum stunt, aster yellows, and tomato aspermy viruses.

Similar species: although the flowers of scentless chamomile (*Tripleurospermum inodorum* [L.] Sch. Bip., p. 142) resemble those of oxeye daisy, the leaves are quite different. The leaves of scentless chamomile are feathery or fernlike, while those of oxeye daisy are broad with few divisions. The white ray florets of scentless chamomile are slightly shorter (8–16 mm long) than those of oxeye daisy. A widely grown ornamental plant, shasta daisy (*L. maximum* L.) is easily mistaken for oxeye daisy. The stem leaves of shasta daisy are 5–12 cm long and 8–22 mm wide with unlobed toothed margins. The flower heads have 21–34 ray florets, each 2–3 cm long.

Also known as: rayless
mayweed, rayless dog
fennel, disc mayweed

Scientific Synonym(s):
Chamomilla suaveolens
(Pursh) Rydb., *M. matri-
carioides* (Less.) Porter,
Tanacetum suaveolens
(Pursh) Hook., *Santolina
suaveolens* Pursh, *M. sua-
veolens* (Pursh) Buch.,
non L., *Lepidotheca
suaveolens* (Pursh) Nutt.,
Lepidanthus suaveolens
(Pursh) Nutt., *Artemisia
matricarioides* auct. non
Less

QUICK ID: plants with a
pineapple scent; flower
heads cone-shaped;
leaves fernlike

Origin: native to western
North America

Life Cycle: Annual
reproducing by seed only.

Weed Designation: CAN: PQ

DESCRIPTION

Seed: Egg-shaped, light brown, 1.3–1.6 mm long
and 0.5–0.7 mm wide, often five-angled, surface
glossy with a few dark orange to brown ribs;
pappus absent.
Viable up to 12 years in soil.
Germination occurs with soil temperatures
between 15°C and 25°C.

Seedling: With spatula-shaped cotyledons, 2–3
mm long and 1 mm wide, short-stalked; first
leaves appearing opposite, margin with a pair
of pointed lobes near the base; later leaves with
several pairs of lobes; leaves alternate after
fourth leaf stage.

Leaves: Alternate, 1–5 cm long, 2–20 mm wide,
divided several times into narrow segments;
segments, 1–2 mm wide, hairless; foliage with a
pineapple scent when crushed.

Flower: Heads cone-shaped, 5–10 mm across; stalks
2–25 mm long; 125–535 disc florets, yellowish-
green, about 1 mm across; 29–47 bracts,
yellowish-green, margins translucent and papery,
borne in three overlapping rows.

Plant: Plants 2–40 cm tall; stems branched and leafy,
hairless; stem and foliage with a pineapple scent
when crushed.

Fruit: Achene.

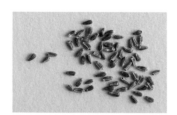

REASONS FOR CONCERN

Pineappleweed is often found growing on compacted soil in farmyards, waste areas, and roadsides. It is also host for raspberry Scottish leaf curl virus.

Onopordum acanthium L.
SCOTCH THISTLE

Also known as: cotton thistle, white thistle, woolly thistle, winged thistle, jackass thistle, heraldic thistle

Scientific Synonym(s): none

QUICK ID: plants grayish-white; leaves prickly; flower heads purplish-pink, globe-shaped

Origin: native to the eastern Mediterranean region of Europe; introduced as an ornamental garden plant

Life Cycle: Biennial (occasionally annual) reproducing by seed only; large plants may produce up to 40,000 seeds.

Weed Designation: USA: AR, AZ, CA, CO, CT, ID, MO, NV, NM, OK, OR, UT, WA, WY; CAN: BC, ON

DESCRIPTION

Seed: Egg-shaped, mottled dark brown, 4–5 mm long, 2–3 mm wide; surface with 15 grayish-brown wrinkles on each face, area between wrinkles yellowish-brown; pappus of numerous pale yellow to reddish barbed bristles, 7–9 mm long.

Viable in the soil for at least 7 years, possibly 20–30 years.

Germination is enhanced by presence of light, nitrogen, and fluctuating temperatures; exposure to 8 hours of daylight is optimal for germination; greatest germination occurs in fall coinciding with increased rainfall; emergence from soil depths of 4.5 cm (0.5 cm is optimal); 8–14% of seeds are nondormant, remaining seeds have various degrees of dormancy.

Seedling: With oval to oblong cotyledons, 11–21 mm long, 5–12 mm wide, somewhat fleshy; first leaves lance-shaped, margins with several weak prickles, leaf stalks with large hairs of various sizes.

Leaves: Alternate, oblong to rectangular, 6–60 cm long, 3–20 cm wide, both surfaces covered with gray velvetlike hairs, margins with eight to 10 pairs of coarse triangular spiny lobes, leaf bases continuing down the stem giving it a winged appearance; leaves of basal rosette up to 60 cm long and 30 cm wide.

Flower: Heads globe-shaped, 2.5–5 cm across, borne singly or in groups of two to seven at the ends of branches; disc florets numerous, violet to reddish, 20–25 mm long; bracts linear to lance-shaped, 2–2.5 mm wide at base, covered in short hairs and cobweblike hairs, spine-tipped, spines 2–6 mm long, borne in eight to 10 overlapping rows.

Plant: Plants 50–400 cm tall; stems branched,

covered in white woolly hairs; leaf bases winged
(2–3 cm wide) and spiny; taproot large and fleshy.
Fruit: Achene; 100–140 seeds per head.

REASONS FOR CONCERN

Scotch thistle populations tend to increase in
pastures and rangeland. Livestock avoid grazing in
areas where plants grow due to their prickly nature.

Similar species: Illyrian thistle (*O. illyricum* L.),
native to Europe, invades pastures and rangeland. It
is distinguished by winged stems 5–20 mm across,
purple disc florets 25–35 mm long, and bracts with
bases 3–8 mm wide. Weed designation: USA: AR, CA.
Another southern European species, Taurian thistle
(*O. tauricum* Willd.), is also known to infest pastures
and rangeland. It is distinguished by green stems
covered in sticky hairs, and spiny bracts 3–4 mm
wide at the base. The spines are 1–4 mm long. Weed
designation: USA: AR, CA, CO.

Rhaponticum repens (L.) Hidalgo
RUSSIAN KNAPWEED

Also known as: Turkestan knapweed, Russian centaurea, Turkestan thistle, hardheads

Scientific Synonym(s): *Acroptilon repens* (L.) DC., *Centaurea repens* L., *C. picris* Pall. ex Willd

QUICK ID: floral bracts with hairy tips; flowers purple to pink; rhizome black and scaly

Origin: introduced from Turkestan into Canada and the United States in the early 1900s as a contaminant in alfalfa seed

Life Cycle: Perennial reproducing by seeds, creeping rhizomes, and root fragments; large plants may produce over 1,200 seeds; colony at Indian Head, Saskatchewan, is reported to be 75 years old.

Weed Designation: USA: AK, AZ, CA, CO, HI, IA, ID, KS, MT, NM, ND, NV, OR, SC, SD, UT, WA, WY; CAN: federal, AB, BC, MB, ON, SK; MEX: federal

DESCRIPTION

Seed: Club-shaped, white to grayish-brown, 2–2.4 mm long and 0.6–0.7 mm wide; surface with several ridges and numerous soft white hairs; pappus consists of white to pale yellow feathery bristles, each 6–11 mm long.
Viable in soil for 2–3 years.
Germination occurs between 0.5°C and 35°C (optimal 20°–30°C); light not required for germination; dormancy broken by alternating temperatures.

Seedling: With lance-shaped cotyledons, 2.5–11 mm long and 2–7.5 mm wide; stem below cotyledons is distinctly curved; first leaves appear opposite and have smooth margins; later leaves alternate with shallowly toothed margins; roots reach 2–2.5 m depth the first growing season.

Leaves: Alternate, 4–15 cm long and 4 cm wide, margins with several deep lobes or teeth; reduced in size upward, 1–3 cm long, margins smooth; young leaves covered in white woolly hairs and becoming hairless with age; rosette leaves 5–10 cm long and 1–2.5 cm wide.

Flower: Heads, 10–15 mm across; 15–36 disc florets, purplish to pink, 11–14 mm long, borne in corymbs at the ends of branches; bracts borne in six to eight overlapping rows; outer bracts about 1 cm long, undivided, green with white margins; inner bracts tipped with fine hairs; flower heads silvery green in bud.

Plant: Plants 40–90 cm tall; stems with numerous branches, covered in cobwebby hairs when young, hairless at maturity; rhizome black and scaly, 2.5–6 m deep, lateral roots 7–20 cm below the surface; new shoots emerge when soil temperature at 5 cm reaches 14°C; limit of northern range in North America about 54° latitude.

Fruit: Achene.

REASONS FOR CONCERN

Russian knapweed is suspected of being poisonous to horses and sheep and may cause a neurological disorder. The roots also produce chemicals that inhibit other plant species from growing in the immediate area. This adaptation has allowed Russian knapweed to infest over 1.2 million acres in Washington, Idaho, Wyoming, and Colorado.

Similar species: spotted knapweed (*C. stoebe L.* ssp. [*Gugler*] *Hayek*, p. 88) is distinguished by floral bracts that are fringed with black hairs and leaf surfaces with translucent dots.

Senecio vulgaris L.
GROUNDSEL

Also known as: grimsel,
simson, birdseed,
ragwort, chickenweed,
old-man-in-the-spring
Scientific Synonym(s): none

QUICK ID: flower heads
yellow composed of disc
florets; stems hollow;
stems, leaves, and flower
heads hairless
Origin: native to Europe;
introduced in 1620 by the
Pilgrims as a treatment
for the early stages of
cholera
Life Cycle: Annual, winter
annual, or biennial
reproducing by seed;
large plants may produce
over 1,700 seeds. Plants
are capable of producing
seeds within 5 weeks
of germination. With
a possibility of four
generations in one
growing season, one
seed has the potential
to produce over 1 billion
seeds.
Weed Designation: USA:
WA;CAN: MB, PQ

DESCRIPTION
Seed: Spindle-shaped, tan, 2–3 mm long; pappus a
tuft of short white bristles, 3–4 mm long.
Viable for 1 year in soil.
Germination enhanced by exposure to
light; maximum germination occurs when
temperatures are between 9°C and 18°C.
Seedling: With oblong cotyledons, 3–11 mm long, 1–3
mm wide, hairless, stalks prominently grooved;
first leaves irregularly toothed and pointing
toward the tip; a few hairs may be found at the
base of the leaf stalk.
Leaves: Alternate, oblong, 5–15 cm long, 1–4 cm
wide, somewhat fleshy, underside often purplish-
green and prominently veined; lower leaf
margins wavy to deeply lobed; upper leaves
stalkless and clasping the stem.
Flower: Heads 5–10 mm across, borne in loose
clusters of eight to 20; five to 80 disc florets,
yellow, 5–8 mm long; 20–23 bracts, yellowish-
green with black tips, 4–6 mm long, borne in a
single row.
Plant: Plant 5–60 cm tall; stems hollow and
somewhat succulent, branched.
Fruit: Achene.

REASONS FOR CONCERN

A prolific seed producer with the ability to flower at temperatures below 0°C, common groundsel is a serious weed of cultivated crops and gardens. It is also host for beet mildew yellowing, beet western yellow, aster yellows, beet curly top, beet ring spot, and lettuce mosaic viruses. Common groundsel contains senecionine, an alkaloid that causes irreversible liver damage in livestock that feed on plants over an extended period of time. The highest levels of toxins are present in flowers and the lowest in the roots.

Silybum marianum (L.) Gaertn.
MILK THISTLE

Also known as: blessed thistle, lady's thistle, Marian thistle, Holy thistle, St. Mary's thistle

Scientific Synonym(s): *Mariana mariana* B. & B., Small., *Carduus marianus* L.

QUICK ID: leaves mottled with white veins, prickly; flower heads, large, purple; outer floral bracts about 4 cm long and 1 cm wide

Origin: native to the Mediterranean region of southern Europe and North Africa; believed to have been introduced in cattle feed or as a herbal treatment for liver disorders

Life Cycle: Annual, winter annual, or biennial reproducing by seed only; plants may produce up to 10,000 seeds.

Weed Designation: USA: AR, OR, WA; CAN: BC; MEX: federal

DESCRIPTION

Seed: Dark brown, flattened, 6–7 mm long, shiny; pappus of tan-colored bristles up to 15 mm long. Viable in soil for up to 10 years. Germination occurs mostly in fall; optimal germination is between 10°C and 15°C; emergence from 8 cm depth; surface germination is decreased by exposure to light; seedling survival is greatest in disturbed areas.

Seedling: With oval to spatula-shaped cotyledons, 2–3 cm long and 1–2 cm wide, somewhat thick, succulent, light green; first leaves conspicuously white-netted along the veins, margins with short yellowish prickles; rapid growth produces rosettes up to 1 m across.

Leaves: Alternate, 10–60 cm long, 15–30 cm wide, pinnately lobed, margins with yellow spines, upper surface dark green and mottled with light green or white veins, lower surface dull; lower leaves stalked while upper are stalkless and clasp the stem; reduced in size upward; rosettes up to 1 m wide produced in the first growing season.

Flower: Heads globe-shaped, 3–6 cm across; disc florets purple, 50–200 per head; bracts numerous, spiny-tipped and -margined, borne in four to six overlapping series; outer bracts leaflike, about 4 cm long and 1 cm wide; spines 2–3.5 cm long; flowering lasts 5 days, and mature seeds shed 17 days later.

Plant: Plants 1–3 m tall with 10–50 heads; stems hollow, vertically ribbed, hairless.

Fruit: Achene.

REASONS FOR CONCERN

Milk thistle is suspected of accumulating nitrates, which may cause livestock (cattle, horses, and sheep) poisoning. Affected animals will show signs of respiratory distress prior to death. Milk thistle is a serious weed of pastures.

Sonchus arvensis ssp. arvensis L.
PERENNIAL SOW THISTLE

Also known as: creeping sow thistle, field sow thistle, field milk thistle, gutweed, swine thistle, marsh sow thistle

Scientific Synonym(s): *Hieracium arvense* Scop., *S. hispidus* Gilib.

QUICK ID: stems with milky juice; flower heads yellow, greater than 2.5 cm across; stems below flower heads with yellow hairs

Origin: perennial introduced from Europe and the Caucasus region of Asia; introduced into Pennsylvania in 1814 as a seed contaminant

Life Cycle: Perennial reproducing by rhizomes and seed; plants may produce up to 13,000 seeds.

Weed Designation: USA: AK, AZ, CA, CO, HI, ID, IL, IA, MI, MN, NV, SC, WA, WY; CAN: federal, AB, BC, ON, PQ, SK

DESCRIPTION

Seed: Rectangular, reddish-brown, 2.5–3.0 mm long, 1.3–1.5 mm wide; surface with 12–15 prominent ribs; pappus of numerous white bristles, 8–10 mm long, deciduous.
Viability in soil 3–6 years.
Germination occurs in top 3 cm of soil; optimal temperature 25°–30°C, germination poor below 20°C and above 30°C.

Seedling: With oval cotyledons, 4–8 mm long, 1–5 mm wide, indented at tip, somewhat fleshy, contain milky juice; first leaves spatula-shaped, margin with irregular downward-pointing teeth and tipped by a weak prickle; cotyledons wither soon after the first leaves appear; at fourth leaf stage lateral roots begin to develop; 1-month-old seedlings with seven to eight leaves and roots 10–15 cm long and 1.5 mm thick; at 4 months roots are 60–100 cm long and up to 50 cm deep.

Leaves: Alternate, lance-shaped, 6–40 cm long, 2–15 cm wide; lower leaves deeply lobed, margins with soft prickles, stalks winged; upper leaves with two to seven backward-pointing lobes per side, margins with spine-tipped teeth, stalkless, and clasping the stem, the basal lobes rounded.

Flower: Heads numerous, 3–5 cm wide, borne on long stalks covered with yellow hairs; 30–250 ray florets, bright yellow; 27–50 bracts, dark green, 14–25 mm long, borne in three to five rows, covered in yellow hairs; seeds viable 5 days after pollination.

Plant: Plants 50–200 cm tall; stems succulent and hollow, branched near the top, exuding milky juice when broken; rhizome creeping and extensive, 2.5–5 mm in diameter, up to 3 m depth, easily broken and producing new plants from buds at 60 cm depth; rhizome spreading 50–280 cm per year.

Fruit: Achene; averaging 30 per head.

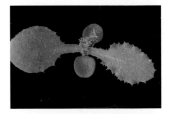

REASONS FOR CONCERN

It has been reported that light infestations of perennial sow thistle can drastically reduce crop yields. Perennial sow thistle competes with the crop for moisture, nutrients (primarily nitrogen), light, and space. Perennial sow thistle is host for aster yellows and beet mosaic viruses.

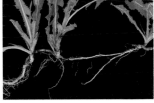

Similar species: perennial sow thistle (*S. arvensis* ssp. *uliginosus* Bieb.) is often mistaken for perennial sow thistle (*S. arvensis* L.), as both subspecies occur throughout the range. *Sonchus arvensis* ssp. *uliginosus* Bieb. does not have yellow hairs on the floral bracts or stem. The floral bracts are green with white margins. Weed designation: CAN: federal, AB, BC, MB, ON, PQ, SK.

Also known as: spiny-leaved sow thistle, spiny annual sow thistle, spiny milk thistle, prickly sow thistle, sharp-fringed sow thistle

Scientific Synonym(s): *S. fallax* Wall., *S. spinosus* Lam.

QUICK ID: stems with milky juice; flower heads yellow, less than 2.5 cm across; leaves with a large, round, basal lobe

Origin: native to Europe and North Africa; first collected in Ontario, Canada, in 1877

Life Cycle: Annual reproducing by seed; large plants producing up to 26,000 seeds.

Weed Designation: CAN: PQ, SK

DESCRIPTION

Seed: Elliptical, reddish-brown, 2.5–3 mm long, 1.3–1.5 mm wide, surface with six prominent ribs; pappus of soft white hairs 6–9 mm long. Viable in soil for up to 8 years. Germination occurs in the top 1 cm of soil.

Seedling: With oval cotyledons, 5 mm long and 2 mm wide, underside with a shiny midvein; first leaf spatula-shaped, margin with a few pointed lobes; later leaves with downward-pointing soft prickles; milky juice present by the fourth leaf stage.

Leaves: Alternate, 6–30 cm long, 1–15 cm wide, dark green above and pale green below, margins shallowly lobed and spiny, often with five to 11 lobes per side, stalkless with large rounded basal lobes, reduced in size upward; becoming stiff and prickly and turning purplish-green in late summer.

Flower: Heads round to pear-shaped, 9–12 mm long, 15–25 cm wide; 30–250 ray florets, light yellow; 27–50 bracts, borne in three to five rows, each 9–16 mm long.

Plant: Plants 20–150 cm tall; stems hollow and leafy, hairless, often reddish-green; taproot short; all parts of the plant exuding milky sap when broken.

Fruit: Achene.

REASONS FOR CONCERN

Annual sow thistle is a problem weed in cultivated crops, gardens, and roadsides. It is host for beet curly top, tobacco streak, cucumber mosaic, lettuce necrotic yellow, lucerne dwarf, and aster yellows viruses.

Similar species: annual sow thistle (*S. oleraceus* L., p. 132) is often confused with the previous species of annual sow thistle (*S. asper* L.). Annual sow thistle (*S. oleraceus* L.) is distinguished by the shape of the base of the leaf and by the number of ribs on the seed. The base of the leaves are pointed. The seed has three to five faint ribs on the surface.

Sonchus oleraceus L.
ANNUAL SOW THISTLE

Also known as: common sow thistle, hare's lettuce, milk thistle, sow thistle

Scientific Synonym(s): *Endiuia agrestis* Hard., *S. ciliatus* Lam., *S. laevis* Vill., *S. levis* Gars., *S. roseus* Bess.

QUICK ID: stems with milky juice; flower heads yellow, about 2 cm across; leaves with pointed basal lobes

Origin: native to Europe, western Asia, and North Africa

Life Cycle: Annual or biennial reproducing by seed; large plants producing up to 35,000 seeds.

Weed Designation: CAN: BC, MB, PQ, SK

DESCRIPTION

Seed: Oblong to oval, 2.5–3 mm long, surface with two to four transverse ribs on each face; pappus of white bristles 5–8 mm long.
Viable for at least 8 years in soil.
Germination occurs in both light and dark; optimal temperatures 10°–25°C; germination does not occur below 7°C and above 35°C; no seedling emergence from 2 cm depth or more.

Seedling: With egg-shaped cotyledons, 3–8 mm long, 1.5–4 mm wide, stalked; first leaves alternate, egg-shaped with irregular toothed margins; later leaves spatula-shaped.

Leaves: Alternate, bluish-green, hairless, upper leaves stalkless and clasping the stem; basal lobes earlike, pointed with weakly spined margins; leaf margins with one to three lobes per side, prickly and toothed; basal leaves oblong to lance-shaped, 6–50 cm long, 1–15 cm wide, margins pinnately lobed and spine-toothed, stalks winged; leaves mostly on lower half of stem.

Flower: Heads about 2 cm wide, borne at the end of stems; 30–250 ray florets; 27–50 bracts in three to five rows.

Plant: Plants 50–100 cm tall; stems hollow, unbranched or with a few branches, hairless; taproot short; stem and foliage exuding milky juice when broken.

Fruit: Achene.

REASONS FOR CONCERN

Annual sow thistle is a common weed of gardens, nursery crops, orchards, grain fields, and cultivated crops. It is a principal weed of rubber and sugarcane in Mexico. It has been documented as a serious weed in 56 countries.

Similar species: annual sow thistle (*S. asper* [L.] Hill, p. 130) is distinguished by large rounded basal lobes on the stem leaves.

Tanacetum vulgare L.
TANSY

Also known as: golden buttons, garden tansy, bitter buttons, hindhead, parsley fern, ginger plant

Scientific Synonym(s): *Chrysanthemum uliginosum* Pers., *C. vulgare* (L.) Bernh.

QUICK ID: leaves aromatic when crushed; stems somewhat woody; flower heads yellow and buttonlike

Origin: native to Europe; introduced as a medicinal garden plant

Life Cycle: Perennial reproducing by rhizome and seeds; plants capable of producing over 50,000 seeds.

Weed Designation: USA: CO, MT, WA, WY; CAN: AB, BC, MB, SK

DESCRIPTION

Seed: Oblong, gray to tan, about 1.5 mm long; surface with longitudinal ribs; pappus reduced to a short five-lobed crown.
Viable in soil for up to 25 years.
Germination occurs in the top 2 cm of soil.

Seedling: With oval cotyledons, 2.3–4 mm long, 1–2 mm wide; first two or three leaves oval, margin with small teeth; later leaves deeply lobed into narrow segments; star-shaped hairs on leaves evident from the first to fourth leaf stage, becoming hairless at the fifth leaf stage.

Leaves: Alternate, 5–25 cm long, 4–10 cm wide; divided twice into narrow-toothed segments; surface covered in numerous small glands.

Flower: Heads buttonlike, 5–10 mm wide, borne in a flat-topped cluster of 20–200 heads; 60–300 disc florets, yellow; ray florets if present, very small; 30–60 bracts, greenish-brown with papery tips, borne in two to five rows.

Plant: Plants 30–180 cm tall, aromatic; stems somewhat woody, purplish-red near the base, hairless; rhizome stout.

Fruit: Achene.

REASONS FOR CONCERN

Tansy is not a problem in cultivated crops because it cannot withstand cultivation. However, it is a problem in pastures where it tends to increase because livestock find it unpalatable. Some reports indicate that common tansy is somewhat poisonous to livestock, although it is eaten when no other food source is available. Tansy is also host for chrysanthemum stunt virus.

Similar species: tansy ragwort (*Jacobaea vulgaris* Gaertn., p. 112), an introduced biennial or perennial from Europe, resembles common tansy. It is distinguished by flower heads with yellow ray florets and seeds with pappus. Leaves and stem are similar in appearance.

Also known as: blowball, faceclock, common dandelion, pee-a-bed, wet-a-bed, lion's tooth, cankerwort, Irish daisy, pissenlit, dent de lion

Scientific Synonym(s): *Lentodon taraxacum* L., *T. dens-leonis* Desf., *T. retroflexum* Lindb. f., *T. taraxacum* Karst., *T. vulgare* Schrank

QUICK ID: plants with milky juice; leaves basal; flowering stalks leafless and hollow

Origin: native to Europe and Asia; introduced as a garden vegetable

Life Cycle: Perennial reproducing by seed only; plants may produce up to 23,400 seeds.

Weed Designation: CAN: MB, PQ, SK

DESCRIPTION

Seed: Elliptical, olive green, 3–4 mm long, 0.6–1 mm wide, surface with five to 10 prominent ribs; pappus of fine white barbed hairs about 5 mm long.
Viable in soil for up to 5 years.
Germination occurs in the top 2 cm of soil.

Seedling: With oval cotyledons, 4.5–10 mm long, 2–3.5 mm long, hairless; first leaves with margins ranging from wavy to deeply lobed, lobes point toward the base of the leaf.

Leaves: Basal, 5–50 cm long, 1–10 cm wide, margins coarsely lobed with triangular-shaped lobes; terminal lobe is always largest; midrib often hollow and winged near the base.

Flower: Heads rising from the base on hollow stalks 5–75 cm tall; heads 2–5 cm across; ray florets yellow, 100–300 per head; 13–18 bracts, green, borne in two overlapping rows; outer row often rolled backward; heads opening in the morning between 6 and 7 A.M.

Plant: With one or more stems per crown; stems leafy, up to 75 cm tall; taproot fleshy, up to 2.5 m deep; all parts of the plant exuding milky juice when broken.

Fruit: Achene; fruiting head 3.5–5 cm across.

REASONS FOR CONCERN

A serious weed of lawns and pastures, dandelion has begun to appear in minimal- and zero-till cultivated fields. In all cases, dandelion competes with the crop for moisture and nutrients. It is also host for aster yellows, beet ring spot, tobacco streak, and raspberry yellow dwarf viruses.

Similar species: red-seeded dandelion (*T. laevigatum* [Willd.] DC.) is common in waste areas, roadsides, and lawns. The leaves of red-seeded dandelion are deeply lobed to the midrib with all the lobes being about the same size. The seeds are reddish at maturity. Stems rarely exceed 30 cm tall. Two other introduced species, spotted catsear (*Hypochoeris radicata* L., p. 110) and fall hawkbit (*Leontodon autumnalis* L., p. 150), are easily confused with dandelion. Spotted catsear is distinguished from dandelion by flower heads that appear in groups of three or four at the ends of stems. Fall hawkbit has a few scale-like leaves on its flowering stem, distinguishing it from dandelion.

Tragopogon dubius Scop.
GOAT'S BEARD

Also known as: western salsify, yellow salsify, salsify, wild oyster plant

Scientific Synonym(s): *T. major* Jacq.

QUICK ID: plants with milky juice; leaves grasslike; flowers yellow; fruiting head 7–10 cm across

Origin: native to Europe; first North American report from Colorado

Life Cycle: Biennial or perennial reproducing by seed; plants capable of producing up to 500 seeds.

Weed Designation: CAN: MB, ON, PQ, SK

DESCRIPTION

Seed: Spindle-shaped, dull brown, 25–35 mm long, 1.5–1.6 mm thick, surface with five to 10 prominent ribs; pappus of 12–20 featherlike webbed hairs, 2.3–3 cm long, brownish-white. Viability unknown; some reports indicate less than 1 year, while others state more than 5 years. Germination occurs in the top 5 cm of soil.

Seedling: With grasslike cotyledons, 4–12 cm long and less than 5 mm wide, hairless, often arched; first leaves resembling grass; seedling often mistaken for grass seedlings.

Leaves: Alternate, narrow and grasslike, 5–30 cm long, somewhat fleshy, bluish-green, hairless, stalkless, and clasping the stem.

Flower: Heads borne singly on long stalks; 3–6 cm across; 30–180 ray florets, yellow; 10–14 bracts, 2–4 cm long, borne in a single row; stem below flower head swollen and hollow; flowers remaining closed on cloudy days, opening on sunny mornings and closing at midafternoon.

Plant: Plants 50–100 cm tall; stems pale green and hairless; taproot deep and fleshy; all parts of the plant exuding milky sap when broken; young plants often overlooked as they resemble grass.

Fruit: Achene; fruiting head globe-shaped, 7–10 cm across.

REASONS FOR CONCERN

Goat's beard is a common weed of cultivated crops, roadsides, and waste areas. It is a problem in fall seeded crops where the biennial phase produces a basal rosette of leaves.

Similar species: another goat's beard (*T. pratensis* L.) is often found alongside goat's beard (*T. dubius* Scop.). *Tragopogon pratensis* L. does not have a swollen stem below the flower head and has eight or nine floral bracts below the flower. These bracts are about the same length as the flowers. Seeds are 15–25 mm long with a feathery pappus. Weed designation: CAN: MB.

goat's beard
(*Tragopogon pratensis*)

Tridax procumbens L.
COATBUTTONS

Also known as: tridax, wild daisy, Mexican daisy

Scientific Synonym(s): none

QUICK ID: leaves opposite, hairy, and toothed; floral bracts in three rows; flowers pale yellow

Origin: native to tropical North and South America

Life Cycle: Annual or short-lived perennial reproducing by seed only; plants producing up to 2,500 seeds.

Weed Designation: USA: federal, AL, CA, FL, MA, MN, NC, OR, SC, VT

DESCRIPTION

Seed: Black, 2 mm long and 1 mm wide, surface covered in fine grayish-brown hairs; pappus of about 20 bristles, 5–6 mm long.
Viable in soil for up to 2 years.
Germination of fresh seeds requires daylight, while older seeds need darkness. Seeds below 2 cm do not emerge and few emerge from 1 cm depth. Optimal germination at 30°C.

Seedling: With round cotyledons, long-stalked, margins minutely haired; first leaves opposite, elliptical to lance-shaped, margins smooth, surface densely haired.

Leaves: Opposite, oval to lance-shaped, 1–7 cm long, 0.5–2 cm wide, margins irregularly toothed, base wedge-shaped, stalks 1–30 mm long, both surfaces hairy.

Flower: Heads 1–2 cm across, borne on stalks 10–25 cm long; three to eight ray florets, pale yellow, three-toothed; 20–80 disc florets, yellow to brownish-yellow, five-toothed; 11–15 bracts, 5–6 mm long, borne in two or three overlapping rows; outer bracts shorter than inner; flowers appearing 35–55 days after germination; seeds maturing 21 days after pollination.

Plant: Plants 30–50 cm tall, shade sensitive; stems somewhat prostrate, branched, sparsely to densely haired; root slender and wavy.

Fruit: Achene; those produced by disc florets generally smaller than those produced by ray florets.

REASONS FOR CONCERN

Coatbuttons are a common weed of maize, cotton, rice, sugarcane, avocado, mango, and bananas. It is also found on roadsides, pastures, and waste land, lawns and nurseries. Hand pulled or mechanically controlled plants often resprout as the stem breaks off at lower nodes. It is also host for branched broomrape (*Orobanche ramosa* L., p. 502), a parasitic plant of tomato crops.

Tripleurospermum inodorum (L.) Schultz-Bip.
SCENTLESS CHAMOMILE

Also known as: mayweed, false chamomile, scentless mayweed

Scientific Synonym(s): *Matricaria perforata* Merat, *M. maritima* L., *M. inodora* L., *T. perforata* (Merat) M. Lainz, *Chamomilla inodora* (L.) Gilib.

QUICK ID: flowers daisy-like; leaves divided into narrow segments, odorless when crushed

Origin: native to northern Europe and western Asia; introduced into New York as a seed contaminant in 1878; first Canadian report from New Brunswick in 1876.

Life Cycle: Annual, biennial, or short-lived perennial reproducing by seed only; large plants capable of producing up to 1 million seeds.

Weed Designation: USA: CO, WA; CAN: federal, AB, SK; MEX: federal

DESCRIPTION

Seed: Rectangular, dark brown, about 2 mm long, surface with three prominent ribs; pappus of small scales or absent.
Viable for 10–12 years in dry storage. Germination requires light; low germination at 10°–20°C; optimal temperature is 20°–30°C.

Seedling: With oval cotyledons, 2–3 mm long, stalkless; first leaves divided into narrow segments; later leaves divided into several branched segments.

Leaves: Alternate, 2–8 cm long, divided into numerous narrow or threadlike branched segments, hairless, short-stalked or stalkless; odorless when crushed.

Flower: Heads resembling daisies, 3–4 cm wide; 12–20 ray florets, white, 8–16 mm long; 300–500 disc florets, yellow, numerous, 2–3 mm long; 28–60 bracts, borne in two or five rows; large plants may have over 3,200 heads.

Plant: Plants 10–100 cm tall, odorless, often flowering in its first year of growth; stems branched, hairless; roots fibrous.

Fruit: Achene; heads producing 345–533 seeds.

REASONS FOR CONCERN

Scentless chamomile, an aggressive weed in forage and noncropland areas, has recently appeared in cultivated crops. Extreme care should be taken to ensure that plants do not set seed.

Similar species: oxeye daisy (*Leucanthemum vulgare* L., p. 116) is often confused with scentless chamomile. While flowers are similar in appearance, the leaves are very different. The leaves of oxeye daisy are lobed and not narrowly dissected like those of scentless chamomile. Other species—stinking mayweed (*Anthemis cotula* L.) and wild chamomile (*M. recutita* L.)—are similar in appearance but have a strong odor when crushed. Stinking mayweed, as the common name implies, is ill-scented, while wild chamomile is pineapple-scented.

Tussilago farfara L.
COLTSFOOT

Also known as: coughwort, gingerroot, clayweed, dovedock, horsehoof, ass's foot, bullsfoot, sowfoot, butterbur

Scientific Synonym(s): none

QUICK ID: flower heads yellow, appearing in early spring; flowering stem leaves scalelike; basal leaves dark green, angular to heart-shaped

Origin: introduced from northern Europe and Asia as a cold remedy

Life Cycle: Perennial reproducing by spreading rhizomes and seed; plants capable of producing over 3,500 seeds.

Weed Designation: USA: AL, CT, MA, OR; CAN: ON, PQ

DESCRIPTION

Seed: Prism-shaped or cylindrical, yellow to reddish-brown, 3–4 mm long, and 0.5 mm wide, glossy, surface with 5–10 ribs; pappus of numerous capillary bristles 8–12 mm long, deciduous and leaving a crown.
Viable for less than 1 year in soil.
Germination occurs in the top 5 mm of soil.

Seedling: With large round cotyledons; first leaves round with narrow stalk; later leaves heart-shaped, long-stalked, woolly haired below.

Leaves: Basal, angular to heart-shaped, 5–20 cm long, 10–20 cm wide, dark green above and white woolly below, palmately veined, long-stalked, margins irregularly toothed to lobed, teeth often purplish; appear in late spring near the end of the flowering period; flowering stem leaves numerous and scalelike, about 1 cm long.

Flower: Heads solitary and terminal, 2–3.5 cm across; 100–200 ray florets, yellow, female, 4–10 mm long; 30–40 disc florets, yellow, 10–12 mm long, perfect but sterile; about 21 bracts, 8–15 mm long, purplish-green, unequal in sizes and borne in one or two rows; flowering stems erect and woolly, the leaves bractlike; flowering in early spring.

Plant: Plants 5–50 cm tall, preferring moist clayey soils; flowering stems covered in woolly hairs and purplish-green bracts; leaves emerge when flower heads have reached maturity; rhizomes creeping and horizontal.

Fruit: Achene; produced by ray florets only; fruiting heads nodding.

REASONS FOR CONCERN

Coltsfoot is host for beet ring spot virus. It is also a problematic weed in low-lying fields where the aggressive nature of the rhizomes allows it to spread rapidly.

Similar species: native species of coltsfoot (*Petasites* Mill.) may be confused with *Tussilago farfara* L. Although these species flower in early spring and produce large basal leaves later in early summer, native coltsfoot species are easily identified by white to pinkish flowers.

Xanthium strumarium L.
COCKLEBUR

Also known as: broadleaved cocklebur, clotbur, sheepbur, ditchbur, buttonbur, noogoora bur

Scientific Synonym(s):
X. macrocarpum DC.,
X. natalense Widder,
X. pungens Wallr.,
X. vulgare Hill

QUICK ID: plants of moist habitats; leaves with a sandpaper texture, oval to triangular; fruit bur–like, two-seeded

Origin: native to North America

Life Cycle: Annual, biennial, or short-lived perennial reproducing by seed only; plants are capable of producing over 10,800 seeds.

Weed Designation: USA: AR, IA; CAN: MB, PQ; MEX: federal

DESCRIPTION

Seed: Elliptical, light brown to black, 8–15 mm long, 5–7 mm wide, surface covered in a silver-black papery membrane; pappus is absent. Viable for about 30 months in soil. Germination dependent on seed type—one type germinating shortly after having been shed, while the other has a delayed germination period; optimal depth is 1–8 cm deep; no germination on soil surface or below 15 cm depth; optimal temperature 20°–30°C.

Seedling: With lance-shaped cotyledons, 6–7.5 mm long, hairless, somewhat fleshy; cotyledons photosynthetically active and enable quick growth of young plants; cotyledons usually present at maturity; stem below the cotyledons purplish-green; first pair of leaves (about day 14) appear opposite, triangular, toothed, prominently three-veined; later leaves opposite and alternate.

Leaves: Opposite below and alternate above, oval to triangular, 2–12 cm long, 3–10 cm wide, palmately three- to five-lobed, base heart-shaped; surface with a sandpaper texture; margins wavy to toothed; stalks 2–14 cm long.

Flower: Heads male or female, green; male heads borne at the tips of branches, 5–8 mm across; 15–20 disc florets; six to 16 bracts, borne in one or two rows; female heads borne in leaf axils, two-flowered; two bracts, becoming the wall of the spiny bur.

Plant: Plants 30–150 cm tall, ill-scented if crushed; stems coarse, purplish-green with purplish-black spots, rough-haired, ridged; roots up to 1.2 m deep, allowing the plant to overwinter in warmer climates.

Fruit: Achene; borne in a woody bur, 22–28 mm long, 14–17 mm wide; bur light brown, covered in hooked spines 2–5 mm long; two-seeded; large plants may produce as many as 5,400 burs.

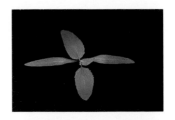

REASONS FOR CONCERN

Cocklebur, a weed of cultivated land, pastures, and waste areas, has been reported to reduce crop yields drastically. Cocklebur seeds and seedlings are poisonous to livestock and man. Seeds are rarely eaten because of the spiny bur that surrounds them. Young pigs are often poisoned because they seek out and eat the succulent seedlings. A lethal dose is reported to be 0.75–1.5% of body weight or about 100 seedlings. Mature plants are not poisonous. It is reported to be weedy in at least 30 countries.

Similar species: spiny cocklebur (*X. spinosum* L.) may be confused with cocklebur. It is distinguished by nodal spines 15–30 mm long, lance-shaped leaves 4–8 cm long and 1–3 cm wide, and burs 10–12 mm long. Weed designation: USA: OR, WA.

OTHER SPECIES OF CONCERN IN THE ASTER FAMILY

AGERATINA ADENOPHORA (SPRENGEL) R. M. KING & H. ROBINSON — CROFTON WEED, STICKY SNAKEROOT

Native to Mexico; plants shrubby, 50–220 cm tall; stems often purplish-green when young, erect, sticky-haired; leaves opposite, stalks 10–25 mm, blades oval to lance-shaped or triangular, 2.5–5.5 cm long, 1.5–4 cm wide, margins toothed; heads clustered; stalks 5–12 mm long, sticky-haired; 10–60 disc florets, white or pinkish; eight to 30 bracts, borne in two series; fruit an achene. Weed designation: USA: federal, AL, CA, FL, HI, MA, MN, NC, OR, SC, VT.

CARTHAMUS OXYACANTHUS BIEB. — JEWELED DISTAFF THISTLE

Native to western Asia; plants 30–100 cm tall; stems hairless; leaves elliptical to oval or lance-shaped, 2–8.5 cm long, margins spiny toothed and tipped; heads borne singly or in cymes; 20–30 disc florets, yellow to red; bracts many, borne in four or five rows; outer bracts spreading to reflexed, spiny toothed and tipped; achenes 4–5 mm long; pappus absent. Weed designation: USA: federal, AL, CA, FL, MA, MN, NC, OR, SC, VT; MEX: federal.

CRUPINA VULGARIS PERSOON EX CASSINI — BEARDED CREEPER, COMMON CUPINA

Native to Europe, Asia, and North Africa; plants 20–80 cm tall; stems leafy, branched; leaves basal, absent at time of flowering, oblong to oval, margins smooth to toothed or deeply divided; stem leaves 1–3.5 cm long, divided one or two times into narrow lobes; heads borne singly or clustered at ends of branches; stalks 0.5–8 cm long; usually three disc florets, purple, about 14 mm long; outer two to four florets sterile; inner one or two fertile; bracts borne in four to six overlapping rows; achenes one per head, barrel-shaped, 3–6 mm long, 1.5–3.5 mm wide; pappus of blackish-brown bristles, unequal in size, borne in two rows. Weed designation: USA: federal,

AL, CA, CO, FL, ID, MA, MN, MT, NV, NC, OR, SC, SD, VT, WA; CAN: federal, AB, BC, SK; MEX: federal.

HELIANTHUS CILIARIS DE CANDOLLE — TEXAS BLUEWEED

Native to southwestern United States and Mexico; plants 40–70 cm tall, spreading by creeping rhizomes; stems hairless; leaves mostly opposite, linear to lance-shaped, 3–7.5 cm long, 0.5–2.2 cm wide, stalkless, often bluish-green; heads 1–5 per stem; stalks 3–13 cm long; 10–18 ray florets, yellow, 8–9 mm long; 35 or more disc florets, reddish, 4–6 mm long; 16–19 bracts, borne in two or three rows, 3–8 mm long, 2–3.5 mm wide, margins hairy; achenes 3–3.5 mm long; pappus of two awn-tipped scales 1.2–1.5 mm long. Weed designation: USA: AZ, AR, CA, OR, SC, WA.

HELIANTHUS TUBEROSUS L. — JERUSALEM ARTICHOKE

Origin uncertain; plants 50–300 cm tall, rhizomes producing tubers in late summer; stems erect, bristly haired; leaves opposite or alternate, oval to lance-shaped, 10–23 cm long, 7–15 cm wide, stalks 2–8 cm long, margins smooth to toothed, surfaces with a sandpapery texture; three to 15 heads; stalks 1–15 cm long; 10–20 ray florets, yellow, 25–40 mm long; 60 or more disc florets, yellow, 6–7 mm long; 22–35 bracts, borne in two or three rows; 8.5–15 mm long, 2–4 mm wide; achene 5–7 mm long; pappus of two awn-tipped scales 1.9–3 mm long. Weed designation: none.

Helianthus tuberosus

HIERACIUM PILOSELLA L. — MOUSE-EARED HAWKWEED

Native to Europe; plants 10–25 cm tall; stems bristly haired; two to 10 basal leaves, stem leaves generally absent (though occasionally up to three), elliptical to oblong or lance-shaped, 10–45 cm long, 5–12 mm wide; margins smooth; surfaces bristly haired; heads borne singly (rarely two or three); stalks bristly haired; 60–120 ray florets, yellow, 8–13 mm long; 20–34 bracts, borne in two rows, bristly haired; achenes 1.5–2 mm long; pappus of over 30 white bristles, 4–5 mm long. Weed designation: USA: OR, WA; CAN: AB.

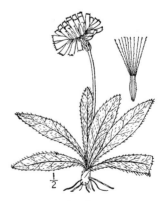

Hieracium pilosella

Inula helenium

INULA HELENIUM L. — ELECAMPE, ALANT, VELVET DOCK

Native to southern Europe and western Asia; perennial reproducing by seed, resembling sunflowers; taproot yellow, large and branched, stems 50–200 cm tall, covered in fine hairs; leaves primarily basal, 25–50 cm long, 10–20 cm wide, long-stalked, underside covered in dense velvety hairs; stem leaves oval, clasping the stem, 10–30 cm long, 4.5–12 cm wide, margins toothed; flower heads 5–10 cm wide, borne singly on long stalks; 15–150 ray florets, yellow, female, 20–30 mm long; 50–250 disc florets, yellow, perfect, 9–11 mm long; bracts leaflike, short-haired, borne in four to 70 overlapping rows; outer bracts 12–20 mm long, 6–8 mm wide; inner bracts narrower and hairless; fruit an achene, 3–4 mm long, four- to five-ribbed; pappus of 40–60 capillary bristles fused at the base; bristles 6–10 mm long. Weed designation: none.

LEONTODON AUTUMNALIS L. — FALL HAWKBIT

Native to Europe and Asia; perennial; flowering stems 10–80 cm tall, often up to 20 stems per crown, woolly haired below the heads; leaves primarily basal, oblong to lance-shaped, 4–35 cm long, 5–40 mm wide, margins smooth to deeply lobed; heads one to five per corymb, 7–13 mm long, stalks with bracts below each head; 18–20 bracts, 10–12 mm long, borne in two rows; 20–30 ray florets, yellow, 13–16 mm long; fruit an achene, 4–7 mm long; pappus yellowish-white to tan, 5–8 mm long, composed of feathery bristles. Weed designation: none.

Leontodon autumnalis

MIKANIA MICRANTHA KUNTH — BITTERVINE

Native to Central and South America; fast-growing perennial vine; stems slender, branched, often rooting at lower nodes; leaves opposite, triangular to heart-shaped, 4–13 cm long, 2–9 cm wide, stalks 2–8 cm long, margins coarsely toothed, surface with three to seven nerves from base; heads 4.5–6 mm long, borne in corymbs; flowers white to greenish-white; four disc florets, 3–4 mm long; four bracts, 2–4 mm long; fruit an achene, 1.5–2 mm long, black, five-angled; pappus of 32–38 white bristles, each 2–4 mm long. Weed designation: USA: federal, AL, CA, FL, HI, MA, MN, NC, OR, SC, VT.

MIKANIA SCANDENS (L.) WILLD. — CLIMBING HEMPVINE

Native to the Bahamas; perennial vine; stems somewhat six-angled; internodes 8–15 cm long; leaves opposite, triangular to heart-shaped, 3–15 cm long and 2–11 cm wide; stalks 20–50 mm long; margins smooth to wavy or toothed; heads borne in compact corymbs, 12–15 cm long and wide; heads 6–7 mm long; four disc florets, pinkish to purple, occasionally white, 3–5.4 mm long; bracts green to purplish, 5–6 mm long; fruit an achene, dark brown to black, 1.8–2.2 mm long; pappus of 30–37 white to pinkish-purple bristles, 4–4.5 mm long. Weed designation: USA: HI.

Mikania scandens

TAGETES MINUTA L. — MEXICAN MARIGOLD, MUSTER JOHN HENRY, AZTEC MARIGOLD, TAGETTE, CHINCHILLA ENANA, HIERBA PUDENT

Native to South America; annual plant producing up to 29,000 seeds; stems 50–200 cm tall, aromatic, pale brown to reddish-green with longitudinal grooves; leaves opposite below, alternate above, 5–20 cm long, deeply divided into nine- to 17-toothed lobes; lobes elliptical to lance-shaped, 1.2–2.5 cm long, 4–7 mm wide, the margins with scattered oil glands; heads borne in terminal clusters of 20–80; heads up to 12 mm long and 2 mm wide; stalks 1–5 mm long; ray florets 1–2 mm long, yellowish-white, one to five per head; disc florets 3–4 mm long, yellow, three to five per head; three to 21 bracts, borne in one or two series; fruit a black achene, spindle-shaped and flattened, 5–8 mm long, tip with four-pointed scales; pappus a few awnlike scales 2–3 mm long and three to five lance-shaped scales less than 1 mm long. Weed designation: USA: CA.

Balsaminaceae

TOUCH-ME-NOT OR JEWELWEED FAMILY

Herbs with succulent stems, often transparent, juice watery; leaves alternate or opposite; stipules absent; inflorescence of solitary flowers or umbels; flowers perfect, irregular; five sepals, the lowest spurred; five petals, the lateral pair united; five stamens; one pistil; ovary superior; fruit an explosive capsule. Worldwide distribution: 4 genera/600 species; predominantly growing in Europe, North America and Africa.

Distinguishing characteristics: plants succulent; flowers irregular; spurs present.

A few species of impatiens are grown as ornamental annuals.

spotted jewelweed (*Impatiens noli-tangere*)

Impatiens glandulifera Royle
POLICEMAN'S HELMET

Also known as: Himalayan impatiens, ornamental jewelweed, Indian balsam, purple jewelweed, touch-me-not

Scientific Synonym(s): *I. roylei* Walp.

QUICK ID: plants succulent; flowers pink, irregular; fruit bursting when touched

Origin: Native to the Himalayan region of Asia; introduced into England in 1839 and spreading throughout the British Isles by 1970; introduced into North America as a garden plant

Life Cycle: annual reproducing by seed; large plants may produce up to 4,000 seeds

Weed Designation: USA: CT, OR, WA;CAN: AB

DESCRIPTION

Seed: Somewhat globe-shaped, brown to black, 2–5 mm across, surface rough.
Viable in soil for 2 years.
Germination can occur underwater; seeds float.

Seedling: With oval to round cotyledons, succulent; roots form 12 days after germination; photosynthesis begins 4 weeks after germination.

Leaves: Opposite or in whorls of three or four, oblong to egg-shaped, 5–23 cm long, 1–7 cm wide, stalks 3–3.5 cm long, margins with 20 or more teeth per side.
Inflorescence a cyme, borne on elongated axillary stalks up to 9 cm long, two- to 14-flowered; bracts elliptical to oval or lance-shaped, 7–8 mm long.

Flower: Pink, white, or purple, irregular, 2–3.5 cm long; three sepals, two fused and the lower with a spur 5–6 mm long; five petals, two fused; five stamens; one pistil; first seeds produced 13 weeks after flowering begins.

Plant: Plants 1–3 m tall, frost-sensitive, partially shade-tolerant, requiring soils with high moisture content; stems succulent and hollow, green tinged with red or purple, hairless, often angular; roots shallow, extending to 6 inches deep, adventitious roots forming near the soil surface.

Fruit: Club-shaped capsule, 14–30 mm long and 4–8 mm wide, nodding, four- to 16-seeded, exploding at maturity, ejecting seeds up to 7 m from parent.

REASONS FOR CONCERN

In Great Britain this species is listed as one of the top 20 worst weeds. A prolific seed producer, policeman's helmet invades moist open areas and forests and displaces native vegetation.

OTHER SPECIES OF CONCERN IN
THE TOUCH-ME-NOT OR JEWELWEED FAMILY

IMPATIENS BALSAMINA L. — BALSAM

Native to India and southeast Asia; annual or perennial; stems 30–60 cm tall, succulent; leaves alternate, elliptical to oval or lance-shaped, 6–15 cm long, margins toothed; inflorescence a solitary flower or raceme (two- or three-flowered); flowers white, pink, red, or purple; three sepals; three sepals; five petals, 25–35 mm long; five stamens; one pistil; spur 15–25 mm long; fruit a capsule, 15–20 cm long, covered in soft hairs. Weed designation: none.

IMPATIENS PALLIDA NUTT. —
PALE TOUCH-ME-NOT

Impatiens pallida

Native to eastern North America; annual; stems 30–150 cm tall, succulent and translucent, juice watery; leaves alternate, oval, 2–10 cm long, 1.5–4 cm wide; stalked; margins coarsely toothed; flowers borne singly on drooping stalks from leaf axils; flowers pale yellow, irregular, 25–40 mm long; spur bent parallel to the sac; three sepals (one large spurred sac and two small green appendages), spur 5–8 mm long; three petals (two fused and one free); five stamens, partly united; one pistil; fruit a capsule 20–25 mm long, explosive, ejecting seeds with force, hence the name touch-me-not. Weed designation: none.

IMPATIENS WALLERIANA HOOK. F. —
BALSAM, JAPANESE BALSAM

Native to eastern Africa; annual or perennial; stems 30–100 cm tall, succulent; leaves alternate, reddish-green to green, elliptical to oval, 4–12 cm long, margins toothed, long-stalked; inflorescence a solitary flower or raceme (two- or three-flowered); flowers red, orange, pink, purple, or white; three sepals; five petals, 18–30 mm long; five stamens; one pistil; spur 2.5–5 cm long; fruit a capsule, 15–25 mm long. An escaped ornamental inhabiting moist shady areas in warm climates, such as Hawaii. Weed designation: none.

Berberidaceae
BARBERRY FAMILY

Shrubs and herbs; leaves alternate, simple or compound; stipules mostly absent; inflorescence of solitary flowers, racemes, and cymes; flowers perfect, regular; four to six sepals; four to six petals; four to 18 stamens, attached to the petals and springing free to dust an insect with pollen; one pistil; ovary superior; fruit a berry. Worldwide distribution: nine genera/600 species; throughout northern hemisphere and South America.

Distinguishing characteristics: shrubs; leaves simple, alternate; flowers yellow.

Members of the barberry family are economically important in positive and negative aspects. Some species are grown for their ornamental value, while others provide food for wildlife. Common barberry (*Berberis vulgaris* L.), once widely planted as an ornamental, is a host for wheat rust, a fungal disease of cultivated grains and grasses.

creeping mahonia (*Berberis repens*)

Berberis vulgaris L.
COMMON BARBERRY

Also known as: European barberry, woodsour, jaundice-tree, pepperidge-bush

Scientific Synonym(s): *B. acutifolia* Prantl

QUICK ID: shrub up to 3 m tall; flowers yellow, in drooping clusters; berries red

Origin: native to Asia; introduced into Europe as an ornamental plant and, later, into North America in the 1600s; proclaimed a weed by Massachusetts in 1754.

Life Cycle: perennial reproducing by seed

Weed Designation: USA: CT, MA, MI, NH; CAN: AB, MB, ON, SK

DESCRIPTION

Seed: Oblong, dark brown, 5–7 mm long; surface shiny and wrinkled.
Viable for at least 4 years at cold storage (1°–3°C). Germination occurs in the top 2 cm of soil; optimal temperature between 13°C and 30°C.

Seedling: With oblong cotyledons two to three times longer than wide; first leaf round, long-stalked, margin bristly haired; stipules prominent.

Leaves: Alternate, oval to elliptical, 2–6 cm long, 9–28 mm wide; stalks 2–8 mm long; margins finely toothed; spines simple or with three branches, located at base of leaf; leaves appearing clustered on the stem; margins with 16–30 teeth.
Inflorescence a raceme, 2–6 cm long, 10–20-flowered, drooping, borne leaf axils.

Flower: Yellow, about 6 mm across; six sepals, yellowish-green; six petals, two glandular spots at base; six stamens; one pistil; sepals and petals falling soon after the flower opens.

Plant: Bushy shrub, 1–3 m tall, hardy to –35°C; stems several from base; bark light gray to yellowish-gray; wood yellowish.

Fruit: Red oblong berry, 8–11 mm long, one- to three-seeded; sour-tasting; remaining attached to branches through winter.

REASONS FOR CONCERN

Common barberry is host for wheat rust (*Puccinia graminis*), which affects wheat, oats, barley, and other grasses. The fungus overwinters on the leaves of common barberry and spreads to cereal crops, causing severe reductions in crop yield. In 1916 a major wheat rust epidemic broke out in the Midwest states, leading to 19 states implementing a major barberry eradication program. By 1975 many of the programs were discontinued as established populations had been eradicated. Since that time, several populations have been discovered in areas where barberry had not existed since the 1940s.

OTHER SPECIES OF CONCERN IN THE BARBERRY FAMILY

BERBERIS THUNBERGII DC. — JAPANESE BARBERRY

Native to Japan; deciduous shrubs 30–300 cm tall; stem bark purple or brown; spines simple or three-branched; leaves alternate, oblong to oval or spatula-shaped; 1.2–2.4 cm long, 0.3–1.8 cm wide; stalks less than 8 mm long; margins smooth; inflorescence an umbel, one- to five-flowered; bracts membranous; six sepals, falling soon after the flower opens; six petals, yellow; six stamens; one pistil; fruit a red berry, elliptical, 7–10 mm long, juicy. Weed designation: USA: CT, MA, MI. Note: Some strains of Japanese barberry are suspected to be susceptible to wheat rust (*Puccinia graminis*) and have been outlawed for growth in Canada. Weed designation: USA: CT, MA.

NANDINA DOMESTICA THUNB. —
HEAVENLY BAMBOO

Native to China, Japan, and India; evergreen shrub
up to 2 m tall; wood bright yellow; leaves alternate,
50–100 cm long, compound with nine to 81 leaflets;
leaflets often reddish-green, elliptical to oval or
lance-shaped, 4–11 cm long, 1.5–3 cm wide; nearly
stalkless, margins smooth; inflorescence a terminal
or axillary panicle of hundreds of flowers; flowers,
10–20 mm long, 5–7 mm across, fragrant, creamy
to white; sepals and petals similar in appearance,
27–36; six stamens; one pistil; fruit a round
berry, 6–9 mm in diameter, red to purple. Weed
designation: none. Considered invasive in Florida
as it develops dense thickets that displace native
vegetation.

Boraginaceae
BORAGE FAMILY

Herbs, occasionally shrubs; leaves alternate, simple; stipules absent; inflorescence a cyme; flowers perfect, regular; five sepals, united at the base; five petals, united into a tube, funnel, or bell; five stamens, borne on the petals; one pistil; ovary superior; fruit a nutlet. Worldwide distribution: 100 genera/2,000 species; tropical and temperate regions well represented.

Distinguishing characteristics: plants bristly haired; flowers borne in coiled cymes; style borne at base of ovary.

The borage family is important horticulturally and include borage, bluebells, and forget-me-nots.

common bugloss (*Anchusa officinalis*)

scorpioid cyme

Anchusa officinalis L.
COMMON BUGLOSS

Also known as: alkanet, bugloss, oxtongue

Scientific Synonym(s):
A. procera Besser & Link.

QUICK ID: plants bristly haired; flowers blue; inflorescence resembling a scorpion's tail

Origin: native to Mediterranean region of Europe; introduced as an ornamental/medicinal plant

Life Cycle: perennial reproducing by seed; average plants producing over 900 seeds

Weed Designation: USA: OR, WA; CAN: BC

DESCRIPTION

Seed: Angular, grayish-brown, 2–3 mm long, tip curved.
Viability unknown.
Germination enhanced by alternating day and night temperatures.

Seedling: With oval to oblong or lance-shaped cotyledons, 10–20 mm long, 7–10 mm wide; first leaf oval to lance-shaped, bristly haired, midrib distinct.

Leaves: Alternate, linear to lance-shaped, 5–10 mm wide, somewhat fleshy, short-stalked, surface with stiff hairs; basal leaves stalked, oblong. Inflorescence a scorpioid cyme; bracts present.

Flower: Purple to dark blue with white centers; five sepals, 5–7 mm long (8–12 mm long in fruit); five petals, 6–12 mm wide, tube 5–10 mm long; five stamens; one pistil.

Plant: With deep taproot; stems angular 30–70 cm tall, coarsely haired.

Fruit: Nutlet, 2–3 mm long; four produced per flower.

REASONS FOR CONCERN

Common bugloss populations often increase in pastures because the bristly nature of the plant makes it unpalatable to livestock.

Cynoglossum officinale L.
HOUNDSTONGUE

Also known as: gypsy flower, sheep bur, dog bur, sheep lice, common bur, glovewort, woolmat, rosenoble, toryweed

Scientific Synonym(s): *C. hybridum* Thuill.

QUICK ID: plants densely haired; flowers red to reddish-purple; upper leaves clasping the stem

Origin: native to Europe and Asia; probably introduced as a cereal contaminant; first Canadian report from Ontario in 1859

Life Cycle: biennial reproducing by seed; large plants producing up to 2,000 seeds

Weed Designation: USA: CO, MT, NV, OR, WA, WY; CAN: AB, BC, SK

DESCRIPTION

Seed: Light brown, angular, 5–8 mm long, 5–6.5 mm wide, surface covered in hooked bristles. Viable for 2–3 years in dry storage; viability reduced when buried in soil. Germination occurs in the top 1 cm of soil, no emergence from 5 cm depth; germination on surface inhibited by light.

Seedling: With oval cotyledons, 15–25 mm long, 8–11 mm wide, thick; upper surface with short stiff hairs pointing toward tip, lower surface smooth; stem below cotyledons very short, barely above ground.

Leaves: Alternate, elliptical to lance-shaped, 2–5 cm long, reduced in size upward, upper leaves stalkless and clasping the stem, surfaces covered with soft velvety hairs; basal leaves oblong to lance-shaped, 10–30 cm long and 2–5 cm wide, long-stalked. Inflorescence an axillary raceme, 10–20 cm long; five- to 35-flowered.

Flower: Dull red to reddish-purple, occasionally white, funnel-shaped, 7–10 mm long; five sepals, united, covered in soft hairs; five petals, united; five stamens; one pistil.

Plant: Plants 30–120 cm tall; stem covered in fine hairs; taproot thick, black, up to 100 cm deep.

Fruit: Nutlet, covered in bristles and assisting in dispersal; four nutlets per flower.

REASONS FOR CONCERN

The seeds of houndstongue become entangled in sheep wool, thereby reducing wool value. It has been reported that the basal leaves (fresh or in cured hay) contain poisonous alkaloids. Livestock losses, primarily in horses and cattle, have been documented. Symptoms of poisoning are diarrhea, dizziness, cardiac problems, respiratory failure, and dark green feces.

Echium vulgare L.
BLUEWEED

Also known as: viper's bugloss, blue thistle, blue devil, snake flower, viper's grass, cat's tails

Scientific Synonym(s): *E. elegans* Noe & Nym.

QUICK ID: plants bristly haired throughout, the hairs with swollen bases; flowers blue, reddish-purple in bud

Origin: native to North Africa; introduced as a garden plant and as an antidote for snake bites

Life Cycle: biennial reproducing by seed only; large plants may produce up to 2,800 seeds

Weed Designation: USA: WA; CAN: AB, BC, MB; MEX: federal

DESCRIPTION

Seed: Grayish-brown, angular, 2–4 mm long. Viable for at least 33 months when buried at 15 cm depth; seeds on soil surface viable less than 12 months.
Germination occurs from the top 3 cm of soil.

Seedling: With elliptical cotyledons, 9–14 mm long, 4.5–7 mm wide, surface covered in fine needle-like hairs; stem below cotyledons also covered in needle-like hairs; first leaf elliptical, bristly haired.

Leaves: Alternate, lance-shaped, 1–15 cm long, surface covered in long stiff hairs; hairs with an enlarged red or black bases; leaves reduced in size upward; basal rosette with long narrow bristly leaves.
Inflorescence a scorpioid cyme; flowers appearing on one side of the stem; bristly haired throughout.

Flower: Blue (reddish-purple in bud), funnel-shaped, about 2 cm across; five sepals, united; five petals, united; five stamens; one pistil; stalks shorter than the sepals.

Plant: Plants 10–90 cm tall; stems reddish-green, several from the base, bristly haired throughout; taproot stout and black.

Fruit: Nutlet; four per flower.

REASONS FOR CONCERN

Blueweed is unpalatable to livestock and tends to increase in overgrazed pastures, thereby replacing many desirable plants. It is also host for cabbage black ring spot and tobacco mosaic viruses.

Lappula squarrosa (Retz.) Dumort.
BLUEBUR

Also known as: sticktight, beggar-ticks, stickseed, sheepbur, European sticktight, bur forget-me-not, European stickweed

Scientific Synonym(s): *L. echinata* Gilib., *L. erecta* A. Nels., *L. fremontii* (Torr.) Greene, *L. lappula* (L.) Karst., *L. myosotis* Moench

QUICK ID: plant with a mousy odor; flowers blue; seeds with two rows of hooked bristles

Origin: native to the eastern Mediterranean region of Europe; first reported in Maryland in 1698, Quebec by 1792

Life Cycle: annual or winter annual reproducing by seed; annual phase producing up to 2,000 seeds, while winter annuals may produce up to 40,000 seeds

Weed Designation: CAN: MB

DESCRIPTION

Seed: Oval, grayish-brown, 3.5–4 mm long; surface with two rows of hooked prickles (assisting in dispersal).
Viable in soil for up to 5 years.
Germination occurs in the top 2 cm of soil.

Seedling: With oval cotyledons, 5–7 mm long, 2.5–4 mm wide, hairless; first leaves similar in appearance but with a distinct crease along the midrib; first two leaves may appear opposite.

Leaves: Alternate, 2–7 cm long, reduced in size upward, both surfaces covered in stiff white hairs; lower leaves oblong, stalked, blunt-tipped; upper leaves stalkless.
Inflorescence a scorpioid cyme.

Flower: Blue with a yellow throat, 3–4 mm wide; five sepals, united; five petals, united; five stamens; one pistil.

Plant: Plants 3–60 cm tall, mousy-scented; stems profusely branched, covered in stiff white hairs that lie flat against the stem and leaves.

Fruit: Nutlet; surface with two rows of hooked prickles with star-shaped tips; fruiting stalks straight and erect.

REASONS FOR CONCERN

Bluebur is not a major weed in cultivated crops but may increase in overgrazed pastures. It is commonly found on roadsides and waste areas. The hooked bristles on the seed assist in the spread of the weed.

OTHER SPECIES OF CONCERN IN THE BORAGE FAMILY

Anchusa arvensis

Echium plantagineum

Heliotropium indicum

ANCHUSA ARVENSIS (L.) BIEB.—
SMALL BUGLOSS

Native to Europe and Asia; annual; stems erect, 50–90 cm tall, leafy; leaves alternate, oblong to oval or lance-shaped, 10–20 mm wide, lower stalked, upper stalkless and somewhat clasping, surface bristly haired and warty, margins crinkly and wavy; inflorescence a scorpioid cyme; flowers sky blue, tube curved and unequal; five sepals, 4–5 mm long in flower and 8–10 mm long in fruit; five petals, 4–6 mm wide and 4–7 mm long; five stamens; one pistil; fruit a nutlet, 2–3 mm long, brown and shiny, tip erect. Weed designation: USA: WA.

ECHIUM PLANTAGINEUM L.—
SALVATION JANE, PATERSON'S CURSE

Native to Mediterranean region of Europe; annual or biennial; taproot stout; stems 30–120 cm tall, coarsely haired throughout; leaves alternate, oblong to lance-shaped, 2–10 cm long, dark green, bristly haired, lateral veins prominent, stalkless, reduced in size upward; basal leaves 5–20 cm long, 1–1.5 cm wide, short-stalked; inflorescence a scorpioid cyme, 5–15 cm long, often with two to eight branches; flowers bluish-purple, 2–3 cm long, irregular, funnel or trumpet-shaped; five sepals, 7–10 mm long at flowering and 8–15 mm long in fruit; five petals; five stamens, two or three long and exceeding the petals; one pistil; fruit a nutlet, grayish-brown, 2–3 mm long, surface wrinkled. Weed designation: USA: OR.

HELIOTROPIUM INDICUM L.—
INDIAN HELIOTROPE

Native to South America; annual with a woody base; stems erect, 30–80 cm tall, bristly haired; leaves alternate to subopposite, oval to triangular, 2.5–10 cm long, 1–5 cm wide, stalks to 5 cm long, surface rough-textured, margin wavy to crinkly; lateral veins 8–14; inflorescence a scorpioid cymelike spike, 8–20 cm long, one-sided; flowers pale violet to blue or whitish with a yellow throat, 3–4 mm wide, 4–5 mm long, trumpet to funnel-shaped; five sepals; five petals; five stamens; one pistil; fruit a nutlet, 2.5–4 mm long. Weed designation: none.

LITHOSPERMUM OFFICINALE L.— GROMWELL, EUROPEAN STONESEED

Native to southern Europe and western Asia; perennial reproducing by seed; taproot thick; stems erect 20–100 cm tall, branched near the top, densely haired throughout; leaves alternate, oval to lance-shaped, 3–8 cm long, 5–15 mm wide, prominently veined, short-stalked, margins smooth; inflorescence a leafy cyme, up to 15 cm long at fruiting; bracts resembling leaves, but smaller; flowers white to pale yellow, tube-shaped; five sepals, 5–7 mm long at fruiting; five petals, 4–6 mm long; five stamens; one pistil; fruit a nutlet, oval, white or yellowish-brown, 3–3.5 mm long, smooth and shiny. Weed designation: none.

Lithospermum officinale

MYOSOTIS SCORPIOIDES L.— FORGET-ME-NOT

Native to northern Europe and Asia; perennial of wetlands and stream banks; stems 15–30 cm tall, hairy throughout, erect to sprawling, often rooting from lower nodes; leaves alternate, oblong to lance-shaped, 2.5–8 cm long, surface hairy, stalkless; inflorescence a scorpioidlike raceme, up to 20 cm long; bracts absent; flowers blue with a yellow throat, 5–9 mm wide; five sepals, 3–5 mm long at fruiting, bristly haired; five petals; five stamens; one pistil; fruit a nutlet. Weed designation: USA: CT, MA.

Myosotis scorpioides

SYMPHYTUM ASPERUM LEPECHIN— PRICKLY COMFREY

Native to Europe and Asia; perennial; taproot thick and black; stems erect, angular and hollow, 40–100 cm tall, covered in bristly hairs; leaves alternate, oval to lance-shaped, 5–30 cm long, reduced in size upward, upper leaves stalkless, surface with short bristly hairs, base not extending down the stem; inflorescence a cyme; bracts absent; flower buds pink and turning blue or purple on opening, 15–20 mm long, tube-shaped; five sepals, 2–4 mm long; five petals; five stamens; one pistil; fruit a nutlet. Weed designation: USA: CA.

Symphytum asperum

Brassicaceae

(Formerly Cruciferae)

MUSTARD FAMILY

Herbs, often with bitter juices; leaves alternate, simple, the surface often with branched hairs; stipules absent; inflorescence a raceme; flowers perfect, regular; four sepals; four petals; six stamens (four long, two short); one pistil; ovary superior; fruit a silicle or silique. Worldwide distribution: 375 genera/3,200 species; greatest diversity occurs in the north temperate regions.

Distinguishing characteristics: herbs; flowers with four petals; fruit podlike (silicle or silique).

The mustard family is economically important. Many species grown as food crops include radish, canola, broccoli, cabbage, rutabaga, and kohlrabi. Many species are troublesome weeds in cultivated fields.

silique (*Erysimum cheiranthoides*)

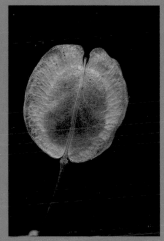

silicle (*Thlaspi urvense*)

Alliaria petiolata (Bieb.) Cavara & Grande
GARLIC MUSTARD

Also known as: herb garlic, hedge garlic, sauce-alone, Jack-by-the-hedge, poor man's mustard, Jack-in-the-bush, garlic root, garlicwort, mustard root

Scientific Synonym(s): *A. alliaria* B. & B., *A. officinalis* Andrz. ex Bieb., *Sisymbrium alliaria* L., *Erysimum alliaria* L.

QUICK ID: plant with a strong garlic odor; flowers white and appearing in spring; leaves triangular

Origin: native to Europe; introduced as a medicinal herb and salad green; first U.S. report from New York in 1868; first Canadian report from Toronto in 1879, spreading to British Columbia by 1948

Life Cycle: annual, winter annual, or biennial reproducing by seed only; large plants producing up to 7,900 seeds

Weed Designation: USA: AL, CT, MA, MN, NH, OR, VT, WA; CAN: AB, BC, SK

DESCRIPTION

Seed: Brownish-black, cylindrical, 2.5–3.8 mm long and 1 mm wide; surface with longitudinal ridges. Viable in soil for up to 6 years. Germination occurs in light or dark after dormancy has been broken; seeds dormant at maturity, requiring 50–150 days of cold stratification; in some climates requiring 1.5–2 years to break dormancy.

Seedling: With oblong to elliptical cotyledons about 6 mm long, stalks 5–7 mm long; stem below cotyledons up to 2 cm long; first leaves alternate, 1–5 cm across, margins coarsely toothed, stalks hairy; rosettes sensitive to summer drought, averaging 4–10 cm across, often purplish-green, continuing to grow during warm winter periods; survival rate from seedling to adult about 42%, often influenced by dry summers.

Leaves: Alternate, triangular to deltoid, 3–8 cm long, stalkless, margins coarsely toothed, reduced in size upward; basal leaves dark green, kidney-shaped, 6–10 cm across, margins with rounded teeth, stalks 1–5 cm long, hairy; young leaves with distinctive garlic odor when crushed. Inflorescence a raceme, elongating to 25 cm in fruit.

Flower: White, 6–10 mm wide; four sepals, green; four petals, white, 3 to 6 mm long; six stamens; one pistil; flowering in early spring.

Plant: Plants 10–100 cm tall, with strong garlic odor, somewhat shade-tolerant; stems rarely branched, hairless; up to 12 stems per rosette, although one to two is common.

Fruit: Linear silique (pod), 2.5–6 cm long and 2 mm wide, stalks spreading, about 5 mm long; six- to 22-seeded (average 17); average plants producing four to 16 seeds/fruit; large plants reporting to produce up to 422 fruits.

REASONS FOR CONCERN

Cattle who feed on garlic mustard are reported to produce milk with a disagreeable odor. Garlic mustard's ability to tolerate semishade has allowed it to spread throughout woodlands of northeastern North America, where it has displaced many of the native spring wildflowers. It is suspected of producing phytotoxic chemicals that interfere with the growth and establishment of native plant species.

Barbarea vulgaris Ait. f.
YELLOW ROCKET

Also known as: winter cress, St. Barbara's cress, bitter cress, rocket cress, yellow weed, water mustard, potherb, herb barbara, wound rocket

Scientific Synonym(s): *B. arcuata* (Opiz ex J. & K. Presl) Reichenb., *Campe barbarea* (L.) W. Wight ex Piper

QUICK ID: flowers yellow, appearing early spring; pods 2–5 cm long; leaves with a large terminal lobe

Origin: native to Europe and Asia; introduced into North America about 1800

Life Cycle: biennial or short-lived perennial reproducing by seed only; large plants may produce up to 116,000 seeds

Weed Designation: CAN: federal, ON, PQ

DESCRIPTION

Seed: Oblong, light yellow to dull grayish-brown, 1.4–1.7 mm long and less than 1.2 mm wide; surface smooth with a slight sheen, honeycomb or meshlike at high magnification.
Viable in soil for up to 20 years.
Germination occurs to a depth of 1 cm; increase germination with alternating temperatures and light intensity; optimal is 15°C in the dark followed by 25°C in bright light.

Seedling: With oval to round cotyledons, 4–9 mm long, 2–4 mm wide, pale green; first leaves spatula-shaped, margins smooth to wavy, underside with shiny veins; producing a rosette in first season's growth.

Leaves: Alternate, 1–2.5 cm long, one to four pairs per stem, margins wavy to lobed, stalkless and partially clasping the stem, reduced in size upward; lower leaves long-stalked; basal leaves dark green, 5–25 cm long, margins with two to several lateral lobes and a large terminal lobe. Inflorescence a pyramidal spikelike raceme.

Flower: Golden yellow, 10–16 mm across; four sepals; four petals, 5–8 mm long and 2–3 mm wide; six stamens; one pistil; appearing in early spring.

Plant: Plants 30–90 cm tall; stems hairless, branched near the top; roots fibrous, occasionally producing new shoots on existing crowns.

Fruit: Silique (pod), 2–5 cm long and about 1.5 mm across, nearly square in cross section, short-stalked and overlapping the pods above; beak 1.8–3 mm long; four- to 20-seeded.

REASONS FOR CONCERN

Yellow rocket is a troublesome weed of winter cereals, as it flowers and sets seed early in the growing season. It is host for the following viral diseases: beet curly top, cabbage black ring spot, cauliflower mosaic, cucumber mosaic, potato yellow dwarf, turnip crinkle, turnip mosaic, and turnip yellow mosaic.

Berteroa incana (L.) DC.
HOARY ALYSSUM

Also known as: false alyssum, hoary false madwort

Scientific Synonym(s): *Alyssum incanum* L.

QUICK ID: plants densely haired; flowers white; fruit a round pod

Origin: native to the Lake Baykal region of Russia

Life Cycle: annual, biennial, or short-lived perennial reproducing by seed; plants capable of producing up to 2,700 seeds

Weed Designation: USA: MI; CAN: AB, BC, SK

DESCRIPTION

Seed: Round to lens-shaped, dark reddish-brown, 1.3–1.8 mm long, 1.3–1.4 mm wide; an orange translucent wing seen at higher magnifications. Viability in soil at least 9 years. Germination occurs at 1–4 cm depth; optimal temperature 20°C.

Seedling: With round to club-shaped cotyledons, 3–8 mm long, 1.5–3.5 mm wide, surface with a whitish bloom; first leaves spatula-shaped, upper surface with numerous star-shaped hairs, lower surface densely covered in star-shaped hairs.

Leaves: Alternate, 1–4 cm long, stalkless and do not clasp the stem; basal leaves oblong to lance-shaped, long-stalked, margins smooth; all leaves covered in whitish-gray star-shaped hairs. Inflorescence a raceme, 5–25 cm long, dense at flowering and elongating in fruit.

Flower: White, numerous, 2–3 mm across; four sepals, densely haired; four petals, deeply notched; six stamens; one pistil; flowering until the ground freezes; stalks 7–8 mm long.

Plant: Plants 20–90 cm tall, leafy; stems grayish-green due to the presence of gray star-shaped hairs, often purplish and branched above; several stems rising from a single rootstalk.

Fruit: Oval silicle (pod), 5–8 mm long, 3–4 mm wide, surface with whitish-gray star-shaped hairs, four- to 12-seeded; borne on erect stalks.

REASONS FOR CONCERN

Hoary alyssum is a common weed of roadsides, waste areas, and the edge of cultivated fields. It is host for cabbage ring necrosis and turnip mosaic viruses.

Similar species: desert alyssum (*Alyssum desertorum* Stapf), an introduced weedy species found in waste sandy areas, resembles hoary alyssum. It is distinguished by yellow flowers and round, hairless fruit, about 3 mm across.

INDIAN MUSTARD

Also known as: Chinese mustard, leaf mustard, brown mustard

Scientific Synonym(s): *Sinapis juncea* L., *B. integrifolia* Rupr., *B. japonica* Thunb., *B. willdenowii* Boiss.

QUICK ID: plants hairless; flowers with four yellow petals; fruit 1.5–5 cm long, podlike and ascending

Origin: native to Asia; introduced as a crop for oilseed

Life Cycle: annual reproducing by seed only

Weed Designation: USA: MI, RI

DESCRIPTION

Seed: Round, dark brown to brownish-red or yellow, 1.6–2 mm long, 1.3–1.6 mm wide, surface conspicuously netted at high magnification. Viability unknown.
Germination occurs over a wide range of temperatures, 8°–37°C; optimal temperature 22°–23°C.

Seedling: With kidney- to heart-shaped cotyledons, hairless; first leaves oblong in shape with shallow lobes.

Leaves: Alternate, lower leaves fiddle-shaped and pinnately lobed, 10–20 cm long, long-stalked; upper leaves linear, margins smooth, stalkless. Inflorescence a raceme.

Flower: Pale yellow flowers, 10–15 mm across; four sepals; four petals, 7–9 mm long; six stamens; one pistil.

Plant: Plant 40–120 cm tall; stems light green, hairless, branched above; taproot white.

Fruit: Silique (pod), 1.5–5 cm long, borne on ascending stalks 8–15 mm long, beak 6–12 mm long.

REASONS FOR CONCERN

Indian mustard may be a problem weed in other mustard crops and canola, as it may go unnoticed during its development. It is a known host for the following viruses: cabbage black ring spot, cabbage ring necrosis, cauliflower mosaic, cucumber mosaic, radish mosaic, rape savoy, turnip mosaic, turnip yellows mosaic, and several others.

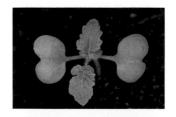

Similar species: white mustard (*Sinapis alba* L., p. 208), an introduced annual, is occasionally cultivated for its seed. The stems, up to 70 cm tall, have fiddle-shaped leaves up to 20 cm long. The stems leaves are reduced in size and less lobed or toothed. The yellow flowers, about 15 mm wide, give rise to pods 10–15 mm long and with four to eight seeds.

Camelina microcarpa DC.
SMALL-SEEDED FALSE FLAX

Also known as: false flax, Dutch flax, western flax, Siberian oilseed, littlepod false flax

Scientific Synonym(s): *C. sativa* ssp. *microcarpa* (DC.) F. Schmid

QUICK ID: flowers pale yellow; fruit round to pear-shaped; leaves alternate with simple to branched hairs

Origin: native to Europe and Asia

Life Cycle: annual or winter annual reproducing by seed

Weed Designation: CAN: federal

DESCRIPTION

Seed: Oblong to oval, rounded to triangular in cross-section, reddish-brown, 0.8–1.4 mm long, 0.5–0.9 mm wide; surface smooth and slightly glossy. Viability unknown.
Germination occurs when soil temperatures are between 20°C and 30°C.

Seedling: With oval cotyledons, 3.5–6 mm long, 2–3 mm wide, pale green; stalks with two or three short hairs; first leaves spatula-shaped, surface with simple and branched hairs.

Leaves: Alternate, 9–15 mm long, 4–6 mm wide, stalkless and clasping the stem with broad arrow-shaped bases; lower leaves covered in gray star-shaped hairs about 2 mm long; basal leaves spatula-shaped, bristly haired, margins smooth. Inflorescence a terminal raceme.

Flower: Pale yellow; four sepals, erect; four petals, spatula-shaped, 4–7 mm long and 4–5 mm wide, dark-veined; six stamens; one pistil.

Plant: Plants 20–100 cm tall, often shedding seeds while still in flower; stems erect and branched, grayish-green due to presence of simple and star-shaped hairs.

Fruit: Round to pear-shaped silique, 4–6 mm long, 10-or-more-seeded, beak less than 2 mm long; stalks spreading, 6–25 mm long; fruiting stem usually over 20 cm long.

REASONS FOR CONCERN

Small-seeded false flax is a common weed of cultivated and abandoned fields, roadsides, and waste areas. It is often a contaminant in livestock feed. Small-seeded false flax is host for cabbage black ring spot, cauliflower mosaic, turnip crinkle, and turnip rosette viruses.

Similar species: large-seeded false flax (*C. sativa* [L.] Crantz), also called gold of pleasure, is another introduced species that closely resembles small-seeded false flax. It has pods 6–9 mm long, containing four to 10 yellowish-brown seeds, each about 2 mm long. Plants are usually hairless. Weed designation: CAN: federal.

Capsella bursa-pastoris (L.) Medik.
SHEPHERD'S PURSE

Also known as: pepper plant, shepherd's pouch, pick pocket, mother's-heart, St. James weed, caseweed, pickpurse, witches' pouches, toothwort, shovel-plant

Scientific Synonym(s): *Thlaspi bursa-pastoris* L., *Bursa bursa-pastoris* (L.) Britt., *Bursa gracilis* Gren., *C. rubella* Reut.

QUICK ID: flowers white; pods triangular; leaves with wavy to deeply lobed margins

Origin: native to southern Europe; reported in North America prior to 1672

Life Cycle: annual or winter annual reproducing by seed; large plants producing up to 90,000 seeds; average plants produce about 4,500 seeds

Weed Designation: CAN: MB, PQ

DESCRIPTION

Seed: Round to oblong, dull orange, 0.9–1 mm long and 0.5 mm wide; surface smooth with a slight sheen.
Viable in soil for up to 35 years, although most seeds cease to be viable after 6.5 years. Germination enhanced by exposure to light, although emergence from the top 2 cm of soil is possible; optimal temperature is 20°–30°C.

Seedling: With oval to egg-shaped cotyledons, 2.5–5 mm long, 1–2 mm wide, long-stalked, surface grainy; stem below cotyledons dull green and stained with purple; first four leaves opposite, covered in numerous simple and star-shaped hairs.

Leaves: Alternate, oblong to lance-shaped, reduced in size upward, stalkless, and clasping the stem; margins smooth to lobed; basal leaves, 3–15 cm long and up to 4 cm wide, varying greatly between plants as some plants have leaves with wavy margins while others are deeply lobed. Inflorescence a raceme; flowering stem elongating with age.

Flower: White, 4–8 mm across; four sepals, green; four petals, 2–4 mm long; six stamens; one pistil.

Plant: Plant 5–80 cm tall, often remaining green during mild winters; stems erect and branched, surface smooth to somewhat hairy; taproot thin and branched; leaf margins of rosettes highly variable.

Fruit: Triangular silicle (pod), 5–8 mm long; two-chambered, 10–20-seeded; fruiting stem elongates with maturity.

REASONS FOR CONCERN

Shepherd's purse is a serious weed of cultivated fields, row crops, gardens, and waste areas. It can drastically reduce crop yields. Shepherd's purse is host for several viral diseases that include aster yellows, beet curly top, beet mosaic, beet ring spot, beet yellow, cabbage black ring spot, cabbage ring necrosis, cauliflower mosaic, cucumber mosaic, radish mosaic, and tobacco mosaic. It also harbors fungi that attack cabbage, turnips, and other members of the mustard family. It is a serious weed in at least 50 countries.

Also known as: flaxweed, tansymustard, herb Sophia, pinnate tansymustard
Scientific Synonym(s): *Sophia multifida* Gilib., *S. sophia* (L.) Britt., *Sisymbrium sophia* L.

QUICK ID: plants grayish-green; flowers yellow; leaves divided into narrow segments
Origin: native to Europe, Asia, and North Africa; first observed in North America at Montreal, Quebec, in 1821; first U.S. report from North Dakota in 1910
Life Cycle: annual, winter annual, or biennial reproducing by seed; large plants producing up to 75,600 seeds; average plants producing 75–650 seeds.
Weed Designation: CAN: MB

DESCRIPTION
Seed: Oblong, dull orange, 1–1.3 mm long, 0.4–0.6 mm wide; surface smooth; some seeds with a thin transparent wing on the rounded end. Viability is about 10% after 5 years in soil. Germination is at or near the surface; optimal temperatures 10°–30°C.
Seedling: With club-shaped cotyledons, 4–12 mm long, 1–1.5 mm wide, long-stalked, densely haired; stem below cotyledons with a few star-shaped hairs; first leaves opposite, three-lobed, covered in star-shaped hairs; later leaves alternate, divided into narrow segments.
Leaves: Alternate, 2–10 cm long, divided two or three times into narrow segments, stalked, surface with fine gray star-shaped hairs; basal rosette produced in the winter annual and biennial life cycles.
Inflorescence a terminal raceme; elongating with age.
Flower: Pale yellow, 2–4 mm across; four sepals; four petals; six stamens; one pistil; continuing to flower as lower fruits mature.
Plant: Plants 20–100 cm tall, grayish-green due to presence of star-shaped hairs; stems branched; taproot slender.
Fruit: Narrow silique (pod), 15–30 mm long and 1 mm wide, 20–40-seeded; stalks 6–12 mm long, making a wide angle with the stem.

REASONS FOR CONCERN

Flixweed is a serious weed that robs the crop of moisture and nutrients. It has been reported to reduce crop yields drastically in winter wheat and fall rye. Flixweed is a prolific seed producer.

Similar species: gray tansymustard (*D. incana* ssp. *incana* [Bernh. ex Fisch. & C.A. Mey.] Dorn) is easily distinguished from flixweed. The pale yellow flowers of gray tansymustard produce pods (5–10 mm long) that are held close to the stem. The leaves of gray tansymustard are divided once into narrow segments. Weed designation: CAN: MB.

Erucastrum gallicum (Willd.) O. F. Schulz
DOG MUSTARD

Also known as: rocketweed

Scientific Synonym(s):
Brassica erucastrum L.

QUICK ID: flowers pale yellow; leaves deeply lobed with a large terminal lobe; lower stem with stiff downward pointing hairs

Origin: native to Europe; introduced into United States in 1903; first Canadian report from Manitoba in 1922

Life Cycle: annual or winter annual reproducing by seed

Weed Designation: CAN: federal, AB, MB

DESCRIPTION

Seed: Oval, reddish to light brown, 1.2 mm long and 0.8 mm wide.
Viability unknown.
Germination occurs between 10°C and 30°C.

Seedling: Round to club-shaped cotyledons, 4–5 mm across, notched at the tip, long-stalked; first two or three leaves with wavy to shallowly lobed margins; later leaves divided into several lobes.

Leaves: Alternate, oblong to lance-shaped, 3–25 cm long, 1–6 cm wide, deeply lobed often to the midrib; basal and lower leaves with a distinct lobe at base; all leaves covered with stiff, simple hairs.
Inflorescence a terminal raceme; elongating with age.

Flower: Pale yellow, 4–8 mm across; four sepals; four petals, 4–7 mm long, veins greenish, six stamens; one pistil.

Plant: Plants 20–100 cm tall; stems branched, lower part with stiff downward-pointing hairs; taproot slender.

Fruit: Four-sided silique (pod), 2–5 cm long; stalks 6–10 mm long, curved upward; beak about 3 mm long.

REASONS FOR CONCERN

Dog mustard has the potential to be a serious weed in canola crops. It is a common weed of orchards, railways, roadsides, and disturbed ground.

Similar species: garden rocket (*Eruca vesicaria* ssp. *sativa* [P. Mill.] Thellung), an introduced plant that has escaped cultivation, is easily distinguished from dog mustard. Garden rocket has whitish-yellow petals, 12–20 mm long that are conspicuously veined. Pods are 12–25 mm long and 3–5 mm wide. Leaves, 5–15 cm long, are fiddle-shaped and deeply lobed.

Erysimum cheiranthoides L.
WORMSEED MUSTARD

Also known as: treacle
   mustard, treacle
   erysimum, wallflower
   mustard
Scientific Synonym(s):
   Cheirinia cheiranthoides
   (L.) Link

QUICK ID: plant covered
   with branched hairs; leaf
   margins wavy, not lobed;
   flowers yellow
Origin: native to North
   America; introduced
   subspecies may be
   present
Life Cycle: annual or winter
   annual reproducing
   by seed; large plants
   producing up to 3,500
   seeds
Weed Designation: CAN: PQ

DESCRIPTION
Seed: Oblong to elliptical; light orange to brown,
   1–1.2 mm long, 0.5–0.6 mm wide; surface slightly
   rough and somewhat shiny.
   Viability unknown.
   Germination occurs in the top 2 cm of soil;
   optimal temperature 20°–30°C.
Seedling: With spatula-shaped cotyledons, 1.5–5
   mm long, 1–2 mm wide, pale green, tips slightly
   indented; first two leaves opposite, angularly
   lobed, surface with a few simple to star-shaped
   hairs.
Leaves: Alternate, oblong to lance-shaped, 1–8 cm
   long and 2 cm wide, dark green, margins smooth
   to wavy, surface covered in numerous forked
   hairs; upper leaves stalkless, distinctly toothed.
   Inflorescence a terminal raceme up to 3 cm
   across; elongating with age.
Flower: Yellow, numerous, 6–10 mm across; four
   sepals, 2–3.5 mm long; four petals, 3.5–5.5 mm
   long; six stamens; one pistil.
Plant: Plants 20–100 cm tall; stems erect, purplish-
   green, occasionally branched, covered in short
   branched hairs; taproot short.
Fruit: Four-angled silique (pod), 1.5–3 cm long; beak
   short; stalks 6–12 mm long, erect, parallel to the
   stem.

REASONS FOR CONCERN

Wormseed mustard is not a concern in most crops, but it has the potential to be weedy in canola where it is not easily controlled. It is a common weed in gardens, lawns, pastures, and roadsides. The seeds of wormseed mustard are bitter tasting and pigs refuse to eat any feed that is contaminated with it. It is also host for cucumber mosaic, turnip crinkle, turnip rosette, and turnip yellow mosaic viruses.

Hesperis matronalis L.
DAME'S ROCKET

Also known as: dame's violet, mother-of-the-evening

Scientific Synonym(s): none

QUICK ID: flowers purple, fragrant; leaves alternate and hairy; fruit 2–14 cm long, constricted between seeds

Origin: native to Europe; introduced as a garden flower

Life Cycle: perennial reproducing by seed only; large plants producing up to 20,000 seeds

Weed Designation: USA: CO, CT, MA CAN: AB, SK

DESCRIPTION

Seed: Round, dark reddish-brown, 3–4 mm long. Viability 94% in dry storage after 1 year. Germination occurs at temperatures 20°–30°C; less than 10% germinate below 10°C.

Seedling: With oval cotyledons, 8–15 mm long, 5–8 mm wide, dull green, surface with short white branched hairs; first leaves opposite, oval; later leaves are alternate.

Leaves: Alternate, lance-shaped, 5–15 cm long; stalkless, margins shallowly toothed, both surfaces hairy.
Inflorescence a terminal raceme.

Flower: Showy and fragrant, pinkish-purple to white, 1.4–2.5 cm across; four sepals; four petals, 15–25 mm long; six stamens; one pistil; flowers reported to be more fragrant during the evening hours.

Plant: Plants 30–90 cm tall; stems branched, covered in simple to forked hairs; taproot slender.

Fruit: Cylindrical silique (pod), 2.5–14 cm long, slightly constricted between the seeds, 20–35-seeded, appearing in a single row; stalks erect to ascending.

REASONS FOR CONCERN

Dame's rocket is a common weed of ditches, fence lines, open forests, and orchards. In northeastern North America, it has invaded woodlands and displaced native vegetation. It is host for the following viral diseases: beet mosaic, cauliflower mosaic, radish mosaic, squash mosaic, and turnip mosaic.

Isatis tinctoria L.
DYER'S WOAD

Also known as: woad, Marlahan mustard

Scientific Synonym(s): none

QUICK ID: leaves bluish-green, hairless and clasping; flowers yellow; fruit blue to black

Origin: native to southeastern Europe or western Asia; once cultivated as a source of blue dye

Life Cycle: biennial or short-lived perennial reproducing by seed; average plants producing 383 seeds

Weed Designation: USA: AZ, CA, CO, ID, MT, NV, NM, OR, UT, WA, WY; CAN: AB

DESCRIPTION

Seed: Oblong, round in cross-section, dull yellowish- to orangish-brown, 3–4 mm long.
Viable up to 3 years in soil.
Germination occurs in fall and early spring; fall-germinating plants produce more seeds than spring-germinating plants; seeds removed from the fruit lack a dormant period, as the fruit wall contains water-soluble inhibitors that prevent germination until leached away.

Seedling: With oval cotyledons, 15–20 mm long, tip slightly indented, stalk 5–12 mm long; first leaves alternate, elliptical to oblong or oval, 15–20 mm long, margins smooth, stalks 4–10 mm long.

Leaves: Alternate, lance-shaped, 2–5 cm long, bluish-green, stalkless and clasping the stem with arrowhead-shaped bases, margins smooth, hairless; basal leaves elliptical to oblong or lance-shaped, 3–18 cm long, 1–4 cm wide, long-stalked, margins weakly toothed to wavy.
Inflorescence a large terminal flat-topped or umbrella-like panicle.

Flower: Bright yellow, 5–7 mm across; four sepals, spreading and unequal; four petals, 3–4 mm long; six stamens; one pistil; seeds maturing about 8 weeks after flowering begins.

Plant: Plants 20–120 cm tall; stems branched above, gray to purplish, hairless; taproot of rosettes to 1 m depth, lateral root growth in top 30 cm occur during second year's growth.

Fruit: Silicle (pod), 8–25 mm long, 5–7 mm wide, blue to purplish-black, flattened, and winged; one-seeded; borne on drooping stalks; fruit not opening at maturity.

REASONS FOR CONCERN

Dyer's woad is primarily a noxious weed of rangeland, crops, and undisturbed natural areas. It is a serious weed of in the intermountain region of the northwestern United States.

Lepidium chalepense L.
LENS-PODDED HOARY CRESS

Also known as: creeping hoary cress, white top, perennial peppergrass, whiteweed

Scientific Synonym(s): *Cardaria chalapensis* (L.) Handel-Mazzetti, *C. draba* (L.) Desv. var. *repens* (Schrenk) O.E. Schulz, *L. repens* (Schrenk.) Boiss.

QUICK ID: flowers white; stem leaves clasping; fruit lens-shaped

Origin: perennial introduced from western Asia (Uzbekistan to Israel) into California in 1918; first Canadian report from Alberta in 1926 as a contaminant in alfalfa seed from Turkestan

Life Cycle: perennial (living up to 8 years) reproducing by creeping rhizomes, root fragments, and seeds; plants are capable of producing 3,400 seeds; fragments 1 cm long at 1.5 cm depth producing new plants

Weed Designation: USA: AZ, CA, OR; CAN: federal, AB

DESCRIPTION

Seed: Oval, reddish-brown, 2 mm long and 1.5 mm wide.
Viability is about 52% after 3 years in soil; 5 years in dry storage.
Germination rate is increased with exposure to sunlight; germination occurs from 0.5°–40°C (optimal 20°–30°C).

Seedling: With club-shaped cotyledons, 7–9 mm long and 2.5 mm wide, dull grayish-green; stem below cotyledons usually brownish-green; first three pairs of leaves opposite, club-shaped; margins of second pair and later leaves with short downward-pointing hairs; lateral roots appear 2–3 weeks after emergence; first seasons growth producing 3.7 m of roots with an average of 455 buds.

Leaves: Alternate, 2–8 cm long, 3–30 mm wide; stalkless and clasping the stem with heart to arrowhead-shaped bases; leaves not reduced in size upward; basal leaf margins variable smooth to irregularly toothed, slightly hairy. Inflorescence a flat-topped raceme.

Flower: White, 4–6 mm across, stalks 5–15 mm long; four sepals, 2–2.5 mm long, green with white margins; four petals, 3–4 mm long; six stamens; one pistil.

Plant: Plants 20–50 cm tall, often forming large colonies in ditches, cultivated fields, and pastures; stems branched above; roots extensive and creeping, bark thick and corklike; most lateral roots found in top 18 cm of soil; roots generally increasing 60–80 cm per year.

Fruit: Lens- to kidney-shaped silique (pod), 2.5–6 mm long, 4–6 mm wide; two- to four-seeded; stems may have up to 850 fruits.

REASONS FOR CONCERN

Lens-podded hoary cress is the most serious weed species of the genus *Lepidium*. It is a serious weed in cultivated crops, forages, and pastures. Once established, it is difficult to eradicate due to the corklike roots.

Similar species: heart-podded hoary cress (*L. draba* L.), a native species of eastern Europe, is closely related to lens-podded hoary cress. As the common name implies, the heart-shaped fruit of heart-podded hoary cress is the distinguishing characteristic between the two species. Weed designation: USA: AK, AZ, CA, CO, HI, IA, ID, KS, MT, NV, NM, OR, SD, UT, WA, WY; CAN: federal, AB. Globe-podded hoary cress (*L. appelianum* Al-Shehbaz), also called Siberian mustard, is distinguished from the other two species by the globe-shaped fruit covered in short simple hairs. Weed designation: USA: AK, AZ, CA, CO, ID, KS, MT, NV, NM, OR, SD, UT, WA, WY; CAN: federal, AB.

heart-podded hoary cress
(*Lepidium draba*)

Lepidium densiflorum Schrad.
PEPPERGRASS

Also known as: common pepperweed, greenflower pepperweed, miner's pepperwort, pepperweed, prairie pepperweed

Scientific Synonym(s): *L. neglectum* Thellung, *L. texanum* Buckl.

QUICK ID: plants bushy; flowers pinkish-white, small; fruit heart-shaped

Origin: native to North America; spread as a contaminant in seed and feed

Life Cycle: annual or winter annual reproducing by seed; plants may produce up to 5,000 seeds

Weed Designation: CAN: PQ

DESCRIPTION

Seed: Oval, orange, 1.4–1.8 mm long, 0.9–1.2 mm wide; surface slightly rough and shiny. Viable in the soil for up to 6 years; majority of seed germinate in the first four years. Germination requires light if seeds are less than 4 months old; older seeds require temperature fluctuations between 15C and 25C with exposure to light at the higher temperature.

Seedling: With oval cotyledons, long-stalked, hairless; first leaves round, long-stalked, hairless; later leaves irregularly lobed.

Leaves: Alternate, stalkless, margins smooth; reduced in size upward; basal leaves 3–10 cm long, 1–2 cm wide, stalked, margins toothed to deeply lobed. Inflorescence a terminal raceme, 7–10 cm long at maturity.

Flower: Pinkish-white, inconspicuous, less than 2 mm wide; four sepals, green; four petals, sometimes absent; six stamens; one pistil; borne on stalks 2–3 mm long.

Plant: Plants 20–50 cm tall, bushy; stems, covered in dense short hairs.

Fruit: Heart-shaped to round silique (pod), 2–3.5 mm long, margin papery, tip notched; two-seeded; an average of nine to 15 pods produced for every 1 cm of flowering stem.

REASONS FOR CONCERN

Common peppergrass is a serious weed of cultivated fields and can drastically reduce crop yield. It is commonly found on roadsides, waste areas, and farmyards.

Neslia paniculata (L.) Desv.
BALL MUSTARD

Also known as: yellow weed, neslia, yellow ball mustard

Scientific Synonym(s): *Myagrum paniculatum* L.

QUICK ID: fruit round, containing one seed; flowers yellow; leaves lance-shaped

Origin: native of southern Europe and North Africa; first Canadian report from Manitoba in 1891

Life Cycle: annual or winter annual reproducing by seed

Weed Designation: CAN: MB

DESCRIPTION

Seed: Oval, yellowish-brown, 2–2.5 mm long. Viability unknown.
Germination requires an after-harvest ripening period; 67% of seeds germinate the following spring, following a period of temperatures between 2°C and 5°C.

Seedling: With round cotyledons, 6–8 mm across, tip indented, hairless; first leaves oblong, tips pointed or rounded, stalked, surface covered in star-shaped hairs.

Leaves: Alternate, lance-shaped, 1–6 cm long; upper leaves, stalkless, and clasping the stem with arrowhead-shaped bases; basal leaves stalked; surfaces with numerous star-shaped hairs. Inflorescence a raceme, elongating with age.

Flower: Bright yellow, 2–3 mm wide; four sepals, green; four petals, six stamens; one pistil; stalks 6–10 mm long.

Plant: Plants 20–100 cm tall; stems yellowish-green, erect, branched; surface with numerous star-shaped hairs.

Fruit: Round silicle (pod), 2–4 mm in diameter, prominently veined; one- or two-seeded; fruit remaining on the stem and not opening at maturity.

REASONS FOR CONCERN

Ball mustard is a weed of concern for canola growers. The pods are similar in size to canola seeds making them difficult to separate, thereby reducing the quality of canola oil. It is also host for cabbage ring necrosis, cauliflower mosaic, turnip crinkle, turnip mosaic, and turnip yellow mosaic viruses.

Raphanus raphanistrum L.
WILD RADISH

Also known as: jointed charlock, jointed radish, wild kale, wild turnip, cadlock, wild rape, runch

Scientific Synonym(s):
R. arvense Mérat,
R. silvestris Lam.,
Rapistrum arvense All.,
Rapistrum raphanistrum Crantz

QUICK ID: flowers yellow, conspicuously veined; leaves alternate, deeply lobed; fruit constricted between seeds

Origin: native to Europe and Asia

Life Cycle: Annual or winter annual reproducing by seed only; average plants produce about 160 seeds, while large plants may produce up to 17,250 seeds; severe infestations may produce up to 61,000 seeds per square meter.

Weed Designation: CAN: federal, PQ

DESCRIPTION

Seed: Egg-shaped, reddish-brown, 4–6 mm long and 2 mm wide.
Viable for up to 6 years; deep burial extends viability.
Germination occurs in the top 2 cm of soil; soil temperature range 5°–35°C, optimal being 20°C.

Seedling: With heart- to kidney-shaped cotyledons, distinctly veined; stem below cotyledons with stiff hairs; first leaves oval, prominently veined; margins toothed, hairy, terminal lobe largest; underside hairy.

Leaves: Alternate, oblong to oval, 1–7.5 cm long, margins wavy to two to five irregularly toothed lobes; reduced in size upward; basal leaves, 5–20 cm long, margins with five to 15 oblong segments, lobes increasing in size toward the leaf tip, terminal segment largest; both surfaces with scattered stiff hairs.
Inflorescence a terminal raceme.

Flower: Bright to pale yellow, 18–20 mm across; four sepals; four petals, 10–20 mm long, often tinged with purple or conspicuously purple-veined; six stamens; one pistil; stalks 10–25 mm long.

Plant: Plants 30–90 cm tall; stems branched and bristly haired at base; taproot thick.

Fruit: Strongly ribbed silique (pod), 3–7.5 cm long, four- to 10-seeded, constricted between the seeds and breaking into barrel-shaped segments, each one-seeded; fruit not opening at maturity; 5–7 months to complete life cycle.

REASONS FOR CONCERN

Wild radish can drastically reduce crop yields, especially in canola where it may go unnoticed. It competes with crops for moisture, nutrients, and sunlight. Wild radish is also a concern in wheat as the pod segments are similar in size to that of the grain, making two difficult to separate. The seeds of wild radish are reported to be poisonous to livestock if eaten in large quantities. It is host for cabbage black ring spot, cauliflower mosaic, turnip crinkle, turnip rosette, and turnip yellow mosaic viruses.

Similar species: wild radish is occasionally confused with wild mustard (*Sinapis arvensis* L., p. 208), another introduced species. Wild mustard has smaller flowers, 10–15 mm across, and a smooth pod that does not break into segments at maturity.

Sinapis arvensis L.
WILD MUSTARD

Also known as: charlock, crunchweed, field kale, krautweed, water cress, yellow-flower, herrick, yellow mustard, corn mustard

Scientific Synonym(s): *Brassica kaber* (DC.) L. C. Wheeler, *B. arvensis* Rabenh., non L., *B. sinapis* Vis, *B. sinapistrum* Boiss, *Caulis sinapiaster* Krause, *Eruca arvensis* Noulet

QUICK ID: stems bristly, purplish at the nodes of branches; flowers yellow; beak of fruit constricted

Origin: native to Europe and Asia; common in New York state by 1748; first Canadian report from Nova Scotia in 1829, and spreading to Manitoba by 1860

Life Cycle: annual or winter annual reproducing by seed; large plants may produce up to 3,500 seeds

Weed Designation: USA: IA, MI, OH; CAN: federal, BC, MB, PQ

DESCRIPTION

Seed: Round, dark brown to black, 1–2 mm in diameter; surface dull and wrinkled under magnification.
Viable in soil for up to 60 years.
Germination occurs to a depth of 2 cm; soil temperatures between 11°C and 30°C required for germination.

Seedling: With kidney- to heart-shaped cotyledons, 6–20 mm long, 6–12 mm wide, hairless; first leaves oblong in shape with shallow lobes, stalks bristly haired; roots up to 87 cm long 5 days after germination, and 12 m by day 21.

Leaves: Alternate; lower leaves stalked, deeply divided into small lateral lobes and a large terminal lobe; upper leaves coarsely lobed, stalkless and not clasping the stem; all leaves prominently veined, somewhat hairy, especially on the veins of the underside.
Inflorescence a long terminal raceme.

Flower: Yellow, 10–15 mm across; four sepals; four petals; six stamens; one pistil; stalks 2–6 cm long; flowers appear 6 weeks after germination.

Plant: Plants 30–180 cm tall; stems with bristly downward-pointing hairs at base, nearly hairless at the top; junction of stem and branches often purplish-green; taproot slender.

Fruit: Purplish-green silique (pods), 2–5 cm long, prominently ribbed, smooth to somewhat hairy, constricted near the beak; 10–18-seeded, the beak one-seeded; stalks about 5 mm long.

REASONS FOR CONCERN

Wild mustard is a strong competitor with crops, especially in canola, where it may go unnoticed. Besides using moisture and nutrients, wild mustard seeds will lower the quality of canola oil.

Similar species: several introduced species of mustard are cultivated in Canada and the United States. Two species, Polish and Argentine canola, are cultivated for the oil content of their seeds. Polish canola (*B. rapa* L.) and Argentine canola (*B. napus* L.) are easily distinguished from wild mustard. The stems of canola are nearly hairless and have clasping stem leaves.

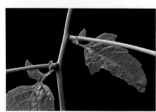

Sisymbrium altissimum L.
TUMBLING MUSTARD

Also known as: Jim Hill mustard, tall sisymbrium, tall hedge mustard, tumbleweed mustard

Scientific Synonym(s): *Norta altissima* (L.) Britt.

QUICK ID: flowers pale yellow; leaves divided into threadlike segments; pods 5–10 cm long

Origin: native to the western Mediterranean region and North Africa

Life Cycle: annual, winter annual, or biennial reproducing by seed; plants may produce up to 2,700 seeds

Weed Designation: CAN: MB

DESCRIPTION

Seed: Oblong, dull orange, 1.1–1.3 mm long, 0.6–0.8 mm wide; surface smooth, semiglossy at high magnification.
Viable for 40 years in dry storage; under field conditions, for between 7 and 17 years. Germination occurs in the top 2 cm of soil; temperature range 0°–20°C.

Seedling: With oval to club-shaped cotyledons, 2–6 mm long, 1–2 mm wide, pale green; first leaves oval with angular lobes, margins bristly haired.

Leaves: Alternate, triangular in outline, divided into threadlike segments; upper leaves smaller with two to five threadlike segments, terminal lobe smaller than lateral lobes; upper and lower leaves pale green, bristly haired; lower leaves often withered by flowering stage; basal and lower leaves 5–30 cm long with five to eight pairs of broad triangular lobes.
Inflorescence a terminal raceme.

Flower: Pale yellow, 6–14 mm wide; four sepals, 4–5 mm long; four petals, 6–8 mm long; six stamens; one pistil.

Plant: Plants 30–120 cm tall, pale green, becoming woody and brittle with age, breaking off at ground level and forming tumbleweeds; stem erect, profusely branched, covered in white hairs at base; roots to 43 cm deep.

Fruit: Silique (pod), 5–10 cm long, wide-spreading and resembling the branches or fruiting stalk; up to 200 seeds per pod.

REASONS FOR CONCERN

Tumbling mustard is a common weed of roadsides, fence lines, and pastures. Recently there have been increased reports of infestations in cultivated fields, coinciding with changes in cultivation practices. Tumbling mustard is host for cabbage ring necrosis, cauliflower mosaic, and turnip mosaic viruses.

Similar species: tall hedge mustard (*S. loeselii* L.) is very similar to tumbling mustard. It is distinguished by shorter petals (5–6 mm long) and shorter pods (1–4 cm long) with a thick stalk. Also, the upper leaf divisions are lance-shaped. Weed designation: CAN: federal.

Thlaspi arvense L.
STINKWEED

Also known as: pennycress, Frenchweed, fanweed, field pennycress, bastardweed, bastard cress, dish mustard, mithridate mustard

Scientific Synonym(s): *Crucifera thlaspi* Krause, *T. collinum* M. Bieb., *Thlaspidea arvensis* Opiz

QUICK ID: fruit heart-shaped and flat; plant with garlic, mustard, or turnip odor; flowers white

Origin: native to the eastern Mediterranean region of Europe and Asia; first report from Detroit in 1701; first Canadian report from Manitoba in 1860

Life Cycle: annual or winter annual reproducing by seed; large plants producing up to 20,000 seeds

Weed Designation: CAN: federal, MB, PQ; MEX: federal

DESCRIPTION

Seed: Oval, purplish-brown, 1.7–2 mm long, 1.1–1.2 mm wide; surface with six fingerprintlike ridges on each face.
Viable in soil for 20 years.
Germination occurs at soil temperatures of 10°–25°C; alternating temperatures and light intensity enhance germination.

Seedling: With oval to spoon-shaped cotyledons, 4–10 mm long, 2–6 mm wide, hairless; stalks up to 7 mm long; first leaves opposite, margins wavy with one to three indentations per side.

Leaves: Alternate, oblong, stalkless, and clasping the stem, margins smooth to toothed; basal leaves oval, 1–10 cm long, widest at the leaf tip, stalked, margins with a few teeth on each side; basal leaves withering soon after the flowering stems emerge.
Inflorescence a terminal cluster; elongating as the plant matures.

Flower: White, 1–3 mm across; four sepals, 1–2 mm long; four petals, 3–4 mm long; six stamens; one pistil; flowering from early spring to late fall and warm periods during the winter.

Plant: Plants 5–80 cm tall, pale green and turning yellowish with age; stems branched at base; crushed plants with a strong turnip odor; studies show that winter annual plants are more prolific than annual plants.

Fruit: Flat heart-shaped silicle (pod), 1–2 cm long, margins broad and papery; four- to 16-seeded.

REASONS FOR CONCERN

Stinkweed is a serious weed of cultivated fields, row crops, gardens, and waste areas. It has been reported to reduce yields drastically in all crops. When eaten by livestock, stinkweed causes off-flavors in milk and meat. Plants are not readily eaten because of bitter mustard oils. Feed containing large amounts of seed may be poisonous to livestock. Research has shown that a lethal dose for cattle is 65 mg/kg of body weight.

Similar species: shepherd's purse (*Capsella bursa-pastoris* [L.] Medic, p. 188) is occasionally confused with stinkweed. Shepherd's purse has triangular fruit, whereas the fruit of stinkweed is heart-shaped.

OTHER SPECIES OF CONCERN IN THE MUSTARD FAMILY

Armoracia rusticana

ARMORACIA RUSTICANA GAERTN. — HORSERADISH

Native to Europe and western Asia; perennial with spindle-shaped woody roots; stems erect, 50–200 cm tall; basal leaves oblong to lance-shaped, 20–45 cm long, 5–12 cm wide; stalks up to 60 cm long, margins toothed; stem leaves linear to lance-shaped, reduced in size upward, margins toothed; inflorescence a raceme; flowers white; four sepals, 2–4 mm long; four petals, 5–8 mm long; six stamens, one pistil; fruit an inflated silicle, 4–6 mm long, borne on ascending stalks 8–20 mm long. Weed designation: none.

Brassica rapa

BRASSICA RAPA VAR. *RAPA* L. (SYNONYM: *B. CAMPESTRIS* L.) — BIRDRAPE, FIELD MUSTARD

Native to Europe; annual, bluish-green in color; stems erect, 20–120 cm tall, branched from the base; leaves alternate, lower leaves 15–20 cm long, 5–7 cm wide, wavy-toothed to deeply two- to four-lobed, stalked; upper leaves reduced in size, oblong to lance-shaped, stalkless, and clasping the stem; inflorescence a raceme; flowers pale yellow, about 1 cm across; four sepals; four petals, 6–11 mm long; six stamens; one pistil; stalks 2 cm long or more; fruit a silique (pod), 3–8 cm long, stalk ascending, 7–25 mm long; seeds dark brown to black, about 1.5 mm wide. Closely related and resembling canola. Weed designation: USA: MI; CAN: PQ.

CHORISPORA TENELLA (PALLAS) DC. — BLUE OR PURPLE MUSTARD, CROSSFLOWER

Native to Russia and southern Asia; winter annual, ill-scented; stems erect to spreading, 15–50 cm tall, leafy, covered in sticky hairs; leaves alternate, sticky-haired, elliptical to oblong or lance-shaped; lower leaves 3–8 cm long, stalked, margins wavy-toothed to lobed; upper leaves reduced in size upward, stalkless, margins smooth to wavy-toothed; inflorescence a raceme; flowers showy, pale purple to blue, 10–15 mm across; four sepals, 6–8 mm long, often purplish; four petals, 10–13 mm long; six stamens; one pistil; fruit a silique, 3–4.5 cm long, 2–4 mm diameter, breaking into segments at maturity; seeds reddish-brown, about 1.5 mm diameter. Weed designation: USA: CA.

CONRINGIA ORIENTALIS (L.) DUMORT—HARE'S EAR MUSTARD, TREACLE MUSTARD

Native to eastern Mediterranean region of Asia; annual or winter annual; stems 30–100 cm tall, slightly succulent, simple to branched, hairless; stem leaves alternate, stalkless, and clasping the stem with heart-shaped bases; basal leaves oval, 5–13 cm long, short-stalked, margins smooth; all leaves bluish-green, hairless, somewhat fleshy; inflorescence a terminal raceme; flowers yellowish- to greenish-white, about 6 mm across; four sepals; four petals, 8–12 mm long; six stamens; one pistil; fruit a four-angled silique (pod), 4–15 cm long and 2 mm wide, slightly twisted or curved; stalks erect, 1–1.5 cm long; seeds oblong, dark reddish-brown, 2.1–3 mm long and less than 1.7 mm wide. Weed designation: MEX: federal.

Conringia orientalis

LEPIDIUM CORONOPUS (L.) AL-SHEHBAZ (SYNONYM: *CORONOPUS SQUAMATUS* [FORSSK.] ASCHERS)—GREATER SWINECRESS

Native to Mediterranean region of western Asia; annual or biennial; stems trailing to spreading, 5–30 cm long, branched; leaves alternate, oblong to oval or lance-shaped, margins deeply lobed and toothed; lower leaves stalked, upper stalkless; inflorescence a dense axillary raceme; flowers white or purplish; four sepals, spreading; four petals (occasionally absent), 1–1.5 mm long; two or four stamens; one pistil; fruit a silicle, 2.5–3 mm long, stalk 1–2 mm long. Weed designation: USA: AZ, CA.

DIPLOTAXIS MURALIS (L.) DC.—STINKING WALLROCKET, WALL MUSTARD

Native to Europe; annual or biennial with an odor resembling rotting turnips; stems 20–50 cm tall, much branched, hairless; leaves primarily basal, 5–10 cm long, oblong to lance-shaped in outline, margins toothed to deeply lobed; stem leaves alternate,

Lepidium coronopus

Diplotaxis muralis

Draba nemorosa

Lepidium latifolium

few; inflorescence a terminal raceme; flowers light yellow, stalks 1–1.5 cm long; four sepals, 3–5 mm long; four petals, 5–10 mm long; six stamens; one pistil; flowering until killed by frost; fruit a silique (pod), 2–3.5 cm long and 2 mm wide; stalks ascending 8–20 mm long; fruiting stem elongating with age; seeds elliptical to oval, light brown, 1.2 mm long and 0.8 mm wide. Weed designation: none.

DRABA NEMOROSA L. — YELLOW WHITLOW GRASS

Native to North America; annual; stems erect, 6–45 cm tall, covered with simple and fork-shaped hairs; basal leaves oblong to oval or lance-shaped, 1–3.5 cm long, 5–15 mm wide, densely haired; three to 12 stem leaves, stalkless, oval to oblong, 5–18 mm long, 3–10 mm wide; inflorescence a raceme, 25–60-flowered; flowers yellow; four sepals, 1–1.6 mm long; four petals, 1.7–2.2 mm long; six stamens; one pistil; fruit a silicle, 5–8 mm long, borne on stalks 7–25 mm long; seeds reddish-brown, less than 1 mm long. Weed designation: none.

LEPIDIUM LATIFOLIUM L. — PERENNIAL PEPPERWEED

Native to southern Europe and western Asia; perennial with creeping rhizomes up to 3 m long; crown woody; stems erect, up to 200 cm tall, branched; leaves alternate, oblong to oval or lance-shaped, margins smooth to toothed; lower leaves up to 30 cm long and 8 cm wide, stalks 3–15 cm long; upper leaves stalkless, 1–4 cm wide, not clasping the stem; inflorescence a panicle; racemes six- to eight-flowered; flowers white; four sepals, green; four petals, spoon-shaped, about 1.5 mm long; six stamens; one pistil; fruit a round silicle, 1.5–2.5 mm across; one seed, reddish-brown. Weed designation: USA: AK, CA, CO, CT, HI, ID, MA, MT, NV, NM, OR, SD, UT, WA, WY; CAN: BC, SK.

NASTURTIUM OFFICINALE W. T. AITON (SYNONYM: *RORIPPA NASTURTIUM-AQUATICUM* [L.] HAYEK) — WATERCRESS

Native to Europe and Asia; aquatic to semiaquatic perennial; roots thin, fibrous; stems 10–60 cm long, hairless, erect to spreading (often floating), often rooting at nodes; leaves 2–15 cm long, compound with three to nine leaflets; each oval to round, terminal segment largest, margins smooth; inflorescence a raceme; flowers white, 4–6 mm wide, borne on stalks 8–20 mm long; four sepals; four petals, 3–4 mm long; six stamens; one pistil; fruit a silique, 10–25 mm long, 1–3 mm wide; seeds about 1 mm diameter. Weed designation: USA: CT.

Nasturtium officinale

RORIPPA AUSTRIACA (CRANTZ.) BESS.— AUSTRIAN YELLOW CRESS
Native to southeastern Europe and western Asia; perennial with thick roots and creeping rhizomes; stems erect, 30–90 cm tall; leaves alternate, 3–10 cm long, linear to oblong, oval, or lance-shaped, bluish-green, hairless, margins smooth to toothed, clasping the stem; inflorescence a panicle; racemes 7–12 cm long, bracts absent; flowers yellow; four sepals, 1–2.5 mm long; four petals, 3–6 mm long; six stamens; one pistil; fruit a silicle, round, 1.5–3 mm wide, borne on spreading stalks 4–15 mm long; seeds reddish-brown, 0.7–0.9 mm. Weed designation: USA: AK, AZ, CA, NV, WA.

RORIPPA SYLVESTRIS (L.) BESS.— CREEPING YELLOW CRESS
Native to western Asia and North Africa; perennial with creeping rhizomes, mat-forming; stems trailing to spreading, hairless, 15–70 cm tall, branched, often purplish-green when exposed to bright light; leaves alternate, 3.5–15 cm long, 1–4.5 cm wide, divided into narrow lobes, stalked, margins toothed, reduced in size upward, often one- to three-lobed; basal leaves 2–10 cm long, 2–2.5 cm wide, three- to six-lobed per side; inflorescence a terminal or axillary raceme 2–10 cm long; flowers light to bright yellow, 4–5 mm wide; four sepals, yellow, 2–2.5 mm long; four petals, spatula-shaped, about 4 mm long and 1.5 mm wide; six stamens; one pistil; stalks spreading, 5–10 mm long; fruit is a silique, 7–20 mm long, 1–1.3 mm wide; seeds reddish-brown. Weed designation: USA: CA, NC, OR.

Rorippa sylvestris

Campanulaceae

HAREBELL OR
BELLFLOWER FAMILY

Herbs (occasionally shrubs), often with milky
juice; leaves alternate, simple; stipules absent;
inflorescence of solitary flowers; flowers
perfect, regular or irregular; five sepals, united;
five petals, united into a bell or funnel, most
often blue; five stamens, borne on the petals;
one pistil; ovary inferior; fruit a capsule.
Worldwide distribution: 70 genera/2,000
species; tropical species often palmlike,
temperate species herbaceous.

Distinguishing characteristics: herbs;
flowers blue, bell or funnel-shaped; fruit a
capsule.

The harebell family is of limited economic
importance. Several species are grown for their
ornamental value.

common harebell (*Campanula rotundifolia*)

Campanula rapunculoides L.
CREEPING BELLFLOWER

Also known as: purple bell, garden harebell, rover bellflower, creeping campanula, creeping bluebell, rampion bellflower

Scientific Synonym(s): none

QUICK ID: flowers purple, bell-shaped; leaves alternate; rhizome white

Origin: native to Europe and Asia; introduced as a garden flower

Life Cycle: perennial reproducing by creeping rhizomes and seeds; plants may produce up to 3,000 seeds

Weed Designation: CAN: AB

DESCRIPTION

Seed: Elliptical, light brown, glossy, 1.5 mm long and 1 mm wide.
Viable for up to 10 years in soil.
Germination occurs in the top 2 cm of the soil.

Seedling: With oval cotyledons, 2–5 mm long, 1–3 mm wide, margins curled upward; first leaves heart-shaped, upper surface and margins slightly hairy.

Leaves: Alternate, 3–7 cm long; lower leaves heart-shaped, long-stalked, margins coarsely toothed; upper leaves lance-shaped, stalkless; reduced in size upward, changing from heart- to lance-shaped.
Inflorescence a one-sided raceme, or solitary from leaf axils.

Flower: Blue to light purple, nodding, bell-shaped, 2–3 cm long; five sepals, united; five petals, united; five stamens; one pistil.

Plant: Plants 20–100 cm tall; stems erect, unbranched, often hairy; rhizomes white and thick, tuberlike, often forming dense clumps.

Fruit: Round capsule, opening by three to five pores.

REASONS FOR CONCERN

Creeping bellflower is a weed of gardens, fence lines, and occasionally cultivated fields. It is shade-tolerant and able to survive in crops. It is a serious weed in lawns, where it competes with the turf, robbing it of moisture and nutrients.

OTHER SPECIES OF CONCERN IN
THE HAREBELL OR BELLFLOWER FAMILY

Lobelia inflata

LOBELIA INFLATA L. — INDIAN TOBACCO
Native to eastern North America; perennial with fibrous roots; stems 20–75 cm tall, winged, milky sap present; leaves alternate, spatula-shaped below and oblong to oval or lance-shaped above, 2–7 cm long, 5–35 mm wide, stalked, margins toothed; inflorescence a raceme, 10–20 cm long; flowers light blue to purplish or white, 6–8 mm long, irregular and two-lipped; five sepals, 3–5 mm long; five petals; five stamens, anthers purple; one pistil; stalks 3–6 mm long; fruit an inflated capsule, up to 8 mm long and 5–6 mm diameter; sepals persistent, 10-nerved. Poisonous. Weed designation: none.

Caprifoliaceae

HONEYSUCKLE FAMILY

Shrubs, woody vines, or occasionally herbs; leaves opposite, simple; stipules absent; inflorescence a cyme; flowers perfect, regular; five sepals, united; five petals, united; five stamens, attached to the petals; one pistil; ovary inferior; fruit a berry, drupe, or capsule. Worldwide distribution: 15 genera/400 species; primarily north temperate.

Distinguishing characteristics: shrubs; leaves opposite; fruit berry-like.

Several members of this family are grown for their ornamental value.

twining honeysuckle (*Lonicera dioica*)

Lonicera tatarica L.
TARTARIAN HONEYSUCKLE

Also known as: bush honeysuckle, fly honeysuckle

Scientific Synonym(s): none

QUICK ID: shrub; leaves opposite; flowers pink to red; fruit a red berry

Origin: native to Central Asia and southern Russia; introduced into North America in 1752 as an ornamental shrub

Life Cycle: perennial reproducing by seed; seed production higher in North American plants than those in the native range

Weed Designation: USA: CT, MA, NH, VT

DESCRIPTION

Seed: Oval, flattened, yellow to orange-red, 2.5–3.5 mm long, 2–2.5 mm wide, glabrous.
Viable up to 5 years in soil.
Germination occurs with soil temperatures between 15 and 25°C.

Seedling: With round to oval cotyledons, dark green, and distinctly veined; first leaves opposite, oval to elliptical, margins smooth and hairy.

Leaves: Opposite, oblong to oval, 2–6.5 cm long, 1–4 cm wide, stalks 1–8 mm long, margins smooth, base slightly heart-shaped.
Inflorescence a cyme, two-flowered, borne in leaf axils; stalks 1.5–2.5 cm long; bracts triangular, 3–8 mm long and 1 mm wide.

Flower: Pink to red (occasionally white), tube-shaped, 15–25 mm long; five sepals; five petals; four stamens; one pistil.

Plant: Plant a shrub 1–5 m tall, hollow; tolerant to shade, periodic flooding, and temperatures from −58°C to +40°C; young branches smooth and green, turning brown to brownish-gray with age.

Fruit: Pair of red, orange, or yellow berries, 5–8 mm diameter; slightly united at the base; two to eight seeds.

REASONS FOR CONCERN

Tartarian honeysuckle outcompetes and displace native vegetation. It is suspected that Tartarian honeysuckle has allelopathic affects on adjacent plants.

OTHER SPECIES OF CONCERN IN THE HONEYSUCKLE FAMILY

Lonicera japonica

LONICERA JAPONICA THUNB. — JAPANESE HONEYSUCKLE

Native to Japan; woody vine, trailing to twining, up to 9 m long; leaves opposite, oval to oblong, 4–8 cm long, 2–4 cm wide, stalks short, margins smooth, base round; inflorescence a cyme, two-flowered, borne in leaf axils, stalks 5–10 mm long; flowers white to cream, fragrant, 25–30 mm long, two-lipped; five sepals; five petals; four stamens; one pistil; fruit a bluish-black berry, 5–7 mm diameter. Weed designation: USA: CT, MA, NH, VT.

LONICERA MAACKII (RUPR.) HERDER — AMUR HONEYSUCKLE

Native to northeastern Asia; shrub up to 7 m tall; bark grayish-brown, scaly; twigs grayish-brown; leaves opposite, 4–9 cm long, long-pointed, margins smooth, surface green above, pale and hairy below; inflorescence a cyme, two-flowered, borne in leaf axils on stalks 1–6 mm long; flowers white, turning yellow with age, 1.8–2.5 cm long, fragrant; five sepals; five petals, the upper four fused; four stamens; one pistil; fruit a red berry, 5–8 mm diameter. Weed designation: USA: CT, MA, VT.

LONICERA MORROWII GRAY — MORROW'S HONEYSUCKLE

Lonicera morrowii

Native to Japan; shrub with scaly bark; stems up to 2.5 m tall; twigs hollow, grayish-brown; leaves opposite, oval, 2.5–6 cm long, bluish-green above and pale below; inflorescence a cyme, two-flowered, borne in leaf axils; flowers white, turning yellow with age, fragrant, 1.8–2.5 cm long; five sepals; five petals, tips free; four stamens; one pistil; fruit a red berry, 5–7 mm diameter. Weed designation: USA: CT, MA, NH, VT.

Caryophyllaceae

PINK OR CARNATION FAMILY

Herbs with swollen stem nodes; leaves opposite, simple; stipules often absent; inflorescence a cyme or solitary-flowered; flowers perfect (occasionally imperfect), regular; five sepals; five petals (occasionally absent); five or 10 stamens; one pistil; two to five styles; ovary superior; fruit a capsule. Worldwide distribution: 80 genera/2,000 species; primarily found in north temperate regions.

Distinguishing characteristics: herbs; leaves opposite; stem nodes swollen.

The pink family contains a number of showy flowers and plants that are often cultivated for their ornamental value. They include carnations, baby's breath, Maltese cross, and bouncing bet. There are several species in this family that are troublesome weeds.

night-flowering catchfly (*Silene noctiflora*)

Agrostemma githago L.
PURPLE COCKLE

Also known as: corn cockle, corn rose, corn campion, crown-of-the-field, corn mullein, old maid's pink, mullein pink

Scientific Synonym(s): *Lychnis githago* L., *A. linicola* Terechov, *Githago segetum* auct.

QUICK ID: stems bristly haired; leaves opposite; flowers purple

Origin: native to Europe and Asia

Life Cycle: annual, winter annual, or biennial reproducing by seed; plants capable of producing over 200 seeds

Weed Designation: USA: AR, SC; MEX: federal

DESCRIPTION

Seed: Kidney-shaped, purplish-black, 3–4 mm long; surface covered in several rows of rounded projections.
Viable for 1 year when buried in the soil. Germination requires a dormant period; fresh seeds germinate at temperatures between 4°C and 20°C.

Seedling: With oval to lance-shaped cotyledons, 16–30 mm long, 5–10 mm wide, dull green above and paler below, hairless; first pair of leaves lance-shaped, densely haired, midrib on underside prominent.

Leaves: Opposite, narrowly oblong, 8–15 cm long, 5–10 mm wide, stalkless, surface with long white hairs; leaves not reduced in size upward. Inflorescence a solitary flower borne at ends of stalks 5–20 cm long.

Flower: Showy, purple, 2.5–5 cm across; five sepals, united, 1–2.5 cm long, prominently 10-ribbed; five petals, often black-spotted; 10 stamens; one pistil; five styles.

Plant: Plants 30–120 cm tall; stems grayish-green, sparingly branched, covered in rough hairs; nodes swollen; taproot shallow.

Fruit: Capsule, 14–24 mm long, enclosed in a 10-ribbed urnlike calyx; seeds numerous.

REASONS FOR CONCERN

The seeds of purple cockle are poisonous to chickens. Wheat crops contaminated with purple cockle seeds produce flours that result in poor-quality breads and unpalatable taste.

Cerastium arvense L.
MOUSE-EAR CHICKWEED

Also known as: chickweed, field chickweed, starry chickweed

Scientific Synonym(s): *Cetunculus arvensis* Scopoli, *Leucodonium arvense* (L.) Opiz

QUICK ID: leaves opposite; flowers white; stem swollen at leaf nodes

Origin: perennial native to North America

Life Cycle: perennial reproducing by seed

Weed Designation: CAN: federal, MB

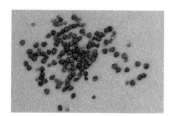

DESCRIPTION

Seed: Triangular, orange to brown, 0.6–1.2 mm long and wide; surface covered with small bumps or projections.
Viable in soil for up to 60 years; 30% germination after 10 years in soil.
Germination greatest when seeds are exposed to light.

Seedling: With oval cotyledons, 2–8 mm long and less than 2 mm wide, upper surface with a few hairs; first pair of leaves spatula-shaped, hairy above and on midveins below.

Leaves: Opposite, narrowly elliptical, 2–7 cm long, 1–15 mm wide, somewhat hairy.
Inflorescence a cyme; one- to 20-flowered.

Flower: White, 18–22 mm across; five sepals, lance-shaped, 5–8 mm long; five petals, 10–16 mm long, deeply notched at the tip; 10 stamens; one pistil; five styles; bracts small and papery.

Plant: Plants 10–60 cm tall; stems erect to ascending, often rooting at lower nodes; numerous branches give a densely matted appearance; surface variable in hairiness, some plants densely haired while others hairless.

Fruit: Cylindrical capsule, 7.5–11.5 mm long, one-chambered, opening by 10 teeth; seeds many.

REASONS FOR CONCERN

Mouse-ear chickweed is not a concern in cultivated fields as it does not survive tillage. However, it is a problem in pastures and rangeland where it invades overgrazed areas. Mouse-ear chickweed is host for strawberry yellow dwarf virus.

Gypsophila paniculata L.
BABY'S BREATH

Also known as: perennial
baby's breath, perennial
gypsophila, tall
gypsophill, maiden's
breath, bachelor's button,
perfoliate baby's breath

Scientific Synonym(s):
Arrostia paniculata (L.)
Raf.

QUICK ID: flowers white;
leaves opposite; stems
and branches bluish-
green

Origin: native to Europe
and Asia; introduced as
an ornamental plant;
first invasive report from
Manitoba in 1887

Life Cycle: perennial
reproducing by seeds;
plants capable of
producing up to 14,000
seeds

Weed Designation: USA: CA,
WA; CAN: AB, MB, SK

DESCRIPTION
Seed: Kidney-shaped, black, 1.5–2 mm long; surface
covered in rows of rounded bumps.
Viable in soil for at least 5 years.
Seeds require little or no dormant period; optimal
depth 2.5 mm; optimal temperature 10°–25°C.

Seedling: With lance-shaped cotyledons, 10–15 mm
long; 3–5 mm wide; first leaves opposite, lance-
shaped, similar in appearance to the cotyledons.

Leaves: Opposite, lance-shaped, 3–10 cm long,
2–10 mm wide; hairless; prominently one-
veined; reduced in size upward; stalkless but not
clasping; upper leaves less than 1 cm long; plants
with very few leaves when in flower.
Inflorescence a cyme.

Flower: Numerous, white, 3–6 mm wide, sweet-
scented; five sepals, about 2 mm long; five petals,
2.5–4 mm long; 10 stamens; one pistil; two styles;
stalks 3–12 mm long.

Plant: Plants 30–100 cm tall, at maturity breaking
at base and becoming a tumbleweed; stems
profusely branched, hairless, covered in a white
powdery film; rootstalk woody, up to 4 m deep,
increasing in size with age; older roots with
sufficient reserves to survive 2 years of drought.

Fruit: Egg-shaped capsule; four-chambered; eight to
20 seeds.

REASONS FOR CONCERN

Baby's breath is a serious weed of roadsides, waste areas, and pastures. It does not withstand any cultivation and is rarely found in crops. Baby's breath is reported to be host for aster yellows and beet curly top viruses.

Saponaria officinalis L.
BOUNCING BET

Also known as: soapwort, Fuller's herb, scourwort, old maid's pink, London pride, hedges pink, wild sweet William, sweet Betty, sea pink, sheepweed, world's wonder, lady's washbowl, wood phlox, mock gillyflower, wild phlox

Scientific Synonym(s): *Lychnis saponaria* Jessen, *L. officinalis* (L.) Scop., *S. hybrida* Mill., *S. vulgaris* Pall.

QUICK ID: plants hairless; leaves opposite, flowers pink

Origin: native to Europe; introduced as an ornamental plant; an extract from the plant (saponin) was used as a detergent

Life Cycle: perennial reproducing by creeping rhizomes and seeds

Weed Designation: USA: CO

DESCRIPTION

Seed: Kidney-shaped, black, 1–3 mm long and 1.5 mm wide; surface covered in small bumps or projections.
Viability unknown.
Germination occurs between 12°C and 25°C.

Seedling: With linear to lance-shaped cotyledons, 15–17 mm long, 2–3 mm wide, dull green; stem below the cotyledons often stained with purple; first pair of leaves lance-shaped; later leaves elliptical to spatula-shaped.

Leaves: Opposite, elliptical, 3–11 cm long, 1.5–4.5 cm wide, grayish-green; hairless; three- to five-nerved; stalks 1–15 mm long.
Inflorescence a cyme up to 15 cm long; densely flowered.

Flower: Showy, white to pink, about 12 mm across; five sepals, united into a tube, 15–20 mm long, 20-veined; five petals, united, clawed; 10 stamens, long-stalked; one pistil; two styles.

Plant: Plants 30–120 cm tall; stems branched, leafy, hairless; rhizomes creeping, somewhat woody, often forming large colonies.

Fruit: Cylindrical capsule, 15–20 mm long; opening by four teeth; seeds numerous.

REASONS FOR CONCERN

Bouncing bet or soapwort, another common name for this plant, contains saponin, a chemical that can be poisonous to livestock. Once established in pastures and rangeland, it is difficult to control due to the woody rhizomes.

Similar species: cow cockle (*Vaccaria hispanica* [P. Mill.] Rauschert, p. 252), another introduced weedy species, is often confused with bouncing bet. It is distinguished from bouncing bet by its unclawed petals and an oval-shaped calyx with five prominent ribs.

Scleranthus annuus L.
KNAWEL

Also known as: annual knawel, German knotwort
Scientific Synonym(s): *Knavel annuum* Scop.

QUICK ID: leaves opposite; flowers green; plants tufted, often forming mats
Origin: native to Europe
Life Cycle: annual or winter annual reproducing by seeds; plants capable of producing up to 100 seeds
Weed Designation: none

DESCRIPTION
Seed: Light brown, oval, about 2 mm long. Viability unknown.
Germination occurs with soil temperatures between 18°C and 20°C.
Seedling: With narrow linear cotyledons, 8–12 mm long, 3–5 mm wide, somewhat fleshy; first leaves 5–20 mm long, 1–2 mm wide.
Leaves: Opposite, awl-shaped, 5–24 mm long, 1–1.5 mm wide, stalkless, margins with a few hairs and transparent papery edges near the base. Inflorescence a cyme.
Flower: Green, 4–5 mm across; five sepals, 3–4 mm long, partially fused; eight to 10 stamens; one pistil; two styles; flowers shorter than the bracts.
Plant: Plants forming mats, prefers acidic soils; stems weak and spreading, 2.5–25 cm long, brownish-green; nodes swollen.
Fruit: Oval capsule, 3.2–5 mm long, enclosed by hardened calyx; one-seeded.

REASONS FOR CONCERN

The roots of knawel are tough and fibrous and can survive light cultivation. Thorough cultivation in the fall is required to control the winter annual plants. Knawel can survive light shade and may be found in some crops.

Similar species: perennial knawel (*S. perennis* L.), native to Europe, is found on roadsides and sandy areas of northeastern North America. It is distinguished by a woody crown, bracts shorter than the flowers, leaves 3–13 mm long, and overlapping sepals 2.5–4 mm long.

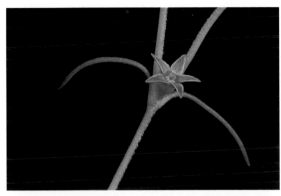

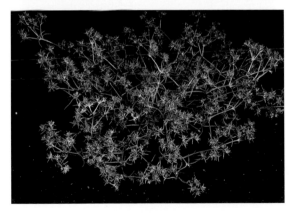

Silene latifolia ssp. *alba* (P. Mill.) Greuter & Burdet
WHITE COCKLE

Also known as: campion, white campion, evening campion, evening lychnis, snake cuckoo, thunder flower, bull rattle, white robin

Scientific Synonym(s): *S. alba* (Mill.) E. H. L. Krause, *S. pratensis* (Raf.) Godr. & Gren., *Lychnis alba* Mill., *L. vespertina* Sibth., *Melandrium album* (Mill.) Garcke

QUICK ID: leaves opposite; calyx with 10 or 20 prominent veins; flowers white, opening at night

Origin: native to Europe; introduced into United States in early 1800s; first Canadian report from Ontario in 1875

Life Cycle: annual, biennial, or short-lived perennial reproducing by seed; plants capable of producing up to 24,000 seeds

Weed Designation: USA: WA; CAN: federal, AB, NS, PQ, SK

DESCRIPTION

Seed: Kidney-shaped, gray to brown, 1–2 mm long; surface covered in rows of warty bumps.
Viable in soil for up to 20 years; immature seeds viable 2–3 weeks after pollination.
Germination requires light; emergence from top 5 cm of soil; optimal temperatures 20°–30°C.

Seedling: With lance-shaped cotyledons, 5–10 mm long, 1–4 mm wide, surface slightly hairy, tip pointed; stem below cotyledons yellowish-green; first leaves oval, surface hairy, margin with long hairs.

Leaves: Opposite, oblong, 2–12 cm long, 6–30 mm wide; lower leaves stalked while the upper stalkless.
Inflorescence a cyme.

Flower: White, fragrant, opening in the evening, and closing in the morning, 2.0–3.5 cm across; male or female; five sepals, united into a tube, 15–20 mm long, males prominently 10-veined, females 20-veined, surface with sticky hairs; five petals, deeply notched; 10 stamens; one pistil; four or five styles.

Plant: Male or female, 20–100 cm tall, somewhat sticky throughout; stems green, leafy; taproot fleshy, up to 120 cm deep; germination to seed maturity in about 40 days.

Fruit: Oval capsule, 10–15 mm long, opening by 10 teeth; up to 500 seeds per fruit.

REASONS FOR CONCERN

White cockle is a serious weed in cultivated crops and pastures. It competes for moisture, nutrients, and sunlight. The seeds, similar in size to clover, are difficult to separate. It is also host for lychnis ring spot virus, which infects sugar beets and spinach crops. Other viral diseases—such as cucumber mosaic, tobacco necrosis, tobacco ring spot, and tobacco streak—are associated with this weed.

Similar species: night-flowering catchfly (*S. noctiflora* L., p. 254), is often confused with white cockle. Night-flowering catchfly has perfect flowers with both stamens and styles present in the same flower.

Also known as: sticky cockle, night-flowering silene, night-flowering cockle, clammy cockle

Scientific Synonym(s): *Melandrium noctiflorum* (L.) Fries

QUICK ID: plants sticky-haired throughout; leaves opposite; calyx prominently 10-veined

Origin: native to Europe; introduced in clover or grass seed; first reported in United States in 1822; first Canadian report Ontario in 1862, spreading to the prairies by 1883

Life Cycle: annual or winter annual reproducing by seed; plants producing up to 2,600 seeds

Weed Designation: CAN: federal, BC, MB, PQ, SK; MEX: federal

DESCRIPTION

Seed: Kidney-shaped, gray, 1.1–1.3 mm long, 0.9–1.2 mm wide; surface covered in several rows of warty bumps.

Viability in soil is about 82% after 5 years. Germination occurs when soil temperatures reach 20°C.

Seedling: With club-shaped cotyledons, 5.5–15 mm long, 3–4.5 mm wide, stalks with short stiff hairs; first pair of leaves spatula-shaped, stalks and margins hairy.

Leaves: Opposite, 2–12 cm long, 3–40 mm wide, sticky-haired, stalkless; reduced in size upward; basal leaves oblong, 4–12 cm long, 2–4 cm wide, stalked.

Inflorescence a cyme; three to 15-flowered.

Flower: White to pale pink, fragrant, and opening at night, 20–25 mm across; five sepals, sticky-haired, about 15 mm long when the flowers open, prominently 10-veined, dark green; five petals, deeply notched, 20–35 mm long; 10 stamens; one pistil; three styles.

Plant: Plants 20–100 cm tall, hairy throughout and trapping small insects; one to three stems, somewhat woody; nodes swollen.

Fruit: Capsule; three-chambered; opening by six teeth, curled backward; containing up to 185 seeds; calyx 25–40 mm long at maturity.

REASONS FOR CONCERN

Night-flowering catchfly is a serious weed in cultivated crops and pastures. It competes for moisture, nutrients, and sunlight. The seeds, similar in size to clover, are difficult to separate. Night-flowering catchfly is host for tobacco streak virus.

Similar species: white cockle (*S. latifolia* ssp. *alba* [P. Mill.] Greuter & Burdet, p. 242), is often confused with night-flowering catchfly. Both species bloom in the evening. White cockle has male and female plants, unlike night-flowering catchfly. Male plants have flowers with 10 stamens and a calyx with 10 prominent veins. Female plants have flowers with four or five styles and a calyx with 20 prominent veins.

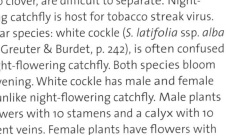

Silene vulgaris (Moench) Garcke
BLADDER CAMPION

Also known as: white bottle,
cowbell, bubble-poppy,
sea pink, maiden's tears,
devil's rattlebox, white
hen, bird's eggs, snappery,
rattleweed, fairy potatoes
Scientific Synonym(s):
S. cucubalus Wibel,
S. inflata Sm., *S. latifolia*
(P. Mill.) Britten & Rundle,
Oberna commutata
(Guss.) S. Ikonnikov

QUICK ID: calyx inflated,
conspicuously veined;
leaves opposite; flowers
white
Origin: perennial introduced
from Europe and Asia
Life Cycle: perennial
reproducing by root
fragments and seeds;
plants producing up to
24,000 seeds
Weed Designation: CAN:
federal, MB, PQ, SK

DESCRIPTION
Seed: Kidney-shaped, brown to gray, 1–1.5 mm long,
surface covered in rows of rounded bumps.
Viable up to 14 years in soil.
Germination occurs at temperatures between
20°C and 25°C; exposure to light produces about
5% more seedlings.
Seedling: With elliptical cotyledons, 3–12 mm long,
1–2.5 mm wide, yellowish-green; stem below the
cotyledons pale green, translucent; first pair of
leaves lance-shaped, yellowish-green; second
pair of leaves with finely toothed margins.
Leaves: Opposite, oval to lance-shaped, 2–8 cm
long, 0.5–3 cm wide; pale green due to a white
powdery film; stalkless.
Inflorescence a cyme; five- to 40-flowered.
Flower: White, 10–20 mm across; five sepals, united,
inflated, 15–20 mm long, prominently 20-veined
(the veins pinkish-white); 10 petals, deeply
notched, 14–16 mm long; 10 stamens; one pistil;
three styles.
Plant: Plants 20–100 cm tall; stems branched at
base, hairless; nodes swollen; taproot deep and
penetrating.
Fruit: Round capsule about 1 cm long; enclosed by
inflated calyx; three-celled; seeds numerous.

REASONS FOR CONCERN

Bladder campion is not a serious weed in cultivated crops, however, it has the potential to be weedy in zero- to minimal-till fields, pastures, roadsides, and gardens. Once established on these sites, bladder campion is difficult to eradicate and often crowds out beneficial plants.

Spergula arvensis L.
CORN SPURREY

Also known as: stickwort, starwort, devil's guts, sandweed, pickpurse, yarr, pinecheat, povertyweed, cowquake

Scientific Synonym(s):
S. sativa Boenn.

QUICK ID: leaves in whorls of 12–16; stems and leaves somewhat fleshy; flowers with five white petals

Origin: native to Europe; introduced as a forage crop into Michigan in 1888

Life Cycle: annual reproducing by seed; plants capable of producing up to 10,000 seeds

Weed Designation: CAN: PQ

DESCRIPTION

Seed: Round to lens-shaped, black, 1–2 mm in diameter; surface often pitted, white ring appearing around the center.
Viable in soil for up to 10 years; seeds excavated from Iron Age dwellings (about 2,000 years ago) have germinated in laboratory experiments. Seeds germinate from 0.5–3 cm soil depth; germination occurs between 20°C and 25°C.

Seedling: With needle-like cotyledons, 10–15 mm long and less than 1 mm wide, tip blunt; first leaves similar in appearance to the cotyledons.

Leaves: Opposite, but appearing whorled, six to eight pairs originating at one node; threadlike leaves, 1.5–5 cm long, somewhat fleshy; slightly round in cross-section.
Inflorescence a loosely branched cyme.

Flower: White, 4–6 mm across, opening only in bright light; five sepals; five petals; five or 10 stamens; one pistil; five styles; ill-scented; flowering occurs within 8 weeks of germination.

Plant: Bushy, readily eaten by livestock; stems spreading up to 60 cm long, fleshy, somewhat sticky; nodes swollen; germination to mature seed in 10 weeks.

Fruit: Round capsule; five-celled; large plants with up to 500 fruits; fruit produced in early summer with up to 25 seeds while those in late fall only have five.

REASONS FOR CONCERN

Corn spurrey is a serious weed in cultivated fields and row crops. It is host for beet yellows, lucerne mosaic, and tobacco streak viruses. It is reported as weedy in 33 countries.

Similar species: knawel (*Scleranthus annuus* L., p. 241) is often confused with corn spurrey because of its growth habit. It is distinguished by awl-shaped leaves that appear in pairs at stem nodes. The flowers of knawel do not have petals and the fruit is one-seeded and surrounded by an enlarged calyx. The seedling of Russian thistle (*Salsola kali* L., p. 267) may be mistaken for the seedling of corn spurrey. The distinguishing feature between the two seedlings is that Russian thistle leaves are spine-tipped.

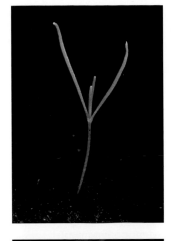

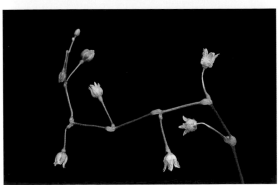

Stellaria media (L.) Vill.
COMMON CHICKWEED

Also known as: chickwhirtles, cluckenweed, mischievous Jack, skirt buttons, tongue grass, starweed, winter weed, satin flower, cyrillo, starwort, bindweed, white bird's-eye

Scientific Synonym(s):
S. apetala Ucria ex Roemer, *Alsine media* L.

QUICK ID: leaves opposite; plants prostrate; flowers with five white petals

Origin: native to Europe; introduced into New England in 1672; common in Quebec and Nova Scotia by 1820s; now found on all continents, including Antarctica

Life Cycle: annual or winter annual reproducing by seed; plants capable of producing up to 15,000 seeds; capable of producing seed 5 to 7 weeks after germination—with four generations possible in one growing season, a single seed can produce over 15 billion seeds

Weed Designation: CAN: federal, MB, PQ

DESCRIPTION

Seed: Oblong, pale yellow to reddish-brown, 1–1.2 mm long; surface covered in rows of small, rounded projections.
Viability is 22% after 10 years in soil.
Germination occurs in the top 3 cm of soil and can occur at temperatures of 2°C, although 12°–20°C is optimal.

Seedling: With oval cotyledons, 1–12 mm long and less than 2 mm wide, tips pointed, stalks with a few clear hairs; first pair of leaves oval with a pointed tip, stalks with a fringe of white hairs.

Leaves: Opposite, oval, 0.5–4 cm long, 2–20 mm wide; stalks of lower leaves 5–20 mm long with a single line of hairs; upper leaves stalkless, about 1 cm wide.
Inflorescence a cyme; five- to many-flowered.

Flower: White, 4–7 mm across; five sepals, green, 4–6 mm long, hairy; five petals (occasionally absent), white, 3–5 mm long, deeply notched; stamens reddish-purple, three to 10, depending on the amount of light present; one pistil; three styles; bracts 1–40 mm long; flowering occurs when temperatures are above 2°C.

Plant: Shade-tolerant, can survive temperatures of −10°C with no damage; stems leafy, weak, and trailing, 10–80 cm long, surface with a line of fine white hairs; roots tough.

Fruit: One-celled capsule, 3–7 mm long; opening by six teeth; eight- to 10-seeded.

REASONS FOR CONCERN

Chickweed prefers soils with high organic matter and increased rainfall. It is able to compete with crops because of its shade tolerance and uses valuable moisture and nutrients. Common chickweed is host for the following viral diseases: beet yellows, beet mild yellowing, cucumber mosaic, beet curly top, beet western yellow, aster yellows, beet mosaic, pea mottle, tobacco mosaic, tomato spotted wilt, and turnip mosaic. It is reported to be a weed of more than 20 crops in 50 countries.

Vaccaria hispanica (P.Mill.) Rauschert
COW COCKLE

Also known as: cow herb,
China cockle, cow
soapwort, spring cockle
Scientific Synonym(s):
V. segetalis Garcke ex
Aschers., *V. pyramidata*
Medik., *V. vaccaria* (L.)
Britt., *V. vulgaris* Host,
Saponaria vaccaria L.

QUICK ID: leaves opposite;
flowers pink; plants
bluish-green, hairless
Origin: native to Europe and
Asia
Life Cycle: annual
reproducing by seed
Weed Designation: CAN:
federal, MB; MEX: federal

DESCRIPTION
Seed: Round, gray to black, 1.7–2.5 mm across;
surface covered in rounded projections.
Viable in soil for up to 3 years; majority of
germination occurs during the first growing
season.
Germination occurs with soil temperatures
between 10°C and 20°C.
Seedling: With oblong cotyledons, 9–32 mm long,
3.5–8 mm wide, somewhat fleshy, hairless,
underside with a prominent midrib; stem below
cotyledons often stained with purple; first leaves
similar in appearance to the cotyledons.
Leaves: Opposite, lance-shaped, 2–10 cm long,
bluish-green, somewhat fleshy, hairless, stalkless
and clasping the stem.
Inflorescence a cyme; 16–50-flowered.
Flower: Bright pink, 10–14 mm across; five sepals,
10–15 mm long, flasklike, five-ribbed; five petals,
about 20 mm long, toothed at tip; 10 stamens;
one pistil; two styles.
Plant: Plants 20–60 cm tall; stems with numerous
branches, bluish-green, hairless; nodes swollen;
taproot slender.
Fruit: Round capsule, 6–8 mm long; opening by four
teeth; often enclosed by the enlarged calyx.

REASONS FOR CONCERN

Cow cockle is an occasional weed of cultivated fields, roadsides, and waste areas. It competes with the crop for moisture and nutrients. The seeds of cow cockle contain saponin, a substance reported to be toxic to livestock. It has been determined that 112 g of seed per 45 kg of body weight can be fatal to livestock.

Similar species: soapwort or bouncing bet (*Saponaria officinalis* L., p. 238), is often confused with cow cockle. The two species are distinguished by two features. The calyx of soapwort is tube-shaped and petals have clawlike appendages.

OTHER SPECIES OF CONCERN IN THE PINK OR CARNATION FAMILY

Dianthus armeria

DIANTHUS ARMERIA L. — DEPTFORD PINK

Native to Europe; plants annual or biennial; stems simple 4–80 cm tall; leaves opposite, linear to oblong or lance-shaped, 2–10.5 cm long, margins smooth, hairy; inflorescence a cyme, three- to six-flowered; four bracts, green, linear to lance-shaped; stalks less than 3 mm long; flowers 11–20 mm across; five sepals, 20–25-nerved, petals reddish to pink, the tips toothed; 10 stamens; one pistil; two styles; capsule 10–16 mm long; seeds 1.1–1.4 mm long. Weed designation: none.

DRYMARIA ARENARIOIDES HUMB. & BONPL. EX J. A. SCHULTES — SANDWORT DRYMARY, ALFOMBRILLA

Native to northwestern Mexico; annual or perennial; stems spreading, up to 20 cm long, glandular hairy; leaves opposite, oblong to elliptical, 5–25 mm long, 1–3 mm wide; stipules about 2 mm long; inflorescence a solitary axillary flower; five sepals; five petals, four- to eight-lobed, 4–6.5 mm long; five stamens; one pistil; two- or three-branched style; fruit an oval capsule, 3–5 mm long; seeds 15–25, 0.4–1.2 mm long, light brown, surface dull. Weed designation: USA: federal, AL, AZ, CA, FL, MA, MN, NM, NC, OR, SC, VT. The toxin, saponin, is very poisonous to livestock, especially cattle, sheep, and goats.

SPERGULARIA SALINA J. PRESL & C. PRESL. — SALTMARSH SAND-SPURREY

Native to Europe, introduced throughout the world; plants less than 25 cm tall; stems erect to spreading; leaves opposite, linear, 1.5–4 cm long, fleshy; inflorescence a cyme or solitary; flowers less than 7 mm across; five sepals, 2.5–4 mm long; five petals, white to pink, about the same length as the sepals; two or three stamens; one pistil; three styles; capsules 2.8–6.4 mm long, greenish to tan; seeds light brown to reddish-brown, less than 0.7 mm long. Often found in saline areas and roadsides where salt has been applied during the winter. Weed designation: none.

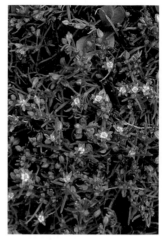

Spergularia salina

Chenopodiaceae
GOOSEFOOT FAMILY

Herbs and shrubs, often found in dry alkaline sites; leaves alternate (opposite in one genus), simple; stipules absent; inflorescence a cyme; flowers perfect or imperfect, regular; bracts present, somewhat fleshy; two to five sepals, united; petals absent; two to five stamens, opposite the sepals; one pistil; ovary superior; fruit an achene, utricle, or nutlet, often surrounded by the calyx. Worldwide distribution: 102 genera/1,500 species; found throughout the world.

Distinguishing characteristics: plants somewhat fleshy; flowers nonshowy; bracts not papery.

Economically important members of this family include spinach, beets, chard, and quinoa. Many species are considered noxious weeds and compete with agricultural crops for light, moisture, and nutrients.

lamb's-quarters (*Chenopodium album*)

Atriplex hortensis L.
GARDEN ORACHE

Also known as: halberleaf
orach, fat hen, lamb's-
quarters, saltbrush,
Hungarian spinach
Scientific Synonym(s):
A. nitens Schkuhr.

QUICK ID: leaves alternate;
flowers borne in dense
clusters; plants reddish-
purple in late summer
Origin: native to Asia;
introduced as a garden
vegetable
Life Cycle: annual
reproducing by seed;
large plants producing
over 6,000 seeds
Weed Designation: none

DESCRIPTION
Seed: Of two types; shiny, black, about 2 mm across
(produced by female flowers with small calyx and
no bracteoles); large, flat, brown, 3–10 mm long
(produced in female flowers with no calyx and
two large bracteoles).
Viable for at least 5 years in dry storage.
Germination of large seeds occurs a few days
after being planted; small seeds remain dormant
for several months.
Seedling: With oblong to lance-shaped cotyledons,
7–20 mm long, 1.2–2.5 mm wide; stem below
cotyledons often tinged with purple; first leaves
alternate, but appearing opposite.
Leaves: Alternate or opposite, varying from lance-
shaped to triangular with heart- to arrowhead-
shaped bases; 5–25 cm long, 8–15 cm wide; stalks
3–40 mm long; green above and whitish-green
below.
Inflorescence a large terminal panicle up to 30
cm long.
Flower: Green; male and female; male flowers with
three to five sepals and three to five stamens;
female flowers of two types: one with no sepals
and two bracts (up to 12 mm long at maturity),
the other with three to five sepals and no bracts.
Plant: Plant 30–200 cm tall; stems yellowish-green
to reddish-purple, erect, branched, hairless.
Fruit: Achene; enclosed by two large bracts.

REASONS FOR CONCERN

Garden orache is a common weed of roadsides and waste areas. It is host for beet mosaic, beet yellows, and beet curly top viruses.

Similar species: at maturity, garden orache could easily be mistaken for curled dock (*Rumex crispus* L., p. 618). Both species have a large reddish fruiting stem and lance-shaped leaves. The distinguishing characteristic between the two species is the shape of the fruit. The fruit of curled dock is three-sided while those of garden orache are flat.

Axyris amaranthoides L.
RUSSIAN PIGWEED

Also known as: axyris,
upright axyris
Scientific Synonym(s): none

QUICK ID: leaves alternate;
flowers in dense clusters;
stems whitish-green
Origin: native to Siberia; first
Canadian report from
Manitoba in 1886
Life Cycle: annual
reproducing by seed
Weed Designation: CAN: MB

DESCRIPTION

Seed: Of two types: oval, dark brown, 2.5–3 mm long,
margin with two-lobed papery wing; or round,
grayish-brown, 1.5–2 mm across, margin wingless.
Viability unknown.
Germination of winged seeds occurs soon after
being shed (18°–22°C); wingless seeds remain
dormant for a longer period of time.

Seedling: With linear to lance-shaped cotyledons,
7–15 mm long, 2–5 mm wide, slightly hairy; first
leaves alternate, dull green, oval, prominently
veined.

Leaves: Alternate, oval to lance-shaped, 5–10 cm
long, 1–3 cm wide, short-stalked, reduced in size
upward; upper surface hairless, the lower with
numerous star-shaped hairs.
Inflorescence a terminal cyme; male flowers
borne at end of branches; female flower borne in
leaf axils.

Flower: Whitish to green; male or female; male
flowers whitish, less than 2 mm long, with three
to five papery sepals and two to five stamens;
female flowers green, with three to five papery
sepals and one pistil.

Plant: Plants 20–120 cm tall, bushy; stems erect,
light green to white, turning straw-colored at
maturity.

Fruit: Achene; enclosed by a papery membrane.

REASONS FOR CONCERN

Russian pigweed is a serious weed of cultivated crops, forages, and pastures. It is commonly found on roadsides, waste areas, and gardens.

Chenopodium album L.
LAMB'S-QUARTERS

Also known as: pigweed, fat hen, white goosefoot, lamb's-quarters goosefoot, netseed lamb's-quarters, mealweed, frostblite, baconweed, wild spinach

Scientific Synonym(s): *C. lanceolatum* Muhl., *C. dacoticum* Standl.; hundreds of subspecies and varieties named due to the variability of the species

QUICK ID: leaves with a mealy texture; stems and leaves with purple blotches; flowers in dense terminal clusters

Origin: native to Europe (native populations may exist); first Canadian report from 1858

Life Cycle: annual reproducing by seed; large plants may produce up to 500,000 seeds; plants in competition with potato and sugar beet crops produce up to 13,000 seeds; life cycle completed within 4 months of germination.

Weed Designation: CAN: MB, PQ

DESCRIPTION

Seed: Seeds disc-shaped, shiny black, 1.1–1.6 mm long and 1 mm wide.
Viable in soil for 30–40 years.
Germination occurs in the top 3 cm of soil; optimal germination occurs between 10°C and 25°C.

Seedling: With oblong to oval cotyledons, 5–15 mm long, and 2 mm wide, fleshy; underside and the stem below cotyledons often pink to purple; first two pairs of leaves opposite, egg-shaped, surface mealy; later leaves alternate.

Leaves: Alternate, green above, mealy white below, 2–12 cm long, 1–8 cm wide, stalks 1–2.5 cm long; margins wavy; leaf shape highly variable: triangular to diamond- or lance-shaped; often with reddish-purple blotches.
Inflorescence a dense terminal or axillary panicle.

Flower: Bluish-green, 1–3 mm wide; five sepals, mealy; two to five stamens; one pistil; two to five styles; wind-pollinated.

Plant: Plants 30–250 cm tall; extremely variable; stems bluish-green, branched, grooved, often splotched with red or purple; taproot short.

Fruit: Achene, surrounded by a papery sac.

REASONS FOR CONCERN

Lamb's-quarters, a rapid-growing plant using large amounts of water, is capable of crowding out cultivated crops and adversely affecting crop yield. Plants are reported to be poisonous to sheep and pigs if eaten in large quantities. Lamb's-quarters is reported a weed of 40 crops in 47 countries.

Kochia scoparia (L.) Schrad.

KOCHIA

Also known as: summer cypress, burning bush, mock cypress, red belvedere, Mexican fireweed, fireball, broom cypress

Scientific Synonym(s):
K. alata Bates,
K. sieversiana (Pallas) C. A. Mey., *K. trichophila* Stapf, *Bassia sieversiana* (Pallas) W. A. Weber, *B. scoparia* (L.) A. J. Scott

QUICK ID: plants pyramid-shaped; plants turning reddish-green in late summer; leaves alternate, hairy

Origin: native to Europe and Asia; introduced into North America as an ornamental garden plant

Life Cycle: annual reproducing by seed; large plants producing up to 50,000 seeds

Weed Designation: USA: CT, OR, WA; CAN: BC, MB, SK

DESCRIPTION

Seed: Wedge- to egg-shaped, dark reddish-brown to black, 1–2 mm long, 1–1.5 mm wide, surface dull and somewhat granular.
Viable less than 1 year at field conditions. Germination occurs at temperatures 3.9°–41°C (optimal at 16°C); dormant period of 2–3 months required.

Seedling: With elliptical cotyledons, 4–5 mm long and 1.5 mm wide, dull green above and bright pink below; first and later leaves lance-shaped, surface with numerous soft hairs; basal rosette produced when the seedling is not in competition with other vegetation; seedlings are tolerant to light frosts.

Leaves: Alternate, numerous, linear to lance-shaped, 5–7 cm long, 2–15 mm wide, three- to five-veined; reduced in size upward, short-stalked; underside and margins with long hairs.
Inflorescence a short spike of two to six flowers borne in leaf axils.

Flower: Green, 2–3 mm across; five sepals, covered in long hairs; three to five stamens, absent in some flowers; one pistil; two styles.

Plant: Bushy, pyramid-shaped, 50–200 cm tall, greenish-yellow turning reddish-purple in autumn, at maturity breaking off and forming a tumbleweed; stems branched, somewhat hairy; roots 1.8–2.4 m deep (up to 4.8 m deep under drought conditions).

Fruit: Achene; often enclosed by the calyx.

REASONS FOR CONCERN

In severe infestations, kochia has been reported to reduce crop yields up to 100%. Seeds germinate in early spring and seedlings are controlled by late spring cultivation. Kochia was once planted as a hay crop on alkaline soils. It is readily grazed by livestock. The nutritional value of kochia is equal to alfalfa. Kochia is host for beet yellows and tobacco mosaic viruses.

Salsola kali L.
RUSSIAN THISTLE

Also known as: Russian tumbleweed, Russian cactus, tumbling Russian thistle, glasswort, burning bush, saltwort, prickly glasswort, wind witch

Scientific Synonym(s): *S. pestifer* A. Nels.

QUICK ID: stems with red stripes; plants bristly at maturity; leaves spine-tipped

Origin: native to Russia; introduced into South Dakota in 1874 as a contaminant in flax seed; first reported from Ontario in 1894

Life Cycle: annual reproducing by seed; large plants capable of producing up to 25,000 seeds

Weed Designation: USA: AR, HI; CAN: BC, MB, ON, SK

DESCRIPTION

Seed: Cone-shaped, dull brown to gray, about 2 mm long.
Viable less than 1 year in soil.
Germination occurs in top 5 cm of soil; no emergence from 8 cm depth; temperature range of −2°–41°C (optimal is 7°–10°C).

Seedling: With narrow linear cotyledons, 10–50 mm long and less than 3 mm wide, fleshy; first pair of leaves opposite, resembling the cotyledons except for a soft spine at the tip; young stem often tinged with purple.

Leaves: Alternate, 2–6 cm long, 1–2 mm wide; lower leaves threadlike; upper leaves 1–2.2 mm long, awl-shaped and spine-tipped, becoming stiff with age.
Inflorescence a solitary flower borne in leaf axils; bracts spine-tipped.

Flower: Green, inconspicuous, less than 2 mm wide; male or female; male flowers with five sepals and five stamens; female flowers with five sepals and one pistil; two bracts, 5–7 mm long, spine-tipped.

Plant: Pyramid-shaped, turning red at maturity and breaking off at the ground and forming a tumbleweed; stems spiny, 30–120 cm tall, branched, often striped with red; taproot to 1 m deep.

Fruit: Coiled, top-shaped utricle, about 2 mm long; single-seeded.

REASONS FOR CONCERN

Russian thistle is a common weed of roadsides, railways, and dry open areas. It establishes itself in areas where there is reduced competition from other plants. The wide temperature range for germination and the mechanism of seed dispersal have assisted in the spread of this plant. It is host for beet yellows, beet curly top, and beet mosaic viruses.

Similar species: the seedling of corn spurrey (*Spergula arvensis* L., p. 249) is often mistaken for the seedling of Russian thistle. The leaf tips of corn spurrey are blunt while those of Russian thistle are spine-tipped. The tumbleweed of kochia (*Kochia scoparia* [L.] Schrad., p. 264) may be mistaken for Russian thistle. The tumbleweed of kochia is slightly hairy unlike that of Russian thistle, which is spiny.

OTHER SPECIES OF CONCERN IN
THE GOOSEFOOT FAMILY

Halogeton glomeratus

Salicornia maritima

HALOGETON GLOMERATUS (BIEB.) C. A. MEY. —
HALOGETON, BARILLA, SALTLOVER
Native to southeastern Russia and northwestern
China; annual or winter annual producing up
to 50,000 seeds; plants spreading, forming a
tumbleweed at maturity, inhabiting alkaline
soils, and growing until killed by frost; stems
stiff, succulent, purplish to reddish-green; leaves
alternate, fleshy, 4–17 mm long, round in cross-
section; upper leaves replaced by spine-tipped floral
bracts; inflorescence a short spike, three- to five-
flowered, often with a single female flower; flowers
green, less than 1 mm across, unisexual or perfect;
five sepals; three to five stamens (fused into two
groups); one pistil; bracts leaflike; fruit a reddish to
yellowish-green utricle surrounded by a five-winged
calyx; seeds of two types, black and winged, less
than 1 mm long, or brown, wingless, 1–2 mm long.
Plants poisonous to livestock. Weed designation:
USA: AZ, CA, CO, HI, NM, OR; CAN: federal, AB, SK.

SALICORNIA MARITIMA WOLFF & JEFFERIES. —
GLASSWORT, SAMPHIRE
Native to North America; plants of saline flats;
stems erect, 5–25 cm tall, fleshy and jointed, green
turning red or purple at maturity, branched; leaves
scalelike, triangular; inflorescence a spike, 1–5 cm
long; flowers perfect or female, borne in groups of
three and sunk into fleshy axis of spike; sepals fleshy,
unlobed; petals absent; one or two stamens; one
pistil; two styles; fruit a utricle surrounded by the
fleshy sepals. Plants may be poisonous due to the
accumulation of sodium and potassium oxalates.
Care should be exercised when livestock graze in
areas where this species exists. Weed designation:
none.

SALSOLA TRAGUS L.—PRICKLY RUSSIAN THISTLE, TUMBLEWEED

Native to Europe and Asia; plants 10–100 cm tall, forming a tumbleweed at maturity; stems erect, profusely branched; leaves alternate, threadlike to linear, less than 1 mm wide, tips with a spine less than 1.5 mm long; inflorescence a solitary flower; bracts spine-tipped; flowers perfect; five sepals; five stamens; one pistil; two or three styles; fruit a utricle, surrounded by a prominent membranous wing, 4–10 mm across. Weed designation: USA: AR, CA, OH.

Salsola tragus

SALSOLA VERMICULATA L.— SHRUBBY RUSSIAN THISTLE

Native to southwestern Asia and North Africa; shrublike perennial; stems 20–70 cm tall, covered in dense brownish hairs when young; leaves alternate, lance-shaped, 5–8 mm long, 0.5–1 mm wide, fleshy; leaf axils with several reduced leaves 1–4 mm long; inflorescence of one to three axillary flower; flowers greenish, bracts distinct; five sepals, becoming winged at maturity (7–10 mm diameter); five stamens; one pistil; fruit a utricle, surrounded by the enlarged sepals. Weed designation: USA: AL, AR, CA, FL, MA, MN, NC, OR, SC, VT; MEX: federal.

SARCOBATUS VERMICULATUS (HOOK.) TORR.— GREASEWOOD, BLACK GREASEWOOD, SEEPWOOD, SALTBUSH

Native to North America; shrub up to 3 m tall; roots to 18 m deep; stems branched, yellowish-white; leaves fleshy, linear, 1.5–4 cm long, hairless, often with a thorn at the base; male flowers borne in spikes 11–40 mm long; female flowers solitary in leaf axils; male flowers consisting of two to four stamens; female flowers with a cuplike perianth and one pistil; fruit an achene, 3–5 mm long, winged. Plants are palatable due to the saline content. Under some conditions, plant material may contain large amounts of sodium and potassium oxalates, compounds poisonous to livestock. Weed designation: none.

Sarcobatus vermiculatus

Commelinaceae
SPIDERWORT FAMILY

Herbs with succulent stems and swollen nodes; leaves alternate, simple, the bases sheathing; inflorescence a coiled or umbel-like cyme; bracts numerous, boat-shaped; flowers perfect, regular; three sepals, green; three petals, often blue; six stamens, the filaments with brightly colored hairs; one pistil; ovary superior; fruit a capsule. Worldwide distribution: 40 genera/600 species; predominantly tropical and subtropical.

Distinguishing characteristics: herbs with succulent stems; leaves basally sheathing; three petals.

Several members of this family are grown as ornamentals, including dayflowers, wandering Jew, and boatflowers.

spreading dayflower (*Commelina diffusa*)

Also known as: Asiatic dayflower

Scientific Synonym(s): none

QUICK ID: plants somewhat fleshy; leaves parallel-veined; two petals, blue

Origin: native to Asia; introduced as an ornamental plant

Life Cycle: annual reproducing by seed and creeping stems

Weed Designation: none

DESCRIPTION

Seed: Brown, 2.5–4.5 mm long; surface rough and pitted.
Viability unknown.
Germination from depths of 1 cm or less; no emergence from 2 cm or deeper.

Seedling: With one oblong to oval cotyledon, tips pointed; later leaves oval to lance-shaped.

Leaves: Alternate, oval to lance-shaped, parallel-veined; 3–12 cm long, 16–40 mm wide; stalkless and clasping the stem; sheath 1–2 cm long, hairy at leaf base; surfaces often hairy; pale green on the underside.
Inflorescence a pair of cymes, enclosed by a spathe borne on a stalk 8–35 mm long and opposite the leaf; spathe 1.5–3 cm long with prominent dark green veins; one cyme with many perfect flowers, the other with one to several male flowers.

Flower: Blue, lasting only 1 day; three sepals, unequal, two partly fused; three petals (two blue and one small white); six stamens (three fertile and three sterile); one pistil; spathe leaflike.

Plant: Susceptible to frost; stems erect to creeping, 20–75 cm long, thick; nodes swollen, smooth to hairy, often rooting when in contact with soil; roots fibrous.

Fruit: Capsule, 4.5–8 mm long; two-celled; four seeds.

REASONS FOR CONCERN

Common dayflower, often found in shady damp areas, is primarily a weed of landscapes and nurseries. Two factors leading to an increased occurrence in agricultural fields are an extended germination period and a relatively high tolerance to glyphosate, an active ingredient in several nonselective herbicides. There are reports of the species spreading into soybean fields in Iowa.

Commelina diffusa Burm. f.
SPREADING DAYFLOWER

Also known as: climbing
dayflower, water grass,
wandering Jew
Scientific Synonym(s):
C. gigas Small,
C. longicaulis Jacq.

QUICK ID: plants
somewhat fleshy; leaves
parallel-veined; flowers
blue
Origin: native to
Mediterranean region of
Europe, Asia, and Africa
Life Cycle: annual or
short-lived perennial
reproducing by seed and
creeping stems
Weed Designation: none

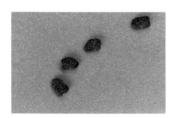

DESCRIPTION
Seed: Brown, 2–2.8 mm long; surface with a network
of ridges.
Viability unknown.
Germination occurs in the top 1 cm of soil.
Seedling: With oval to lance-shaped cotyledon,
sheathing; first leaf lance-shaped; later leaves
similar in appearance.
Leaves: Alternate, lance-shaped, parallel-veined, 1–14
cm long, 0.5–3.3 cm wide.
Inflorescence a pair of cymes, enclosed by a
spathe borne on a stalk 5–20 mm long and
opposite the leaf; spathe 8–40 mm long, 4–14
mm wide, sickle-shaped, not prominently veined;
one cyme with two to four perfect flowers, the
other with one to several male flowers.
Flower: Blue; three sepals, green; three petals, blue
(one smaller than the other two); six stamens
(three fertile and three sterile); one pistil; bract
broadly rounded to shallowly heart-shaped, 2–3
cm long, 1.5–2 cm wide.
Plant: Creeping; stems smooth to sparsely haired,
nodes swollen, often rooting when in contact
with soil.
Fruit: Capsule, 4–6.3 mm long; fewer than five seeds.

REASONS FOR CONCERN

Spreading dayflower is a serious weed in vegetable and cereal fields with high moisture content. It is a principal weed of bananas, beans, citrus, coffee, corn, sugarcane, and cotton in Mexico. In Indonesia, it is a common weed of rice. Spreading dayflower is a serious weed in 26 countries.

Similar species: Carolina dayflower (*C. caroliniana* Walt.), native to India, may be mistaken for spreading dayflower. It is distinguished by a spathe that is not sickle-shaped and a reduced secondary cyme with a single male flower.

OTHER SPECIES OF CONCERN IN THE SPIDERWORT FAMILY

COMMELINA BENGHALENSIS L.—JIO, BENGHAL DAYFLOWER, TROPICAL SPIDERWORT

Origin uncertain; plants annual; rhizomes short; stems spreading to trailing; leaves alternate, oval to elliptical or lance-shaped, 2–9 cm long, 1–3 cm wide, tip rounded, sheath not clasping, red hairs present at tip; inflorescence a pair of cyme borne in a funnel-shaped spathe borne on stalks 1–3.5 mm long; spathes 5–20 mm long; one cyme exserted and one-flowered; flowers of two types, large sterile flowers produced from base of plant, small fertile flowers perfect or male; male flowers with blue petals (two large, one small), three fertile stamens and two or three sterile; fruit a capsule, 4–6 mm long; five seeds, brown or blackish. Weed designation: AL, CA, FL, MA, MN, NC, OR, SC, VT; MEX: federal.

MURDANNIA KEISAK (HASSK.) HAND.-MAZ.— WARTREMOVING HERB

Native to Asia; annual; stems trailing to ascending; leaves alternate, linear to oblong or lance-shaped, 1.5–7 cm long, 2–10 mm wide, hairless; inflorescence a cyme, one-flowered; flowers purple to pink or white, about 1 cm wide; three sepals, 5–6 mm long; three petals, 5–8 mm long; two or three stamens, filaments bearded, three staminodes; one pistil; fruit a capsule, 5–9 mm long; six to 18 seeds. Weed designation: USA: WA.

TRADESCANTIA SPATHACEA SW. — OYSTER PLANT, MOSES-IN-THE-CRADLE, BOATLILY

Native to southern Mexico, Central America, and the West Indies; stems unbranched; leaves alternate (spirally arranged), linear to strap-shaped, 15–30 cm long, 2.5–8 cm wide; upper surface green, lower purple; inflorescence of cymes borne in boat-shaped spathes 3–4 cm long; flowers stalked; three sepals, white, 3–6 mm long; three petals, white; six stamens, the filaments hairy; one pistil; fruit a capsule, 3–4 mm long; two or three seeds. Weed designation: none.

TRADESCANTIA FLUMINENSIS VELL. — GREEN WANDERING JEW, WHITE-FLOWERED WANDERING JEW

Native to Brazil and Argentina; stems spreading, often rooting at nodes; leaves two-ranked, elliptical to oval or lance shaped, 2.5–5 cm long, 1–2 cm wide, hairless, shiny, often tinged with purple; margins hairy; sheaths closed, hairy at top; inflorescence of one or two cymes with leaflike bracts borne opposite the terminal leaf; flower stalks 1–15 cm long, glandular-hairy; three sepals, 5–7 mm long; three petals, 8–9 mm long; six stamens, the filaments white and densely bearded; one pistil; fruit a capsule; six seeds. Weed designation: none.

Convolvulaceae

MORNING-GLORY FAMILY

Herbs, trees, and shrubs, often twining; milky juice often present; leaves alternate, simple; stipules absent; inflorescence a cyme; bracts present; flowers perfect, regular; five sepals; five petals, united into a tube or funnel; five stamens, borne on the petals; one pistil; ovary superior; fruit a capsule, berry, or nut. Worldwide distribution: 50 genera/1,400–1,650 species; distributed throughout tropical and temperate regions.

Distinguishing characteristics: plants climbing; flowers tube- or funnel-shaped; fruit a capsule.

The morning-glory family is of limited economic importance, with some genera grown for their horticultural and ornamental value. A few species are troublesome weeds in agricultural land.

field bindweed (*Convolvulus arvensis*)

Calystegia sepium (L.) R. Br.
HEDGE BINDWEED

Also known as: devil's vine, great bindweed, bracted bindweed, wild morning glory, Rutland beauty, hedge lily, bearbind, devil's gut, hedgebell, old man's nightcap

Scientific Synonym(s): *Convolvulus sepium* L.

QUICK ID: plants climbing or trailing; leaves alternate; flowers white, funnel-shaped

Origin: native to North America; introduced subspecies may be present

Life Cycle: perennial reproducing by creeping rhizomes and seeds

Weed Designation: USA: AR, TX; MEX: federal

DESCRIPTION

Seed: Round to oblong, dark brown to black, 4.6–5.1 mm long, 3.2–3.7 mm wide; surface dull. Viable in soil for at least 50 years. Germination occurs within 2 weeks of being shed; ungerminated seeds can remain dormant for long periods of time.

Seedling: With rectangular cotyledons, 26–50 mm long, 16–22 mm wide, prominently veined on the underside; stem below cotyledons often dull red; first leaf arrowhead-shaped, prominently veined below.

Leaves: Alternate, triangular to arrowhead-shaped, 4–15 cm long, dark green, long-stalked, tips pointed. Inflorescence a solitary flower borne on stalks 5–15 cm long from leaf axils; stalks four-angled.

Flower: Bell- to funnel-shaped, white to pink, 4–8 cm long, 2–8 cm across; five sepals, united; five petals, united; five stamens, 2–3.5 cm long; one pistil; two bracts, heart-shaped, 1–5 cm long.

Plant: Large vine up to 3 m long, shade-intolerant; stems branched; rhizomes creeping and fleshy, buds giving rise to numerous shoots.

Fruit: Globe-shaped capsule, 8–10 mm across, two-celled, each one- or two-seeded.

REASONS FOR CONCERN

Hedge bindweed is a serious weed in orchards and vineyards, where it often climbs over the crop. It is also a weed of cultivated crops, fencerows, and roadsides. Hedge bindweed is host for cucumber mosaic and tobacco streak viruses.

Similar species: field bindweed (*Convolvulus arvensis* L., p. 282), an introduced species, is closely related to hedge bindweed. It is distinguished by smaller pinkish-white flowers, about 2 cm across, and two small bracts below the flower.

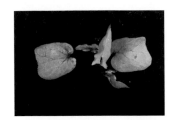

Convolvulus arvensis L.
FIELD BINDWEED

Also known as: European bindweed, small-flowered morning glory, creeping Jenny, European glorybind, cornbine, lesser bindweed, barbine, corn bind, devil's guts, greenvine, corn lily, laplove, hedge bells

Scientific Synonym(s): *Strophocaulos arvensis* (L.) Small., *C. incanus* auct. non Vahl, *C. ambigens* House

QUICK ID: stems twining; leaves arrowhead-shaped; flowers funnel-shaped and white

Origin: native to Europe; introduced into New England in 1739 and spreading to California by 1900; first Canadian report from Ontario in 1879, spreading to western Canada by 1890s

Life Cycle: perennial reproducing by creeping rhizomes and seed; plants capable of producing over 500 seeds; severe infestations capable of producing over 800 kg of seed per acre; 5 cm root fragment can produce 25 shoots within 4 months

Weed Designation: USA: AK, AR, AZ, CA, CO, HI, IA, ID, KS, MI, MN, MO, MT, NM, ND, OR, NJ, SD, TX, UT, WA, WI, WY; CAN: federal, AB, MB, NS, SK; MEX: federal

DESCRIPTION

Seed: Egg-shaped, three-sided, grayish-brown, 3–5 mm long, 0.7–3 mm thick, surface dull and covered in rough bumps.
Viable in soil for at least 50 years.
Germination occurs in the top 2 cm of soil when temperatures are between 5°C and 40°C.

Seedling: With heart- to kidney-shaped cotyledons, 9–22 mm long, 3.5–10 mm wide, long-stalked; stem below cotyledons often red at or near the soil surface; first leaves arrowhead-shaped, stalked, tip rounded; can tolerate frost temperatures of –8°C; 6 weeks after emergence roots are 45–60 cm deep.

Leaves: Alternate, arrowhead-shaped to triangular, 2–6 cm long, 2–3.5 cm wide, long-stalked, tips rounded, margins smooth.
Inflorescence a raceme of one to four flowers borne in leaf axils; stalks four-sided.

Flower: Funnel-shaped, white to pink, 2–3 cm across, 1.5–2 cm long; five sepals, fused, about 3 mm long; five petals, fused, 1.5–3 cm long; five stamens; one pistil; two bracts, small; flowers lasting only 1 day; seeds viable 10 days after pollination.

Plant: Twining or trailing vine up to 7 m long, stems frost-tolerant to –10°C; stems twisting in a counterclockwise direction, hairless; rhizomes creeping and cordlike, white, up to 30 m in length and 9 m deep; under favorable conditions, flowers may be produced within 6 weeks of germination.

Fruit: Egg-shaped capsule, 2–4 mm long, two-celled; one- to four-seeded.

REASONS FOR CONCERN

Field bindweed has been known to reduce crop yields by 50%. It competes with crops for moisture and nutrients, and once established, it is difficult to eradicate. The twining nature of the plant also hampers harvesting of crops. Field bindweed is a serious weed of orchards and vineyards, where it climbs over the crop. It is a serious weed in 42 countries.

Similar species: wild buckwheat (*Fallopia convolvulus* L., p. 608), a member of the buckwheat family (Polygonaceae), is often confused with field bindweed. Wild buckwheat is distinguished by greenish-pink flowers less than 5 mm across and leaves with pointed tips.

Cuscuta ssp.
DODDER

Also known as: goldthread vine, love-tangle, coralvine (other common names refer to specific hosts, i.e., onion dodder, large alfalfa dodder, etc.)

Scientific Synonym(s): none

QUICK ID: stem climbing, parasitic; leaves reduced to small scales or absent; roots absent

Origin: native and introduced species occur in North America

Life Cycle: annual reproducing by seed

Weed Designation: USA: federal, AL, AZ, AR, CA, FL, MA, MI, MN, NC, OR, SC, SD, VT, TX, WA; CAN: federal, BC, MB, ON,; MEX: federal

DESCRIPTION

Seed: Round, less than 1.5 mm in diameter, variety of colors ranging from white to brown; surface dull, covered in small scales.

Viability ranges from 5 to 50 years, depending on the species.

Germination occurs at the soil surface; optimal temperature is 30°C.

Seedling: Seedlings do not produce cotyledons or roots. The slender twining leafless stem must attach itself to a host plant within a short period of time, or the young plant will wither and die. The stem develops suckers and attaches itself by penetrating the tissue of the host plant. Once attached, the seedling loses contact with the soil.

Leaves: Alternate, very small, scalelike.

Inflorescence a compact raceme borne in leaf axils.

Flower: Yellow to white, bell to globe-shaped, 2–5 mm across; four or five sepals, united; four or five petals, united; five stamens; one pistil.

Plant: Parasitic on herbs, occasionally trees and shrubs, spreading rapidly and forming large masses; stems yellow to reddish-brown, threadlike; roots absent; studies show one seed can produce over 720 m of stem in 4 months.

Fruit: Round capsule, four- to eight-seeded.

REASONS FOR CONCERN

Dodder parasitizes agricultural crops and drastically reduces yield. The twining stems also hamper harvesting of the crop. Some species are host for a variety of viral diseases, including beet yellows, cucumber mosaic, and tobacco etch.

Similar species: a tropical species, love vine (*Cassytha filiformis* L.), a parasitic vine of the laurel family (Lauraceae), is often mistaken for dodder. Parasitic on woody plant species, love vine produces yellowish-green to orange leafless stems. Flowers are borne in short spikes.

A key to common *Cuscuta* ssp. of North America can be found in figure 2.

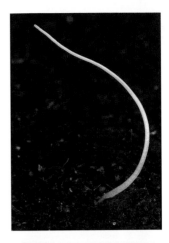

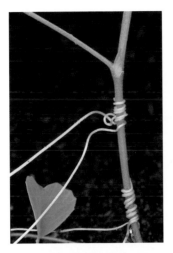

love vine (*Cassytha filiformis*)

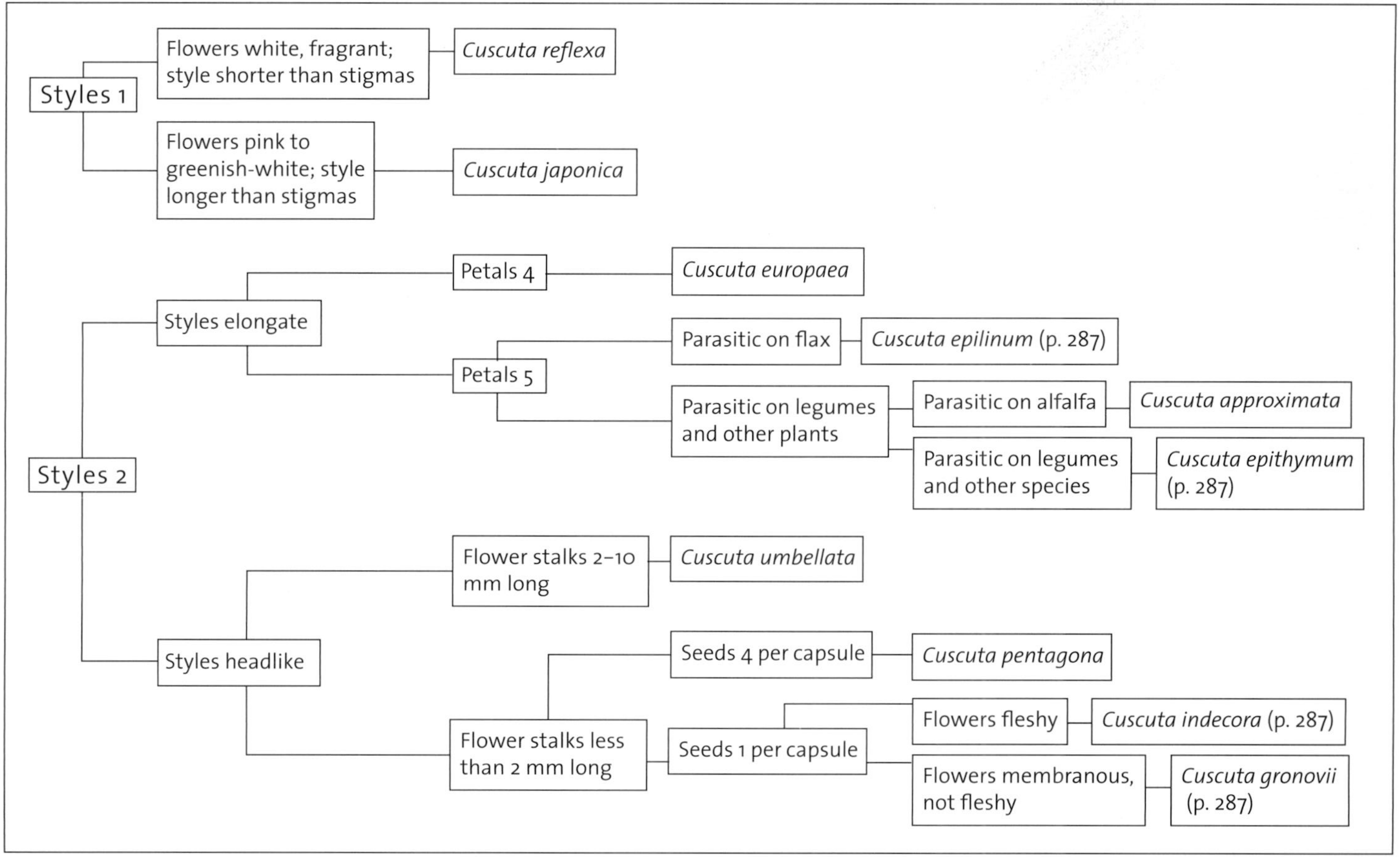

Figure 2. Simplified key to common Cuscuta species in North America

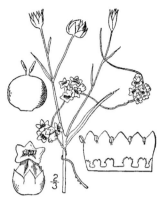

flax dodder (*Cuscuta epilinum*)

scaldweed (*Cuscuta gronovii*)

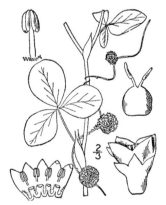

clover dodder
(*Cuscuta epithymum*)

bigseed alfalfa dodder
(*Cuscuta indecora*)

Ipomoea aquatica Forsk.
WATER SPINACH

Also known as: water cabbage, swamp morning glory, Chinese spinach

Scientific Synonym(s): *I. repens* Roth, *I. reptans* Poiret, *Convolvulus repens* Vahl

QUICK ID: aquatic to semiaquatic plant; leaves arrowhead-shaped; flowers white, funnel-shaped

Origin: perennial native to China; introduced as a vegetable for human and livestock consumption

Life Cycle: perennial reproducing by root fragments, cuttings, and seed; plants may produce up to 250 seeds

Weed Designation: USA: federal, AL, AR, AZ, CA, FL, MA, NC, OR, SC, TX, VT

DESCRIPTION

Seed: Elliptical to oval, rounded or three-sided, dark reddish-brown to dark gray, 4.5–5.5 mm long, 3.5–4.5 mm wide; surface often with short hairs. Viability unknown.
Germination occurs within 3 days under optimal conditions; growth is delayed if temperatures are below 25°C.

Seedling: Seedling with deeply lobed irregularly shaped cotyledons, prominently veined; first leave lance-shaped.

Leaves: Alternate, oblong to oval or lance-shaped, 3–15 cm long, 1–10 cm wide, base arrowhead-shaped, heart-shaped, or truncated, stalks 3–20 cm long, margins slightly wavy.
Inflorescence a raceme of one or two flowers borne in leaf axils; stalks 2–10 cm long.

Flower: White to pink or purplish, showy, funnel-shaped, 3–5 cm long, 1.5–3 cm across; five sepals, united, about 8 mm long, green to brownish-pink; five petals, united; five stamens; one pistil; bracts small.

Plant: Aquatic, trailing up to 3 m long, often spreading over the water surface; stems hollow or spongy, hairless, often rooting at nodes; under favorable conditions single plants may have up to 23 m of stem; sap milky.

Fruit: Capsule 6–10 mm long, surrounded by persistent sepals; one- to four-seeded; stems often with three to five fruits per node.

REASONS FOR CONCERN

Water spinach has been reported to impede water flow in irrigation canals and drainage ditches. It is a principal weed of rice paddies in southeast Asia and South America.

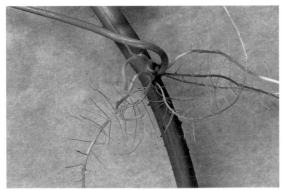

capsule

Ipomoea hederacea Jacq.
IVY-LEAVED MORNING GLORY

Also known as: morning glory, Mexican morning glory, entireleaf morning glory

Scientific Synonym(s): *Pharbitis barbigera* (Sweet) G. Don, *P. hederacea* (Jacq.) Choisy, *I. barbigera* Sweet, *I. desertorum* House, *I. hirsuta* auct. non Jacq. f.

QUICK ID: stems twining; leaves alternate, three- to five-lobed; flowers pale blue, funnel-shaped, turning to pinkish-purple on opening

Origin: native to tropical North America

Life Cycle: annual reproducing by seed

Weed Designation: USA: AR, AZ

DESCRIPTION

Seed: Disc-shaped, dark brown to black, 4.8–5.6 mm long, 3.4–4.2 mm wide; surface dull, somewhat wrinkled.
Viability unknown.
Germination occurs in the top 1 cm of soil; emergence from 20 cm has been documented; soil temperatures of 20°–30°C required.

Seedling: With butterfly-shaped cotyledons, 16–57 mm long, 15–40 mm wide, shiny; stem below cotyledon dull purple and a ridge running from the base of each seed leaf.

Leaves: Alternate, 5–12 cm long and wide, deeply three-lobed (occasionally five-lobed), long-stalked, hairy, bases heart-shaped.
Inflorescence a raceme, one- to three-flowered, borne in leaf axils.

Flower: Pale blue turning pinkish-purple on opening, funnel-shaped, 3–5 cm long; five sepals, lance-shaped, 15–25 mm long, densely haired to bristly; five petals, united; five stamens; one pistil; flowers may appear by the fourth leaf stage.

Plant: Twining or spreading vine up to 2 m long; stems hairy.

Fruit: Egg-shaped capsule, partly surrounded by the calyx, three-celled; four- to six-seeded.

REASONS FOR CONCERN

Ivy-leaved morning glory is a serious weed in orchards and vineyards, climbing over the crop. When found in cultivated fields and row crops, it hampers harvest of the crop. Ivy-leaved morning glory is host for sweet potato internal cork virus.

Similar species: purple morning glory (*I. purpurea* [L.] Roth) closely resembles ivy-leaved morning glory. It is distinguished by shorter sepals (10–15 mm long), larger flowers (4–6 cm long), and heart-shaped leaves.

OTHER SPECIES OF CONCERN IN
THE MORNING-GLORY FAMILY

Ipomoea quamoclit

IPOMOEA QUAMOCLIT L. —
CYPRESSVINE MORNING GLORY

Native to Mexico; annual vine with milky sap; stems smooth; leaves alternate, 23–120 mm long, 17–60 mm wide, pinnately divided into eight to 30 linear segments, each 10–25 mm long and less than 1 mm wide, stalks 7–41 mm long; stipules dissected; inflorescence a cyme, one- to four-flowered; flowers red; five sepals; five petals, 2.7–3.5 cm long; five stamens; one pistil; stalks 15–20 mm long; fruit a capsule, 8–9 mm long; usually four seeds. Weed designation: USA: AR, AZ.

IPOMOEA TRILOBA L. —
LITTLEBELL, PUYUI, CAMPANILLA

Native to tropical America; annual vine with milky
sap; stems twining to trailing, angular; leaves
alternate, heart-shaped, 2–6 cm long, 2–6 cm wide,
stalks 2.5–6 cm long, margins smooth or three-
lobed, hairless; inflorescence an umbel of one to five
flowers, stalks 2.5–10 cm long; flowers pink to pale
purple, funnel-shaped, 1.5–2 cm long; five sepals, 5–8
mm long; five petals; five stamens; one pistil; stalks
5–7 mm long; fruit a globe-shaped capsule, 5–6 mm
long, bristly haired; seeds dark brown, about 3.5 mm
long. Weed designation: USA: AR, AZ, FL, SC.

Cucurbitaceae

GOURD FAMILY

Herbs, sprawling to climbing; leaves alternate, simple, often palmately lobed and veined; tendrils present, branched inflorescence a panicle or solitary flowers; flowers imperfect (male or female), regular; male flowers, five sepals (small), five petals, five stamens; female flowers, five sepals (small); five petals; one pistil; ovary inferior; fruit a pepo (berry with a leathery rind). Worldwide distribution: 100 genera/850 species; found throughout subtropical and warm temperate regions of the world.

Distinguishing characteristics: plants trailing or climbing; flowers male or female; fruit berry-like (pepo).

The gourd family has several economically important members, including cucumbers, melons, squash, and luffa. A few species are considered weedy. Members of this family play an important role serving as reservoirs for viral disearse of cultivated cucurbit crops, particularly the widely transmitted squash vein yellowing virus (SqVYV).

balsam apple (*Momordica charantia*)

Also known as: balsam apple, mock apple, creeping Jenny, four-seeded bur cucumber, mock cucumber

Scientific Synonym(s): *Sicyos lobata* Michx., *Micrampelis lobata* (Michx.) Greene

QUICK ID: climbing vine; leaves alternate; flowers white; fruit green with soft spines

Origin: native to Europe; introduced as an ornamental plant

Life Cycle: annual reproducing by seed; plants capable of producing over 400 seeds

Weed Designation: none

DESCRIPTION

Seed: Elliptical, flattened, 16–20 mm long, 8–10 mm wide, less than 4 mm thick; surface dull brown marbled with patches of dark brown. Viability unknown.
Germination occurs when soil temperatures reach 25°C; a cold period of at least 4 months is required to break dormancy.

Seedling: With oval cotyledons, 3–7 cm long, 1–3 cm wide, dark green above and pale beneath, thick, prominently veined; stem below cotyledons pale green, hollow until the second leaf stage; first leaves kidney-shaped with three to seven prominent lobes, both surfaces rough textured.

Leaves: Alternate, palmately three- to seven-lobed, 5–13 cm across, bright green, long-stalked, base heart-shaped, margins toothed, surface with a sandpapery texture; tendril long and branched, opposite the leaf.
Inflorescence of 2 types; male flowers produced in panicles; female flowers solitary in the axil of the tendril.

Flower: Two types, male or female; male flowers white to greenish-yellow, six sepals, six petals, two or three stamens, yellow; female flowers yellowish-green, six sepals, six petals, one pistil.

Plant: Creeping or climbing vine up to 8 m long; stem bright green, twining, often grooved and angled.

Fruit: Pepo (berry), fleshy, oblong, 2.5–5 cm long; skin thick, mottled pale green, covered in weak prickles; inside fibrous and mesh like; four-seeded; opening at the apex at maturity.

REASONS FOR CONCERN

Wild cucumber is a weed of fence lines, waste areas, and meadows. It is host for cucumber mosaic, cucurbit mosaic, and several Prunus viruses.

Momordica charantia L.
BALSAM APPLE

Also known as: bitter melon, balsam pear, bitter gourd, carilla gourd, cerasee, peria, squirting cucumber

Scientific Synonym(s): none

QUICK ID: climbing vine; flowers yellow and producing orange-ribbed fruit; seeds covered in sticky crimson pulp

Origin: native to Africa; introduced into North America during the slave trade

Life Cycle: annual reproducing by seed; average plants producing up to 1,600 seeds

Weed Designation: none

DESCRIPTION

Seed: Light brown to black, 5–9 mm long, 2.5–6 mm wide; surface ridged or pitted; margin thick and ragged.
Viable at least 3 years in dry storage.
Germination occurs between 20°C and 40°C; optimal temperature is 30°C.

Seedling: With oval to oblong cotyledons, 10–15 mm long, 8–12 mm wide.

Leaves: Alternate; 2.5–10 cm across, five- to seven-lobed, stalked, base heart-shaped, margins notched; tendrils opposite the leaf.
Inflorescence of two types; male flowers borne in panicles; female flowers borne on short stalks.

Flower: Yellow, 15–20 mm long, 8–13 mm wide; male flowers with five sepals, 3–9 mm long, five petals, two stamens; female flowers with five sepals, five petals, and one pistil; bract round, about 1 cm long; open in early morning with all pollen shed by noon; flowering occurs 30–35 days after germination; mature fruit 15–20 days later.

Plant: Creeping or climbing vine, ill-scented; stems 2–3 m long, branched.

Fruit: Egg-shaped pepo (berry), orange to orangish-yellow, surface warty, 2–7 cm long in wild forms, up to 30 cm long in cultivated forms, eight to 10 ridges, 15–20-seeded; seeds covered in a sticky crimson pulp.

REASONS FOR CONCERN

Balsam apple is a serious weed in sugarcane, bananas, citrus, cotton, pineapple, and maize. Plants climb over and shade out native vegetation, thereby reducing the habitat diversity. It is also a host of several viruses for celery, papaya, and melons.

Sicyos angulatus L.
BUR CUCUMBER

Also known as: one-seeded bur cucumber, nimble kate, wall bur cucumber, star cucumber, wild pickle

Scientific Synonym(s): none

QUICK ID: climbing vine; flowers white; fruit single-seeded and bristly

Origin: native to eastern North America

Life Cycle: annual reproducing by seed; large plants capable of producing up to 80,000 seeds; may still produce seed even when germination occurs as late as mid-August

Weed Designation: USA: DE, IN, KY

DESCRIPTION

Seed: Oval, black to dark brown seeds, 8.6–12 mm long and 6 mm wide, surface covered in warts and prickles.

Viability unknown.

Germination occurs in the top 15 cm of soil; dormant period required; delayed germination produces several flushes of seedlings from May to September.

Seedling: With oval cotyledons, thick, prominently veined; stem below cotyledons with downward-pointing hairs; first leaves alternate, hairy, lobed.

Leaves: Alternate, palmately three- to –five-lobed, bases round to heart-shaped; surface rough; tendrils with three forked branches.

Inflorescence of two types; male flowers borne in panicles; female flowers borne in headlike clusters on stalks 5–8 cm long; stalks of both clusters originating in the same axil.

Flower: Greenish-white; male or female; male flowers 8–10 mm across, five sepals, five petals, three stamens; female flowers with five sepals, five petals, and one pistil.

Plant: Climbing vine up to 6 m tall; stems clammy-haired, angular stems.

Fruit: Pepo, dry; single-seeded; often borne in clusters of two to 30.

REASONS FOR CONCERN

Control of bur cucumber is difficult due to its extended germination period and growth habit. It continues to flower and produce seed until killed by frost. Bur cucumber is host for cucurbit mosaic, tobacco streak, and wild cucumber mosaic viruses.

OTHER SPECIES OF CONCERN IN THE GOURD FAMILY

***COCCINIA GRANDIS* (L.) VOIGT—IVY GOURD**
Native to India, southwestern Asia, and North Africa; perennial vine, male or female; rootstock tuber-bearing; stems climbing up to 13 m tall, hairless; leaves alternate, oval, five-lobed, 5–9 cm long, 4–9 cm wide, base with three to eight glands, margins toothed, stalks 1–5 cm long; tendrils simple, originating in leaf axils; inflorescence a solitary axillary flower; flowers white, bell-shaped, 3–4.5 cm long; five sepals, the lobes curled back; five petals, deeply five-lobed; three stamens (as staminodes in female flowers); one pistil; fruit a bright red pepo (berry), oval to elliptical, 2.5–6 cm long, 2.5–3.5 cm wide, surface smooth. Weed designation: USA: HI.

THLADIANTHA DUBIA BUNGE—
GOLDENCREEPER, MANCHU TUBERGOURD

Thladiantha dubia

Native to northeastern China and Korea; perennial vine with underground potato-like tubers, male or female; stems climbing 1–2 m tall; tendrils branched; leaves alternate, heart-shaped, 7.5–15 cm long, margins smooth, surfaces with bristly hairs; flowers yellow, 15–25 mm across, imperfect; male flowers, sepals 5 (small), five petals, five stamens; female flowers, five sepals (small); five petals; one pistil; fruit a pepo (berry). Weed designation: none.

Cyperaceae
SEDGE FAMILY

Herbs, grasslike; stems three-sided, solid; leaves three-ranked; sheaths closed; ligule absent; inflorescence a panicle or umbel of spikelets, or spike; flowers perfect or imperfect; one bract; sepals and petals, when present, reduced to bristles, hairs, or scales; three stamens (occasionally one or six); one pistil; two or three styles; in *Carex*, female flowers surrounded by a hollow saclike structure called a perigynium; ovary superior; fruit an achene or nutlet, often lens-shaped or three-sided. Worldwide distribution: 90 genera/4,000 species; found throughout the world with greatest numbers in the cool temperate and arctic regions.

Distinguishing characteristics: plants grasslike; stems three-sided; flowers with a single bract.

A few species are important economically, including papyrus and Chinese water chestnut. Many provide food and habitat for a variety of wildlife.

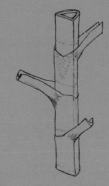

triangular stem

umbrella sedge (*Cyperus iria*)

Cyperus esculentus L.
YELLOW NUTSEDGE

Also known as: chufa, chufa flatsedge, northern nut grass, coco, coco sedge, rush nut, edible galingale, earth almond

Scientific Synonym(s): *C. repens* Ell., *C. tuberosus* Pursh

QUICK ID: stems triangular in cross-section; flower clusters yellowish-brown; leaves grasslike

Origin: perennial native to North America; weedy nature has developed over the last 50 years, coinciding with the use of selective herbicides

Life Cycle: perennial reproducing by tubers and seed; plants capable of producing over 90,000 seeds; field research shows that one tuber can produce 1,900 plants and 7,000 new tubers, in an area 2.1 m in diameter

Weed Designation: USA: CA, CO, HI, OR, WA; CAN: AB, BC, MB, NS, PQ; MEX: federal

DESCRIPTION

Seed: Oval, three-sided, whitish-brown, 1.2–1.8 mm long, 0.3–0.8 mm wide; surface shiny. Viability of tubers in soil is about 3.5 years; 60-year-old seed collected from herbarium sheets has proved viable. Germination in greenhouse experiments, over 51% of the seeds germinated. New plants are formed from the germination of underground tubers. Tubers are rarely found below 23 cm depth and, most commonly, in the top 15 cm of soil. Requires a cold period prior to producing vegetation; seed emergence between 6 and 25 mm depth. Optimal temperatures between 20°C and 30°C.

Seedling: Rarely produced in nature; cotyledons grasslike, shiny, often overlooked.

Leaves: Alternate, three-ranked, grasslike, flat to V-shaped, 20–90 cm long, 4–9 mm wide; waxy and glossy, midrib prominent; tips sharp-pointed; sheath closed, triangular; auricles and ligules absent; usually three to seven leaves per stem. Inflorescence an umbrella-shaped panicle; four to 10 rays, 2–12 cm long; 10–20 spikelets, 1–3 cm long, 12–35 mm wide, many-flowered.

Flower: Yellowish-brown, less than 2 mm across; six to 34 scales, 1.8–2.7 mm long; three stamens; one pistil; three-lobed style; bracts borne in whorl of three to nine, leaflike, 5–30 cm long, 0.5–4 mm wide, longest bract generally longer than the flower cluster; seeds viable 2–3 weeks after flowering.

Plant: Plants 15–90 cm tall; stems triangular, often reddish-brown at base, 3.4–6 mm wide; basal bulb stout; roots creeping, fibrous, producing tubers at root tips; tubers black, 8–19 mm long, woody, each producing over 10 fibrous roots and ultimately basal bulbs and new plants; native of warmer climates and migrating into cooler climatic zones and overwintering as tubers; range climatically controlled, as soil temperatures below −6.5°C kills most tubers; can tolerate −18°C air temperatures for short periods.

Fruit: Achene; rare in nature.

REASONS FOR CONCERN

Yellow nutsedge is a serious weed of cultivated fields and row crops. It is a major pest of corn, beans, and potatoes. Tubers of yellow nutsedge are capable of growing inside potato tubers, reducing the quality of the crop. Yellow nutsedge is also a host for two nematode species that attack cotton. Over 30 countries report yellow nutsedge as a principal weed.

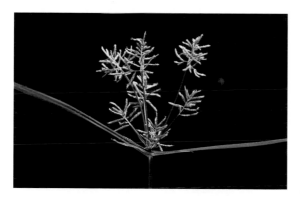

Cyperus iria L.
UMBRELLA SEDGE

Also known as: ricefield
   flatsedge
Scientific Synonym(s): none

QUICK ID: plant grasslike;
   roots yellowish-red;
   lowest bract longer than
   flower cluster
Origin: native to tropical
   regions of Asia and
   Africa; first reported from
   southeastern United
   States in 1840s
Life Cycle: annual
   reproducing by seed;
   large plants producing up
   to 5,000 seeds
Weed Designation: none

DESCRIPTION
Seed: Seeds 1–1.5 mm long, 0.5–0.7 mm wide, brown,
   triangular in cross-section.
   Viability unknown.
   Germination of fresh seed requires a 75-day
   dormant period; light enhances germination;
   optimal temperature 20°C; constant moisture
   required until plant established.
Seedling: With threadlike cotyledon; first leaf
   grasslike, wider than the cotyledon; later leaves
   similar in appearance.
Leaves: Alternate, V-shaped, grasslike, linear to
   lance-shaped, three to four per stem, shorter
   than the stem; 15–40 cm long, 3–6 mm wide,
   margins rough, sheath membranous and closed.
   Inflorescence of two to eight dense spikes, 2–20
   cm long, 3–12 mm wide, simple or compound;
   four to eight rays, 1–11 cm long; 10–30 spikelets,
   six- to 24-flowered, 5–20 mm long, 1.5–2 mm
   wide; six to 26 scales, brown to golden brown,
   1.3–1.8 mm long, 1.2–1.8 mm wide; two or
   three stamens; one pistil; four to seven bracts
   (commonly five), V-shaped, 8–30 cm long, 1–6
   mm wide.
Flower: Yellow; three stamens; one pistil, three-
   lobed style; three to seven bracts, the lowest
   longer than the cluster.
Plant: Tufted; stems triangular, 20–60 cm tall, 1–2.4
   mm wide, hairless; roots yellowish-red and
   fibrous; rhizomes, stolons, and tubers absent.
Fruit: Achene.

REASONS FOR CONCERN

Umbrella sedge is a principal weed of rice fields in Sri Lanka, India, and the Philippines. It is also a common weed of tea, vegetables, pastures, bananas, corn, and sugarcane in subtropical and tropical regions of the world. It is a serious weed of 17 crops in 22 countries.

Cyperus rotundus L.
PURPLE NUTSEDGE

Also known as: coco grass, nutgrass, nut grass, nutsedge

Scientific Synonym(s): *Chlorocyperus rotundus* Palle, *C. purpuro-variegatus* Boeck., *C. stoloniferum pallidus* Boeck., *C. tetrastachyos* Desf., *C. tuberosus* Rottb.

QUICK ID: stems three-sided; leaves dark green; flower clusters reddish- to purplish-brown

Origin: perennial native to India and western Asia

Life Cycle: perennial reproducing by rhizomes, tubers, and seed; each stem may produce up to 260 seeds; infestations may produce over 40,000 kg of tubers per hectare

Weed Designation: USA: AR, CA, OR, WA; CAN: BC; MEX: federal

DESCRIPTION
Seed: Linear to oblong, 1.4–1.7 mm long, 0.8–1.1 mm wide, triangular in cross-section, olive gray to brown or black, surface with a network of gray lines; rare in nature.
Viable up to 7 years.
Germination occurs in top 1 to 1.5 cm of soil; under optimal conditions in native range only 26% of seeds germinate, rarely exceeding 5% outside native range; germination of tubers requires 30–40% soil moisture; optimal soil temperature is 30°–35°C (range 15°–40°C); above 40°C tubers are no longer viable; bulbs exposed to soil temperatures lower than –5°C are killed.

Seedling: With a grasslike cotyledon, short; root development appears at the base of the sheath after emergence of the cotyledon; first two or three leaves emerge simultaneously, blades linear, folded lengthwise, smooth, membranous, light green, overlapping to form a triangular structure in cross section; tubers produced 6–8 weeks after emergence.

Leaves: Primarily basal, V-shaped, grasslike, dark green, 3–8 mm wide, 5–50 cm long, surface smooth and shiny, grooved on the upper surface; sheath closed; ligule absent.
Inflorescence a loose umbel with three to nine spikelets; stalk up to 30 cm long; three to seven rays, 0.2–10 cm long; two to 12 spikelets, 4–35 mm long, 1.3–1.8 mm wide, reddish-brown to purplish-brown, 10–40-flowered; two to five bracts, leaflike, 0.5–10 cm long, 0.5–4 mm wide, about the length of the cluster.

Flower: Reddish-brown to purplish-brown; six to 36 scales, 2–3.5 mm long, midribs green, 1.8–3.4 mm long, 2.2–3 mm wide; three stamens; one pistil;

three-branched style; flowering stem leafless; flowers appear 6–8 weeks after emergence.

Plant: Plants 20–100 cm tall, tolerant of drought and waterlogged conditions, intolerant of shade; stems triangular, 3.4–7 mm wide, hairless; rhizomes long, producing a basal bulb and bearing tubers at intervals of 5–25 cm along the length of the rhizome; tubers white and succulent when young, turning brown or black and fibrous with age, irregular to round, 5–25 mm long, tubers with papery scalelike leaves; roots to a depth of 135 cm, most tubers produced in top 15 cm of soil; does not persist in areas where the average January temperature is lower than 1°C.

Fruit: Achene.

REASONS FOR CONCERN

Purple nutsedge has been designated the world's worst weed, being reported in 52 crops in 92 countries. It is a serious weed of sugarcane, cotton, corn, rice, and vegetables. Rhizomes have also been reported to penetrate root vegetable crops, making the crops unusable.

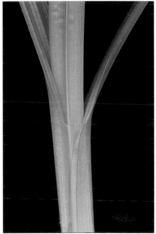

OTHER SPECIES OF CONCERN IN THE SEDGE FAMILY

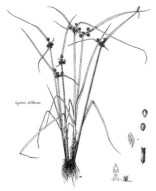

Cyperus difformis

***CYPERUS DIFFORMIS* L. — VARIABLE FLATSEDGE**
Native to southern Europe, Africa, and western Asia; wetland annual plant; stems triangular, one to 15 per plant, 7–30 cm tall, 1.2–2.5 mm wide, hairless; leaves three-ranked, two to seven per stem, 7–22 cm long, 2.2–4 mm wide; inflorescence a dense head, 7–17 mm across; one to five rays, 2–32 mm long; two to four bracts, 1–22 cm long, 0.5–3.5 mm wide, margins sharp; 30–120 spikelets, greenish- to purplish-brown, 3–5 mm long, 0.8–1.2 mm wide; 12–20 scales, straw-colored to purple, midrib green; one or two stamens; one pistil; fruit an achene, 0.6–0.8 mm in diameter. Weed designation: none.

CYPERUS INVOLUCRATUS ROTTB. — UMBRELLA PLANT

Native to Madagascar; plants rhizome-bearing; stems triangular 30–150 cm tall, 1–8 mm wide; leaves reduced to sheaths; inflorescence a panicle; heads 15–30 mm across; 20–22 rays, 5–12 cm long; secondary rays 3–33 mm long; tertiary rays 3–25 mm long when present; 18–22 bracts, 15–27 cm long, 8–12 mm wide, spreading; eight to 20 spikelets, 5–25 mm long, 1.5–2 mm wide; eight to 28 scales, whitish to light brown, 1.6–2.4 mm long, 1.2–1.5 mm wide; three stamens; one pistil; fruit an achene, 0.6–0.8 mm long, 0.4–0.6 mm wide. Weed designation: none.

Cyperus involucratus

Dioscoreaceae

YAM FAMILY

Vines with fleshy rhizomes or tubers; male
or female; stems twining; leaves alternate,
opposite, or whorled, simple, net-veined;
stipules absent; inflorescence solitary, spikes,
panicles, and racemes (often in groups of two
to seven); flowers imperfect; three sepals;
three petals; six stamens, in two whorls
of three (inner whorl reduced or absent);
one pistil; three styles; ovary inferior; fruit
a capsule, berry, or samara. Worldwide
distribution: five to nine genera/620 species;
predominantly found in tropics and subtropics.

Distinguishing characteristics: plant is
a climbing vine; rhizomes or tubers fleshy;
axillary bulblets often present.

A few species are grown for their food value.

Chinese yam (*Dioscorea polystachya*)

Dioscorea polystachya Turcz.
CHINESE YAM

Also known as: cinnamon vine, potato vine, air yam

Scientific Synonym(s): *D. batatas* Decne.

QUICK ID: twining vine; leaves opposite, heart-shaped; bulbils borne in leaf axils

Origin: native to China and southeast Asia; introduced as a food plant into the United States in the 1800s; not recognized in the natural environment until the 1980s

Life Cycle: perennial reproducing by aerial tubers and seeds; average-sized plants capable of producing over 100 bulbils per season.

Weed Designation: none

DESCRIPTION

Seed: Yellowish-brown, about 1 cm wide, with a prominent membranous wing.
Viability of bulbils in not known, but expected to be 3–5 years.
Germination of bulbils occurs within 2 weeks of being shed; small pieces of bulbils capable of producing new plants; seeds germinate within 3 weeks at 20°C.

Seedling: Rarely produced, as female plants are not found in the wild in North America.

Leaves: Opposite below, alternate or in whorls of three above, oval to heart-shaped, 4–8 cm long; distinctly nine- to 13-veined; tip with a long tapered point; base arrowhead-shaped; hairless; stalks longer than the blade; new leaves with a bronze tint.
Inflorescence of two types; male flowers borne in groups of three in panicles; female flowers solitary with two small bracts.

Flower: Greenish-yellow, fragrant with a cinnamon scent; stalkless; male flowers with three sepals, three petals, and three stamens; female flowers with three sepals, three petals, and one pistil with a three-branched style; female plants not found in North America.

Plant: Herbaceous or slightly woody twining vine; hardy to −18°C; male or female; stems up to 4 m long, twining from left to right, branched; root tuberlike, fleshy or woody, up to 1 m long, and weighing 5 pounds; aerial tubers (bulbils) 7–30 mm long, borne in leaf axils.

Fruit: Three-angled capsule, 2–4 cm long and 2 cm wide; three to six seeds.

REASONS FOR CONCERN

Chinese yam is known to climb over and shade out native vegetation. The aerial bulblets are often spread by wildlife. Research has demonstrated that partially eaten bulblets germinate and produce new plants.

OTHER SPECIES OF CONCERN IN
THE YAM FAMILY

DIOSCOREA ALATA L. — WATER YAM, WHITE YAM, GREATER YAM

Native to southeast Asia, widely cultivated for edible yams; plants tuber-bearing; tubers often weighing up to several kilograms; stems twining clockwise, up to 20 m, winged and four-angled; bulblets up to 4 cm across borne in leaf axils, wings purplish; leaves alternate below, opposite above, oval, 6–16 cm long, 4–13 cm wide, five- to seven-veined; stalks 6–16 cm long, winged, clasping; male flowers borne in one or two panicles per leaf axil, sepals and petals whitish, six stamens; female flowers borne in racemes, four- to eight-flowered, 6–35 cm long, sepals and petals pale, sterile six stamens; one pistil; fruit a capsule, 2–3.5 cm long; seeds winged. Weed designation: USA: FL.

DIOSCOREA BULBIFERA L. — AIR YAM, POTATO YAM

Native to tropical Asia and Africa; plants with or without tubers; tubers when present, globe-shaped, less than 1 kg; stems twining counterclockwise, over 20 m long, purple-spotted; axillary bulblets more than 5 cm across; leaves alternate, oval to heart-shaped, 5–25 cm long, 5–25 cm wide, five- to 11-veined; stalks less than 25 cm long, clasping; male inflorescence a panicle up to 70 cm long; male flowers fragrant, white, three sepals, three petals, six stamens; female inflorescence a spike, 6–40 cm long; five- to 50-flowered; female flowers greenish-white, three sepals, three petals, sterile six stamens, one pistil; capsule 1.8–2.8 cm long, 1–1.5 cm wide; seeds winged. Weed designation: USA: AL, FL.

Dioscorea bulbifera

Dipsacaceae

TEASEL FAMILY

Herbs, resembling members of Asteraceae (aster family); leaves opposite; stipules absent; inflorescence a capitulum (head); two bracts, below each flower; flowers perfect, irregular; five sepals, fused into a cuplike structure; four or five petals, united; four stamens, borne on the petals; one pistil; ovary inferior; fruit an achene. Worldwide distribution: 10 genera/270 species; native to Europe and western Asia.

Distinguishing characteristics: leaves opposite; flowers borne in heads, sepals present.

The teasels are of little economic importance.

teasel (*Dipsacus fullonum* ssp. *sylvestris*)

Dipsacus fullonum ssp. *sylvestris* (Huds.) Clapham
TEASEL

Also known as: wild teasel, Fuller's teasel, Venus's cup, prickly back, Adam's flannel, church brooms, gypsy-combs, Indian thistle

Scientific Synonym(s): *D. sylvestris* Huds.

QUICK ID: stems and leaves prickly; flowers, lilac to white, borne in dense heads; bracts below the head prickly

Origin: native to western Russia; introduced for use in carding wool; first reported from Niagara Falls in 1877

Life Cycle: biennial or short-lived perennial reproducing by seed; plants capable of producing up to 22,000 seeds; each head producing between 500 and 1,200 seeds

Weed Designation: USA: CO, IA, MO, NM

DESCRIPTION

Seed: Dull grayish-brown, 2.2–3.7 mm long, 0.8–1.1 mm thick; four-sided, each face with a prominent ridge; surface with stiff white hairs.
Viability in soil after 6 years is about 86%. Germination of teasel seeds is inhibited by the presence of quack grass.

Seedling: With oval cotyledons, 10–13 mm long, 5–6 mm wide, bitter-tasting, hairless; stem below cotyledons pale green: first leaves opposite, one leaf larger than the other, margins crinkled and wavy, midvein below bristly haired; later leaves with toothed margins.

Leaves: Opposite, lance-shaped, stalkless, and clasping the stem, forming a cup with the opposite leaf, margins toothed and prickly, underside with prickles and stiff spines on veins on the leaf underside; basal leaves oblong to lance-shaped, 10–60 cm long, upper surface with stout prickles; veins curving forward and connecting with those above.
Inflorescence an elliptical head, 3–10 cm long, borne on leafless stalks, up to 35 heads per plant (three to nine common); bracts stiff, prickles straight; flowering begins at the middle of the head, progressing upward and downward at the same time.

Flower: Lilac to white; four sepals, fused, about 1 mm long; four petals, fused, 10–15 mm long; four stamens, long-stalked; one pistil.

Plant: Plants 50–200 cm tall; stems coarse, very prickly; taproot up to 75 cm deep; easily recognizable by the tall stiff stems and large flower heads.

Fruit: Achene, 2–5 mm long, surrounded by the grayish-brown calyx; seeds shed within 1.5 m of the parent plant.

REASONS FOR CONCERN

Due to its prickly nature, livestock tend to avoid areas where teasel is growing. Teasel is host for aster yellows, teasel mosaic, and tomato black ring virus.

Similar species: cutleaf teasel (*D. laciniatus* L., p. 326) is distinguished by its stem leaves, which are lobed or divided. The leaves of cutleaf teasel are never prickly but may be bristly haired. Weed designation: USA: CO, IA, MO, OR.

Knautia arvensis (L.) Duby
FIELD SCABIOUS

Also known as: blue buttons, knautia, pincushion, gypsy rose, blue caps

Scientific Synonym(s): *Scabiosa arvensis* L.

QUICK ID: flower heads blue to pale purple; leaves opposite; lower stem bristly haired

Origin: native to Europe; introduced into North America as an ornamental plant

Life Cycle: perennial reproducing by seed; plants may produce up to 2,000 seeds

Weed Designation: CAN: AB, BC, SK

DESCRIPTION

Seed: Rectangular, four-sided, light brown, 5–6 mm long and 2 mm wide; surface covered in long hairs.
Viability is about 16% after 32 months, although 35-year-old seed has been reported to germinate. Germination occurs at 1–2 cm depth; optimal temperature is 20°C; no germination below 10°C.

Seedling: With club-shaped cotyledons, 12–18 mm long, 4–6 mm wide; tip slightly indented; first leaves oval, margins wavy, surface with scattered white hairs; later leaves shallow to deeply lobed.

Leaves: Opposite, deeply divided into five to 15 narrow lobes, 6–25 cm long, 2–6 cm wide, dull grayish-green due to the presence of short stiff hairs; terminal lobe largest; reduced in size upward; lower leaves stalked, margins toothed; upper leaves stalkless, margins toothed to deeply lobed.
Inflorescence a terminal head, 1.5–4 cm across; bracts 10–15 mm long, borne in a single row.

Flower: Blue to purple, tube-shaped, 9–12 mm long; four sepals, each with two or three teeth; four or five petals; four stamens; one pistil.

Plant: Plants 50–130 cm tall, rough-textured; stems erect, often purple-spotted, base with short stiff hairs, upper part hairless.

Fruit: Achene; enclosed by a four-ribbed hairy bract.

REASONS FOR CONCERN

Blue buttons is not a weed of cultivated land but has the potential to be a serious weed of pastures and forage crops. Once established, it is difficult to eradicate.

Similar species: blue buttons is often confused with several blue-flowered members of the aster family (p. 52). It is distinguished from asters by the presence of a calyx and four stamens and the absence of pappus. Members of the aster family have no calyx, five stamens, and pappus consisting of hairs, scales, or awns.

***DIPSACUS LACINIATUS* L.—CUTLEAF TEASEL**
Native to Europe; biennial or short-lived perennial;
taproot stout, 20–60 cm deep, about 2.5 cm
diameter; stems erect 1–3 m tall, surface with
hooked and straight prickles; leaves opposite,
deeply divided into narrow segments, bases fused
and forming a cup; upper stem leaves with smooth
margins; basal rosette 30–100 cm across produced
in first growing season, leaves oval to lance-shaped,
up to 40 cm long and 10 cm wide, pinnately lobed
and coarsely toothed; inflorescence a dense terminal
head 5–10 cm long and 5 cm wide; bracts linear to
lance-shaped, 5–10 cm long, 3–12 mm wide; flowers
white, 10–15 mm long, often fragrant; floret bract
1–2 cm long, stiff, sharp-pointed; four sepals, small
and cuplike; four petals; four stamens, exceeding
the petals; one pistil; fruit a grayish-brown achene,
about 3 mm long. Weed designation: USA: CO, IA,
MO, OR.

DIPSACUS SATIVUS (L.) HONCKNEY—
INDIAN TEASEL

Native to Europe; biennial; stems up to 2 m tall, spiny and rough-haired on ridges; leaves opposite, oblong to lance-shaped, margins wavy to toothed, bases fused and forming a cup, midrib below spiny; inflorescence a dense terminal head, 4–4.5 cm long; bracts curved down, unequal, 1.6–6 cm long; flowers pink to pale lilac or lavender; four sepals, cup-shaped; four petals; four stamens; one pistil; floret bract linear, stiff, spine-tipped; fruit an achene, 6–8 mm long, four-angled, hairy. Weed designation: USA: IA.

Elaeagnaceae

OLEASTER FAMILY

Shrubs and trees; leaves alternate or opposite, often covered in star-shaped or scalelike hairs; stipules absent; inflorescence a raceme or solitary flower; two or four sepals, fused or partly fused; petals absent; four or eight stamens; one pistil; ovary superior or resembling inferior; fruit an achene surrounded by fleshy calyx tube, often berry- or drupelike. Worldwide distribution: three genera/50 species; native to temperate and subtropical regions of the world.

Distinguishing characteristics: shrubs or trees; leaves alternate or opposite; two or four sepals.

A few members of the family are grown as ornamentals.

wolf willow (*Elaeagnus commutata*)

Elaeagnus angustifolia L.
RUSSIAN OLIVE

Also known as: oleaster, trebizond date

Scientific Synonym(s): *E. angustifolia* var. *orientalis* (L.) Kuntze, *E. hortensis* M. Bieb, *E. moorcroftii* Wall. ex Schltdl., *E.orientalis* L.

QUICK ID: trees; leaves alternate, grayish-green to silver; flowers yellow, fragrant

Origin: native to southern Europe and western Asia; introduced as an ornamental shrub in the early 1900s

Life Cycle: small tree or large shrub (living 30–50 years) reproducing by seed and resprouting of damaged crowns

Weed Designation: USA: CO, CT, NM

DESCRIPTION

Seed: Oblong, brown, 6–13 mm long. Viable in soil for up to 3 years. Germination is enhanced by temperatures below 5°C for a period of 90 days.

Seedling: With oval cotyledons, hairless, base with two small lobes; first leaves opposite, surfaces mealy, stem hairy.

Leaves: Alternate, dull grayish-green to silver-colored, oblong to elliptical or lance-shaped, 2–10 cm long, 1–4 cm wide, stalks 6–13 mm long, margins smooth, surface scaly, brown-dotted below.
Inflorescence an axillary raceme (two- or three-flowered) or a solitary flower.

Flower: Yellow, fragrant, 3–12 mm long, bell-shaped, perfect or imperfect; four sepals, silvery gray on the outside and yellow on the inside; petals absent; four stamens; one pistil.

Plant: Plants 5–12 m tall, trunks 10–50 cm diameter at maturity, tolerant of temperatures from −45°C to 46°C, somewhat shade-tolerant; bark dark, smooth; branches reddish-brown, ending in a short spine; seed bearing at 3–5 years of age, with greatest number of fruit produced after 14 years.

Fruit: Achene; berry-like, surrounded by the fleshy calyx, mealy textured, 1–2 cm long, surface covered in scales.

REASONS FOR CONCERN

Russian olive is somewhat shade-tolerant and eventually replaces native poplars along riverbanks. Seedlings of native poplars are shaded by mature Russian olives, thereby limiting the regenerative capacity of the native stands.

Similar species: a native species of western North America, wolf willow or silverberry (*E. commutata* Bernh.) may be confused with Russian olive. It is distinguished by silvery leaves 2–8 cm long and 0.5–4 cm wide, flowers 12–15 mm long, and fruit 8–10 mm long.

OTHER SPECIES OF CONCERN IN THE OLEASTER FAMILY

Elaeagnus umbellata

***ELAEAGNUS UMBELLATA* THUNB. —
AUTUMN OLIVE**

Native to China, Korea, and Japan; deciduous shrub 2–4 m tall; spines, when present, up to 2.6 cm long, young twigs covered in scales; leaves alternate, elliptical to oval or oblong, 4–8 cm long, 1–2 cm wide, underside covered with white scales, stalks 5–10 mm long, white scaly, margins smooth; inflorescence an axillary umbel, one- to seven-flowered, stalks 3–8 mm long, scaly white; flowers yellowish-white, bell-shaped, fragrant, 8–12 mm long and 7 mm across; four sepals; petals absent; four stamens; one pistil; stalks 5–6 mm long; fruit red to pink, drupelike, elliptical to globe-shaped, 3–9 mm long and 5 mm across, dotted with scales; seed 5–8 mm long, 2–3 mm across, saffron yellow. Weed designation: USA: CT, MA, NH, WV; CAN: AB.

Equisetaceae

HORSETAIL OR SCOURING-RUSH FAMILY

Herbs with creeping rhizomes; stems hollow and jointed; branches, when present, originating in leaf axils; leaves reduced and scalelike, borne in whorls at nodes; strobili terminal, conelike; sporangia borne on shieldlike scales attached to the central axis; spores minute, green. Worldwide distribution: one genus/23 species; found throughout the world, except for Australia and southeast Asia.

Distinguishing characteristics: herbs; stems jointed; leaves scalelike, borne in whorls.

Horsetails are of little economic importance. They are toxic to livestock, especially horses. Horsetails, one of the most primitive living plant families, were huge trees that dominated the landscape during the Carboniferous period (from 354 to 290 million years ago).

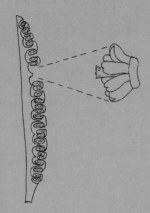

cross-section of strobilus

strobilus of field horsetail (*Equisetum arvense*)

Equisetum arvense L.
FIELD HORSETAIL

Also known as: mares tail,
pipe weed, paddy's pipe,
scouring rush, horsetail
fern, meadow pine,
pine grass, foxtail rush,
bottlebrush, horse pipes,
snake grass, cornfield
horsetail, toadpipes
Scientific Synonym(s):
E. calderi Boivin

QUICK ID: stems jointed;
leaves in whorls of eight
to 12; rhizome black,
feltlike
Origin: native to North
America
Life Cycle: perennial
reproducing by spreading
rhizomes and spores;
rhizome fragments 3 cm
long capable of producing
new plants
Weed Designation: CAN: PQ

DESCRIPTION

Seed: Not produced by horsetails; reproduction by
spores, pale green to yellow, globe-shaped, less
than 0.5–2 mm across and 4 mm tall.
Viability of spores about 48 hours.
Spores will only germinate in damp soil. A small,
flat green structure called a prothallus, less than
1 mm across, will grow into a new plant. Male
prothalli are 2–3 mm across, while the female
are 4–5 mm; mature plant develops from the
fertilized egg.
Seedling: Not produced by horsetails; young
horsetail plants are called sporelings.
Leaves: Appear in whorls of eight to 12, small,
scalelike, brown.
Flower: Not produced by horsetails. Spores produced
in a conelike structure (strobilus), 1–3 cm long,
borne at the top of reproductive stems.
Plant: With two types of stems: reproductive and
vegetative, both types jointed, up to 30 cm tall;
reproductive stems unbranched, flesh-colored,
appearing in spring, eight to 12 leaves per whorl,
5–9 mm long, unbranched, cone-tipped and
withering soon after spores shed; vegetative
stems green, 2–100 cm tall, 12 leaves per node,
internodes 1.5–6 cm long, 2–5 mm thick, branches
appearing in whorls from leaf axils, branches
three- to five-sided, six to eight branches per
node, 10–15 cm long, branches with three or four
teeth per node; rhizome creeping, up to 200 cm
long and 1 m deep, dark brown, feltlike, tuber-
bearing; plants extremely variable in appearance
depending on habitat and climatic conditions.
Spores produced by the millions in small tubelike
appendages inside the conelike structure at the
top of the reproductive stem.

REASONS FOR CONCERN

Common horsetail is an aggressive weed capable of reducing crop yields by 50%. In 1910, it was determined that horsetails are poisonous, especially to young horses. Ingestion of the plant causes vitamin B deficiency in animals feeding on it. Hay containing horsetails can be more toxic than horsetail plants found in pastures. A few sheep deaths have been reported. It is recognized as a serious weed in 25 countries.

reproductive stem

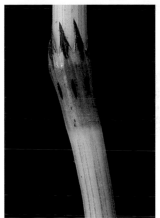

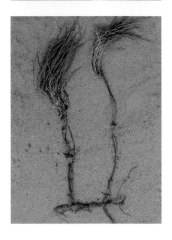

OTHER SPECIES OF CONCERN IN
THE HORSETAIL OR SCOURING-RUSH FAMILY

EQUISETUM TELMATEIA SSP. *BRAUNII*
(J. MILDE) HAUKE — GIANT HORSETAIL

Native to the west coast of North America; plants with two types of stems, reproductive and vegetative; reproductive stems brown, 17–45 cm tall, unbranched, sheaths 15–40 mm long, fleshy; vegetative stems green, branched, sheaths 7–18 mm long, 5–13 mm wide; teeth with green bases and dark tips, 14–30 mm long, 3–12 mm wide; branches in whorls, four- or five-ridged. Weed designation: USA: OR.

Euphorbiaceae
SPURGE FAMILY

Trees, shrubs, and herbs, often cactuslike; milky juice often present; leaves alternate, opposite, or whorled, simple or compound; stipules present, often reduced to hairs, glands, or spines; inflorescence a cyathium, cymes, or raceme; flowers imperfect, regular; zero to eight sepals (commonly five); petals absent (occasionally four to eight); one to many stamens; one pistil; three styles; ovary superior; fruit a capsule. Worldwide distribution: 290 genera/7,500 species (one genus, *Euphorbia*, with over 1,000 species); greatest diversity in tropical Africa and America.

Cyathiums are found predominantly in one genus, *Euphorbia*. This highly modified flower cluster often resembles a single flower. The cuplike involucre consists of four or five fused bracts. Bracts have a pair of nectar-secreting glands. Each cyathia contains several male flowers, each consisting of a single stamen, often in groups of four or five. A single female flower is found in each cyathia.

Distinguishing characteristics: plants with milky juice; flowers often borne in cyathia; fruit an explosive capsule.

The spurge family is a diverse group of economically important plants. Castor oil, rubber, cassava, and tapioca are produced by tropical members of this family. Other species, such as crotons, poinsettias, crown-of-thorns, and castor bean, are grown as ornamental plants. Several species are designated noxious weeds, and many more are considered poisonous.

petty spurge (*Euphorbia peplus*)

cyathium

Also known as: Virginia copperleaf, copperleaf, rhombic copperleaf, wax balls, mercuryweed

Scientific Synonym(s): *A. virginica* var. *rhomboidea* (Raf.) Cooperrider, *A. digynea* Raf.

QUICK ID: leaves diamond-shaped; floral bracts five- to nine-lobed; plants greenish-purple to copper-colored.

Origin: native to North America

Life Cycle: annual reproducing by seeds

Weed Designation: none

DESCRIPTION

Seed: Reddish-brown to grayish-black, 1.6–1.8 mm long, 1–1.3 mm wide; surface smooth; often covered in a thin gray covering.
Viable for up to 50 years.
Germination occurs in the top 2 cm of soil.

Seedling: With oval to round cotyledons, 3–15 mm long, 1.5–8 mm wide, distinctly veined, stalks with short hairs; stem below cotyledons dull reddish-brown, covered in short, stiff, downward-pointing hairs; first leaves opposite, thin, and pale green, often copper-colored at base; later leaves often alternate.

Leaves: Opposite near the base, alternate above, diamond-shaped, 1–9 cm long, 0.5–4 cm wide, green to copper-colored, stalks slender, 1–4 cm long, margins irregular with rounded teeth; leaves turning purple or copper-colored in autumn.
Inflorescence a raceme; male flowers in a short spike, 4–15 mm long, borne in leaf axils; female flowers borne in axillary clusters of one to three.

Flower: Green; male or female; male flowers less than 0.5 mm across, four sepals, eight to 16 stamens; female flowers with three to five sepals and one pistil with three irregularly branched styles; bracts five- to nine-lobed, often covered in glandular hairs.

Plant: Plants 7.5–100 cm tall; stems slender and branched; young plants resemble immature redroot pigweed (*Amaranthus retroflexus*, p. 6) or stinging nettle (*Urtica dioica*, p. 730).

Fruit: Three-lobed capsule; three-seeded.

REASONS FOR CONCERN

Threeseeded mercury is a serious weed in cereal, corn, and soybean fields, where it is reported to reduce crop yields drastically. It is also a host for tobacco broad ring spot virus.

Similar species: Virginia copperleaf (*A. virginica* L.) closely resembles threeseeded mercury. It is distinguished by floral bracts with nine to 15 lobes. The bracts do not have glands like those of threeseeded mercury. Both species are often treated as subspecies of each other.

Virginia copper leaf
(*Acalypha virginica*)

Chamaesyce glyptosperma (Engelm.) Small
THYME-LEAVED SPURGE

Also known as: ridgeseed spurge, ribseed sandmat

Scientific Synonym(s): *Euphorbia glyptosperma* Engelm.

QUICK ID: stems with milky sap; plants prostrate; leaves opposite

Origin: native to North America

Life Cycle: annual reproducing by seed

Weed Designation: none

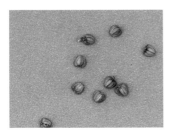

DESCRIPTION

Seed: Prism-shaped, grayish-white to tan, about 1.2 mm long and 0.7 mm wide; surface pitted or wrinkled; very sticky when wet and aiding in dispersal.
Viability unknown.
Germination occurs in the top 1 cm of soil.

Seedling: With oval cotyledons, 4–6 mm long, underside pinkish; first leaves opposite and resembling later leaves.

Leaves: Opposite, narrowly oblong, 3–15 mm long, dark green with a red midrib, margins smooth except for a few small teeth near the tip. Inflorescence a cyathium.

Flower: Greenish-white; male or female; male flowers borne in clusters of one to five, one stamens, one small bract; female flowers with one pistil.

Plant: Prostrate and forming large mats up to 60 cm across; stems freely branched, reddish-green, exuding milky juice when broken.

Fruit: Three-angled capsule, 1.5–2 mm long; three-seeded.

REASONS FOR CONCERN

Thyme-leaved spurge is a common weed of graveled areas, railways, and roadsides. It is occasionally found in gardens and new or poorly maintained lawns.

Similar species: a closely related species, also called thyme-leaved spurge (*C. serpyllifolia* Pers.), is distinguished from *C. glyptosperma* by wider-toothed leaves and clusters of five to 18 male flowers.

thyme-leaved spurge
(*Chamaesyce serpyllifolia*)

Chamaesyce hirta (L.) Millsp.
HAIRY SPURGE

Also known as: pillpod sandmat, milkweed, spurge, asthma plant, red euphorbia, cats hair, pill-bearing spurge, garden spurge, Queensland asthma weed, red milkweed, snakeweed

Scientific Synonym(s): *Euphorbia hirta* L., *E. pilulifera* auct. non L.

QUICK ID: stems with milky sap; stems prostrate; leaves with brown hairs

Origin: annual native to tropical America or Africa

Life Cycle: annual reproducing by seed; plants capable to producing up to 3,000 seeds; life cycle usually completed in less than 30 days

Weed Designation: none

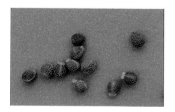

DESCRIPTION

Seed: Reddish-brown, ovoid, three-ridged, 0.5–1 mm long, surface slightly ribbed or wrinkled. Viability unknown.
Germination occurs in the top 1 cm of soil.

Seedling: With oval to round cotyledons; first leaves opposite, oblong to lance-shaped, margins smooth and fine-haired; later leaves minutely toothed and hairy.

Leaves: Opposite, elliptical to oblong or lance-shaped, 1–5 cm long, 5–25 mm wide, dark green above and pale green below, often purple blotched in the middle, densely haired on veins, stalks about 1 mm long, margins minutely toothed; stipules small, comblike.
Inflorescence a cyathium, dense and subglobose, axillary or terminal on stalks 4–15 mm long.

Flower: Greenish-yellow; male or female; bract cup-shaped, four-toothed.

Plant: Prostrate and forming mats up to 40 cm across; stems reddish-green, branched from the base, covered in yellowish-brown crinkled hairs; exuding milky juice when broken.

Fruit: Three-angled capsule, about 1.5 mm long, hairy, explosively breaking into three segments at maturity.

REASONS FOR CONCERN

Hairy spurge is a principal weed of sugarcane, sorghum, corn, and vegetables in Mexico and Hawaii. It is also host for cucumber mosaic virus. It has been reported as a serious weed in 47 countries.

Euphorbia cyparissias L.
CYPRESS SPURGE

Also known as: salvers spurge, quacksalvers grass, graveyard weed, tree-moss, Irish moss, Bonaparte's crown, kiss-me-Dick, welcome-to-our-house

Scientific Synonym(s): *Tithymalus cyparissias* (L.) Hill., *Galarheous cyparissias* (L.) Small ex Rydb.

QUICK ID: stem with milky sap; flowers greenish-yellow; leaves alternate, numerous

Origin: native to central Europe; introduced as an ornamental plant; first Canadian report from Ontario in 1870

Life Cycle: perennial reproducing by seeds and creeping rhizomes; plants producing up to 900 seeds; diploid plants sterile while tetraploids fertile

Weed Designation: USA: CO, CT, MA; CAN: MB, ON, SK

DESCRIPTION

Seed: Egg-shaped, yellow to bluish-gray, 2–3 mm long, 1.5–2 mm wide, surface finely pitted. Viability unknown.
Germination occurs when temperatures alternate between 15°C and 25°C; germination is low if temperature is constant.

Seedling: With oblong to lance-shaped cotyledons, 2.5–9 mm long, 2–6.5 mm wide, often larger than the first few pairs of leaves; stem below the cotyledons dull red; first leaves opposite, lance-shaped; milky juice present at early stages of growth; rhizomes and buds appear by the tenth leaf stage.

Leaves: Alternate, slightly club-shaped, 0.5–3.2 cm long, 1–3 mm wide; dark green; leaves of branches small, bristle-like; leaves dropping early in the season.
Inflorescence an umbrella-shaped terminal cyathium; rays 5–22, 9–40 mm long; bracts below clusters in whorls of 10–18, 5.6–21.2 mm long, 1.2–4.9 mm wide, yellowish-green and turning to bronze or purple with age.

Flower: Greenish-yellow; male or female; male flowers with a single stamen and small bract; female flowers with a single pistil.

Plant: Plant often forming large patches, both sterile and fertile plants in nature; stems 20–80 cm tall, leafy, exuding milky juice when broken; rhizomes with numerous pink buds.

Fruit: Three-lobed capsule, about 3 mm long; one- to three-seeded.

REASONS FOR CONCERN

Cypress spurge is a serious weed of roadsides, waste areas, and pastures in dry open areas. The milky sap of cypress spurge is reported to cause skin irritation in humans with sensitive skin. It is also reported to be somewhat toxic to cattle and horses.

Euphorbia esula L.
LEAFY SPURGE

Also known as: wolf's-milk, euphorbia, spurge, Hungarian spurge, faitours grass

Scientific Synonym(s):
E. discolor Ledeb.,
E. glomerulans Prokh.,
E. gmelinii Steud.,
E. jaxartica Prokh.,
E. kaleniczenkii Czern. ex Trautv.

QUICK ID: stem juice milky; flower clusters greenish-yellow; leaves alternate (whorled below the flower clusters)

Origin: native to Caucasus region of western Asia; introduced as seed impurity in 1827 at Massachusetts, Minnesota in 1890 from Russia; first Canadian report in Ontario in 1889; spreading to British Columbia by 1939

Life Cycle: perennial reproducing by seed and creeping rhizomes; each stem capable of producing 250 seeds

Weed Designation: USA: AK, AZ, CA, CO, CT, HI, ID, IA, KS, MA, MN, MT, NE, NV, NM, ND, OR, SD, UT, WA, WI, WY; CAN: federal, AB, BC, MB, NS, ON, SK; MEX: federal

DESCRIPTION

Seed: Oval to oblong, grayish with yellow and brown flecks, 2–2.4 mm long, 1.7–1.8 mm wide; surface smooth and somewhat glossy; seeds float on water, establishing new colonies on the shores of ditches, canals, and rivers.
Viable for 5–8 years in soil.
Germination occurs when air temperatures reach 26°C; can emerge from 10 cm, but 1.5–5 cm is optimal; seeds float and may germinate on water surface.

Seedling: With narrowly oblong cotyledons, 13–19 mm long, 2–4 mm wide, surface granular; stem below cotyledons pale green, reddish-brown at soil level; first leaves opposite, lance-shaped, stalkless; seedlings contain milky juice.

Leaves: Alternate or opposite, sometimes whorled below the flowering cluster, lance-shaped, 2–7 cm long, 3–5 mm wide, stalkless, margins smooth.
Inflorescence an umbrella-shaped cyathium; male flowers borne in groups of five; female flowers borne in groups of three; bracts opposite, 1.2 cm long and 1 cm wide, four crescent-shaped glands present in each cluster.

Flower: Greenish-yellow; male or female; male flowers with a single stamen; female flowers with a single pistil; seeds produced 30 days after the first flowers appear.

Plant: Plants 30–100 cm tall; stems somewhat woody, branched, hairless, often bluish-green, exuding milky juice when broken; rhizome deep, up to 5 m long, buds numerous, pink, and giving

rise to new stems; research has shown root
fragments from 2.8 m depth produce new plants.
Fruit: Nodding capsule, 4 mm long; three-seeded,
explosive, and scattering seeds up to 5 m away.

REASONS FOR CONCERN

Leafy spurge is a serious weed of pastures,
rangeland, and roadsides. If found in cultivated
fields or row crops it is difficult to eradicate because
of its rhizome and seed dispersal mechanism. In
North America over 2.5 million acres are infested.
Plants are poisonous to most livestock, except sheep
who feed on it without any affect. The milky juice
may cause severe skin rashes in humans. Some
reports indicate that decomposing plant tissue
produces an allelopathic response in sensitive
crops such as legumes and tomatoes.

Ricinus communis L.
CASTOR BEAN

Also known as: castor oil plant, wonder tree, African coffee tree

Scientific Synonym(s): none

QUICK ID: leaves alternate, palmately six- to 11-lobed; flowers inconspicuous; fruit a bristly capsule

Origin: native to northeastern Africa; introduced medicinal crop and garden ornamental

Life Cycle: annual (cool climates), or small tree (warm climates) reproducing by seed; persisting in climates with temperatures above 7°C; persisting as an annual weed in southern Ontario

Weed Designation: none

DESCRIPTION

Seed: Oval, 0.5–1.5 cm long; mottled white, gray, brown, yellow, red, or black; surface shiny. Viable for 2–3 years in soil.
Germination occurs in the top 7.5 cm of soil; optimal temperatures between 20°C and 35°C. Seedling with oval to round cotyledons, prominently veined; first leaves opposite, three- to five-lobed; later leaves alternate.

Leaves: Alternate, round in outline, 10–60 cm wide, palmately six- to 11-lobed, margins toothed; surface hairless.
Inflorescence a panicle up to 50 cm long; male flowers borne at base while females produced at top.

Flower: Male or female; male flowers with three to five sepals, numerous stamens with branched filaments, 5–10 mm long; female flowers with three to five sepals and one pistil with three red styles.

Plant: Plant 7–13 m tall and 2–5 m wide (1–3 m tall in temperate climates), frost-intolerant, requiring 140–180 days to reach maturity; stems 7.5–15 cm diameter, hairless; roots fibrous.

Fruit: Globe-shaped capsule; 2–2.5 cm long and wide; spiny; green turning brown at maturity; three-seeded.

REASONS FOR CONCERN

Castor bean may crowd out native vegetation when growing in dense colonies. Seeds contain the potent cytotoxin, ricin, and when chewed may cause abdominal pain, vomiting, and diarrhea. Death usually occurs within 36 hours. The number of seeds required for a fatal dose varies among species. A toxic dose for children is two to four seeds; adults, fewer than six; livestock, five to seven; dogs, 11; chickens and ducks, about 80.

OTHER SPECIES OF CONCERN IN
THE SPURGE FAMILY

Euphorbia dentata

Euphorbia helioscopia

Euphorbia heterophylla

EUPHORBIA DENTATA MICHX.—
TOOTHED SPURGE

Native to North America; plants annual, sap milky; stems 20–50 cm tall, branched, hairy; leaves opposite, linear to oval, stalked, margins toothed; inflorescence compact; involucres about 3 mm long, fringed, fleshy two-lipped gland present; fruit a capsule, about 5 mm diameter; seeds oval, 2.5–3 mm long, surface rough. Weed designation: USA: ID.

EUPHORBIA HELIOSCOPIA L.—SUN SPURGE

Native to Europe; plants annual; stems 20–50 cm tall, sap milky; leaves alternate, spatula-shaped, 1.5–5 cm long, 1–1.5 cm wide, blunt-tipped, margins finely toothed; leaves below inflorescence broadly elliptical to oblong or oval, in a whorl of five; inflorescence umbel-like; five rays, repeatedly branched; involucre about 2 mm long; fruit a capsule, 2.5–4 mm long; seeds 2–2.5 mm long; surface rough. Weed designation: CAN: PQ.

EUPHORBIA HETEROPHYLLA L.—
MEXICAN FIREPLANT, WILD POINSETTIA

Native to tropical America; annual with milky sap; stems 30–80 cm tall, branched, hollow, somewhat hairless; leaves alternate below, opposite or whorled above, oblong to oval, 5–10 cm long, 3–5 cm wide, green with a red or whitish base, or completely red, stalks 5–30 mm long; inflorescence a funnel-shaped cyathium; flowers male or female; male flowers with a single stamen, borne in groups of 10–12, and surrounding the female flower (a single pistil); fruit a nodding capsule, 5–6 mm across; three seeds, 2–2.5 mm long, dark brown to black. Weed designation: USA: FL, SC; MEX: federal.

EUPHORBIA MYRSINITES L.—MYRTLE SPURGE, DONKEYTAIL SPURGE, CREEPING SPURGE

Native to Mediterranean region of southern Europe; plants perennial, sap milky; stems 10–15 cm tall, but trailing to 45 cm, densely leaved; leaves alternate, egg-shaped, bluish-green and fleshy; inflorescence umbel-like, bracts showy yellowish-green; fruit a capsule.

Weed designation: USA: CO, OR.

Euphorbia myrsinites

TRIADICA SEBIFERA (L.) SMALL—CHINESE TALLOW

Native to China; tree to 16 m tall, bark reddish-brown; sap milky; leaves alternate, broadly oval, 3–8 cm long, 3–6 cm wide, stalks 2–5 cm long, margins smooth; inflorescence a spike up to 20 cm long; flowers male or female, yellowish-green; two or three sepals; two or three stamens; one pistil; three styles; fruit a capsule 1–2 cm long and 2 cm wide, turning brown to black at maturity; three seeds, globe-shaped, dull white. Weed designation: USA: FL, LA, MS, TX.

Fabaceae
(Formerly Leguminosae)

PEA FAMILY

Herbs, vines, shrubs, and trees; roots with bacterial nodules; leaves alternate, simple to pinnately or palmately compound; stipules present; inflorescence a head, raceme, umbel, or panicle; flowers perfect, mostly irregular; five sepals, fused; five petals, the lower two fused; 10 stamens (occasionally four to many); one pistil; ovary superior; fruit a legume or loment. Worldwide distribution: 600 genera/13,000 species; found throughout the world.

Fabaceae is often separated into the subfamilies Caesalpinioideae, Papilionoideae, and Mimosoideae. In the largest subfamily, Papilionoideae, the upper petal is often referred to as the banner or standard and located outside the two lateral petals. The two lower petals are fused and form a keel. In Caesalpinoideae, the banner is located inside the lateral petals. In Mimosoideae, the petals are all free.

Distinguishing characteristics: leaves compound; flowers pealike; fruit a pod (legume or loment).

The pea family is the second most important plant family to agriculture and human civilization, second only to grasses (Poaceae). The family includes significant crop species such as peanuts, beans, peas, and lentils. Other species, including alfalfa and clover, are grown for their crucial forage value. Members of the pea family play a valuable role in crop rotation where they serve to fix nitrogen.

common caragana (*Caragana arborescens*)

dissected flower

legume

loment

Abrus precatorius L.
ROSARY PEA

Also known as: crab's eye,
jumbie bean, licorice
plant, jequerity, precatory
bean

Scientific Synonym(s): *Abrus
abrus* (L.) Wight, *Glycine
abrus* L.

QUICK ID: plants climbing;
flowers white, pealike;
seeds scarlet with a black
base

Origin: native to India and
Guinea in Africa; once
used as balance weights
for measure, rosary beads
for Buddhist monks; still
used today in necklace
jewelry

Life Cycle: perennial
reproducing by seed

Weed Designation: none

DESCRIPTION

Seed: Scarlet with black base, 6–7 mm long, hard,
surface smooth and shiny.
Viability unknown.
Germination occurs once the hard seed coat
has disintegrated; about 61% of seeds will
have germinated between 11 and 182 days after
sowing.

Seedling: With thick and fleshy cotyledons
remaining below the surface of the soil; first leaf
compound with four to six leaflets, each 4–8 mm
long and less than 3 mm wide; later leaves with
eight or more leaflets; after 6 months plants may
be 30 cm tall.

Leaves: Alternate, 5–13 cm long, stalks 6–12 mm
long; compound with 10–40 leaflets; leaflets
oval to oblong, 1–1.8 cm long, margins smooth;
stipules 3.5–6 mm long, deciduous.
Inflorescence a short-stalked dense raceme borne
in leaf axils.

Flower: White to pink or reddish-purple; five sepals,
about 2.5 mm long; five petals; 10 stamens; one
pistil.

Plant: Climbing, twining, or trailing woody vine;
stems 5–10 m tall; older stems gray, younger
stems green; taproot dark reddish-brown, very
deep, lateral roots producing white nodules;
mature plants may grow 2 m or more per year.

Fruit: Short oblong legume, 25–42 mm long; three to
eight seeds, ovoid.

REASONS FOR CONCERN

The seeds of rosary pea contain abrin, a toxin that is extremely poisonous to humans, cattle, and horses. A single well-chewed seed (about 0.5 grams) is fatal for humans. Death usually occurs 3–4 days after severe stomach cramps and vomiting.

Cytisus scoparius (L.) Link.
SCOTCH BROOM

Also known as: broom, English broom, striated broom

Scientific Synonym(s): *Spartium scoparius* L., *Sarothamnus scoparius* (L.) Wimmer ex Koch

QUICK ID: shrub with green angular branches; flowers yellow; fruit a legume

Origin: native to Europe and North Africa; introduced into Hawaii, seeds taken to Victoria, British Columbia, in 1850

Life Cycle: perennial reproducing by seed and resprouting; mature plants may produce up to 18,000 seeds; plants live up to 20 years

Weed Designation: USA: CA, HI, ID, OR, WA

DESCRIPTION

Seed: Greenish-brown to black, ovoid, about 2 mm long, shiny, hard-coated; several species of ants attracted to seed appendages and disperse seeds while foraging.
Viable in soil for 30 years or more.
Germination temperatures 4°–33°C (18°–22°C optimal); emergence from 8 cm depth, optimal depth 2 cm; 50% of seed produced dormant for at least 1 year.

Seedling: With oval to elliptical cotyledons; first leaf compound with three leaflets; leaflets oval.

Leaves: Alternate, compound with three leaflets (occasionally one on new twigs), stalks 2–8 mm long; leaflets oblong to oval, 5–20 mm long, 1.5–8 mm wide, upper surface hairless, lower surface with short flattened hairs; stipules leaflike, 5–10 mm long, deciduous.
Inflorescence a long terminal raceme; one- to three-flowered in leaf axils; stalks less than 12 mm long.

Flower: Bright yellow, often purple-tinged, 20–25 mm long; five sepals, united; five petals, two united and three free; 10 stamens, one pistil; flowers produced in the second or third year of growth.

Plant: Deciduous shrub up to 3 m tall, frost-tolerant, shade-intolerant; stems green (photosynthetic), stiff, branched, sharply five-angled; taproot deep, branched; range controlled by severe winters and inadequate moisture; greatest growth in the first 4–5 years, and declines slowly thereafter.

Fruit: Dark brown to black legume 3–5 cm long, two- to 20-seeded (average five to eight), opening explosively and ejecting the seed; mature plants may produce 2,000–7,000 pods.

REASONS FOR CONCERN

Scotch broom seedlings can outcompete conifer seedlings and prevent reforestation. Scotch broom is known to colonize open disturbed sites, such as logged or burned sites, roadsides, pastures, oak woodlands, and open forests. It does not tolerate heavy shade but can tolerate minimal shade along the edges of forest canopies.

Similar species: another introduced species, French broom (*Genista monspessulana* [L.] L. Johnson, p. 383), is recognized by light yellow flowers less than 1 cm long, borne in compact clusters of three to 10 flowers. Weed designation: USA: CA, OR, HI.

Galega officinalis L.
GOAT'S RUE

Also known as: professor-weed, French lilac

Scientific Synonym(s): none

QUICK ID: plants with hollow stems; flowers white to pink; fruit a legume

Origin: native to central and southern Europe and western Asia; introduced forage crop into Utah in 1891

Life Cycle: perennial reproducing by seed; large plants producing up to 135,000 seeds

Weed Designation: USA: federal, AL, CA, FL, MA, MN, NV, NC, OR, PA, SC, VT, WA; MEX: federal

DESCRIPTION

Seed: Oblong to kidney-shaped, greenish to mustard yellow or reddish-brown, 2.5–4.5 mm long, 1–2 mm wide, surface dull.
Viable in soil for 5–10 years.
Germination enhanced by scarification; optimal soil temperatures between 20°C and 30°C.

Seedling: With oblong cotyledons; first leaf round, long-stalked; second leaf oval to round; later leaves compound.

Leaves: Alternate, compound with 11–17 leaflets; leaflets elliptical to lance-shaped, 1–5 cm long, 4–15 mm wide; stipules arrowhead-shaped, toothed or lobed.
Inflorescence a raceme, 7–10 cm long; 20–50-flowered

Flower: Purple to white, 10–15 mm long; five sepals; five petals; 10 stamens; one pistil.

Plant: With up to 20 stems from the crown, taproot deep; stems 0.5–2 m tall, hollow, branched, hairless.

Fruit: Legume, 2–5 cm long, 2–3 mm wide; one to nine seeds; large plants producing up to 15,000 legumes.

REASONS FOR CONCERN

Leaves contain galegin, a poisonous alkaloid, making it unpalatable for livestock. Plants prefer moist areas, enabling seeds to be spread by water. Goat's rue is known to displace native vegetation.

Similar species: wild licorice (*Glycyrrhiza lepidota* [Nutt.] Push.), a native perennial, may be mistaken for goat's rue. It is distinguished by solid stems, burlike legumes, and glandular-dotted leaflets. The yellowish-white flowers, about 12 mm long, are borne in dense racemes, 3–4 cm long.

wild licorice
(*Glycyrrhiza lepidota*)

Lotus corniculatus L.
BIRDSFOOT TREFOIL

Also known as: yellow trefoil, birdfoot deervetch, bloomfell, cat's clover, crowtoes, ground honeysuckle

Scientific Synonym(s): *L. glaber* Mill.

QUICK ID: plants prostrate to ascending; flowers yellow, borne in umbels; fruit cluster resembling a bird's foot

Origin: native to Europe; introduced as a forage crop; North American introduction occurred prior to 1753

Life Cycle: perennial reproducing by seed; plants producing up to 120 seeds per inflorescence

Weed Designation: none

DESCRIPTION

Seed: Oval, olive to brown or mottled black, 1.7–2.2 mm long, 1.4–1.7 mm wide, surface smooth. Viable up to 11 years in soil.
Germination within 4 days at 20°C, delayed when below 15°C or above 30°C; optimal depth 5–11 mm.

Seedling: Oval to lance-shaped cotyledons; first leaf compound with three leaflets; leaflets oval to elliptical or lance-shaped.

Leaves: Alternate, compound with five leaflets (two near the stem and resembling stipules, three near the tip); leaflets 5–20 mm long, 2–9 mm wide, sharp-pointed, margins finely toothed; stipules absent.
Inflorescence an umbrella-shaped cluster; two- to 10-flowered (commonly six); stalk 3–10 cm long.

Flower: Yellow to reddish-orange, 8–15 mm long; five sepals, united; five petals (two united, three free); 10 stamens; one pistil.

Plant: Prostrate to ascending; stems 30–100 cm long, branched, crown tough; taproot to 1 m deep.

Fruit: Legume 2–4 cm long, brown to black, hanging or spreading, five- to 20-seeded, opening explosively at maturity (about 37 days after pollination); fruiting cluster resembling a bird's foot; seeds mature 7–10 days before being shed.

REASONS FOR CONCERN

Birdsfoot trefoil is rapidly spreading along roadways and invading cultivated fields, lawns, and waste areas.

Similar species: another introduced species, trefoil (*L. pedunculatus* Cav.), is recognized by its hairy sepals, flowers about 12 mm long appearing in clusters of six to 12, and leaflets up to 2 cm long. The hollow stems, over 60 cm long, produce stolons.

Medicago lupulina L.
BLACK MEDIC

Also known as: trefoil, black
clover, none-such, hop
medic, spotted burclover,
blackseed, yellow trefoil,
hop clover
Scientific Synonym(s):
Lupulina aureata Noulet,
M. willdenowii (Merat)
Urb., *Melilotus lupulinus*
Trautv.

QUICK ID: leaves alternate,
compound with three
leaflets; flowers yellow,
borne in globe-shaped
clusters
Origin: native to eastern
Europe, western Asia,
and North Africa; first
reported in Canada in
1792
Life Cycle: annual, winter
annual, biennial, or a
short-lived perennial
reproducing by seed;
each plant capable of
producing over 6,000
seeds
Weed Designation: none

DESCRIPTION
Seed: Oval to kidney-shaped, yellowish-brown with
a green or orange tinge, 1.2–1.6 mm long, 0.9–1.2
mm wide, surface smooth.
Viable in soil for up to 11 years.
Germination does not require a dormant period;
seedlings emerging in fall are usually killed by
winter temperatures.
Seedling: With oval cotyledons, 4–9 mm long,
2–4 mm wide, dull green above, pale beneath;
first leaf simple, spatula-shaped; later leaves
compound with three leaflets.
Leaves: Alternate, compound with three leaflets;
leaflets oval, hairy, 5–30 mm long; terminal
leaflet stalked while laterals nearly stalkless;
stipules lance- or awl-shaped, spine-tipped.
Inflorescence a globe-shaped umbel 10–15
mm across, 20–50-flowered; borne on stalks
originating in leaf axils.
Flower: Yellow, 3–4 mm long; five sepals, united;
five petals (two united, three free); 10 stamens;
one pistil; flowers may appear within 6 weeks of
germination; seeds 9 weeks after germination.
Plant: Prostrate; stems 20–80 cm long; roots thick,
often difficult to pull from the ground.
Fruit: Black kidney-shaped pod, 1.5–3 mm long;
single-seeded.

REASONS FOR CONCERN

Black medic is a weed of lawns, gardens, roadsides, and pastures. It is occasionally found in cultivated crops and is increasing in importance in strawberry production. Black medic is host for bean yellow mosaic, pea mottle, pea wilt, potato yellow dwarf, and red clover mosaic viruses.

Similar species: alfalfa (*M. sativa* L.), another introduced species, is distinguished from black medic by the blue or purple flowers and a spirally coiled pod that contains several seeds. Stems, up to 1 m tall, rise from a deep taproot. Alfalfa was introduced from southeastern Europe as a forage crop in the early 1800s.

alfalfa (*Medicago sativa*)

Melilotus alba Medikus
WHITE SWEET CLOVER

Also known as: white melilot, honey clover, honey lotus, tree clover, white millet

Scientific Synonym(s): *M. albus* Medikus

QUICK ID: flowers white; leaves compound with three leaflets; plants sweet scented

Origin: native to Europe and Asia; introduced as a forage crop in mid 1600s

Life Cycle: annual or biennial reproducing by seed; plants capable of producing up to 350,000 seeds

Weed Designation: CAN: PQ

DESCRIPTION

Seed: Yellow, oval to kidney-shaped, 2–2.5 mm long and 1.5 mm wide.
Viable in the soil for up to 81 years.
Germination does not require a dormant period.

Seedling: With oblong cotyledons, 5–8.5 mm long, 2–2.5 mm wide, whitish-green above, pale green below; first leaf simple, round to kidney-shaped; later leaves compound with three leaflets.

Leaves: Alternate, compound with three leaflets, 1.5–5 cm long; leaflets club-shaped, 12–25 mm long, margins finely toothed; stipules 7–10 mm long. Inflorescence a terminal or axillary raceme, 5–15 cm long, 40–80-flowered.

Flower: Fragrant, white, 4–6 mm long; five sepals, united; five petals (two united, three free); 10 stamens; one pistil; stalks 1.5–2 mm long.

Plant: Sweet-scented; stems 50–250 cm tall, branched and leafy; roots extending to 1.2 m depth.

Fruit: Smooth papery legume, 3–4 mm long, black to dark gray; single-seeded (occasionally two seeds).

368 ■ PEA FAMILY

REASONS FOR CONCERN

White sweet clover is a common weed of roadsides and waste areas. The plant contains coumarin, a compound reported to be toxic to livestock. Seeds are a contaminant in cereal grains and adversely affect flour quality. White sweet clover is associated with over 28 viral diseases, many of which affect vegetable crops.

Similar species: yellow sweet clover (*M. officinalis* [L.] Lam.) is distinguished from white sweet clover by its yellow flowers and wrinkled pod. Each plant is capable of producing over 100,000 seeds. Weed designation: CAN: PQ.

yellow sweet clover
(*Melilotus officinalis*)

Oxytropis monticola A. Gray
LATE YELLOW LOCOWEED

Also known as: crazyweed
Scientific Synonym(s):
  O. campestris (L.) DC.,
  O. gracilis (A. Nels.)
  K. Schum., *O. luteola*
  (Greene) Piper & Beattie,
  O. villosa (Rydb.) K. Schum.

QUICK ID: leaves
  basal, compound with
  17–33 leaflets; flowers
  yellowish-white
Origin: native to North
  America; several
  subspecies occur
  throughout North
  America
Life Cycle: perennial
  reproducing by seed;
  each plant capable of
  producing over 1,500
  seeds
Weed Designation: CAN: MB

DESCRIPTION

Seed: Brown to black, kidney-shaped, 2–3 mm long
  and 2 mm across.
  Viability unknown.
  Germination shows no difference between
  exposure to light or darkness.
Seedling: With oval cotyledons 4–6 mm long, 2–3
  mm wide; stem below cotyledons pale green,
  succulent; first leaf simple, lance-shaped,
  densely haired; second to sixth leaf stage leaves
  compound with three leaflets; later leaves with
  17–33 leaflets.
Leaves: Basal, 6–23 cm long, composed with 17–33
  leaflets; leaflets oblong, 6–25 mm long; surface
  with short silky hairs; stipules present.
  Inflorescence a terminal raceme 5–10 cm long,
  10–30-flowered.
Flower: Creamy white (occasionally pinkish-blue);
  five sepals, united, 5–7 mm long, covered in
  white and black hairs; five petals (two united,
  three free), 12–17 mm long; 10 stamens; one pistil;
  bracts covered in black hairs.
Plant: Of grassland and open forests; flowering
  stems 10–40 cm tall; taproot short and thick.
Fruit: Oblong legume, 16–20 mm long, papery,
  surface with black and white hairs; several-
  seeded.

REASONS FOR CONCERN

Late yellow locoweed contains a poisonous agent that affects the nervous system of animals that ingest it. The common names, locoweed and crazyweed, refer to the behavior of animals that have been poisoned by this plant.

Similar species: early yellow locoweed (*O. sericea* Nutt.) is often confused with late yellow locoweed. Early yellow locoweed blooms in late April to early May. Its leaves are 4–30 cm long with 11–17 leaflets. The fruit is a leathery pod about 20 mm long. Weed designation: CAN: MB. Another native species, showy locoweed (*O. splendens* Dougl. ex Hook.), is often found growing alongside late yellow locoweed. It has blue to reddish-purple flowers, and leaves with 21–60 leaflets. Weed designation: CAN: MB.

showy locoweed
(*Oxytropis splendens*)

showy locoweed
(*Oxytropis splendens*)

Also known as: purple crown vetch, trailing crown vetch, rosenkronill

Scientific Synonym(s): *Coronilla varia* L.

QUICK ID: leaves with 15–29 leaflets; flowers pink, borne in umbels; fruit a loment

Origin: native to Europe and Asia; introduced as an erosion control and soil stabilizer

Life Cycle: perennial reproducing by seed and creeping rhizomes; large plants producing up to 1,000 seeds

Weed Designation: none

DESCRIPTION

Seed: Round to oblong, notched on one side; purplish-red, 3.1–3.7 mm long, 1–1.3 mm wide. Viable in soil for up to 5 years. Germination occurs shortly after being shed as no dormant period is required.

Seedling: With oblong cotyledons, 8–10 mm long; first leaf compound with three triangular leaflets, bristly haired; later leaves pinnately compound with oval leaflets.

Leaves: Alternate, compound with 15–29 leaflets; leaflets oblong to oval or elliptical, hairless, 6–20 mm long, 3–12 mm wide, stalks up to 1.2 mm long; stipules awl-shaped, 1–6 mm long. Inflorescence a headlike umbel; five- to 20-flowered; stalk up to 7 cm long.

Flower: Whitish-pink to deep pink, 9–15 mm long; five sepals; five petals, the keel inflated; 10 stamens; one pistil; stalks 3–4 mm long, hairless.

Plant: With fleshy rhizomes, tolerant of drought, waterlogged periods, and cold temperatures (−33°C), intolerant of shade; stems trailing 20–200 cm long, hollow, branched at the base;

Fruit: Loment, 2–8 cm long, 1.5–2 mm across, three to 14 segments, beak to 5 mm long, leathery; fruiting cluster crownlike.

REASONS FOR CONCERN

Once used extensively for erosion control, crown vetch escaped regeneration sites and invaded natural areas. It forms large monocultures and displaces native vegetation.

Sesbania herbacea (P. Mill.) McVaugh
TALL INDIGO

Also known as: bigpod sesbania, hemp sesbania, peatree, Colorado River hemp, coffeeweed

Scientific Synonym(s): *S. exaltatus* (Raf.) Rydb., *S. macrocarpa* Muhl ex Raf., *S. exaltata* (Raf.) Rydb. ex A. W. Hill., *Darwinia exaltata* Raf.

QUICK ID: plants semiwoody; flowers yellow; fruit a legume

Origin: native to North America

Life Cycle: perennial reproducing by seed

Weed Designation: USA: AR

DESCRIPTION

Seed: Orange to reddish-brown, kidney-shaped. Viability in soil is about 18% after 5.5 years. Germination at temperatures ranging from 15°C to 40°C. Optimal growth occurred from 30°C to 35°C.

Seedling: With oblong to spoon-shaped cotyledons, about two times longer than wide, thick, upper surface green, lower surface grayish-green and hairless; first leaf simple; later leaves with six to eight or more leaflets.

Leaves: Alternate, 10–30 cm long, compound with 20–70 leaflets; leaflets oblong to elliptical, 8–30 mm long, 2–6 mm wide, hairless above and sparingly hairy below, margins smooth; stipules deciduous.
Inflorescence an axillary raceme; two- to six-flowered.

Flower: Yellow, often purple-streaked or -spotted; 1.5–2.0 cm long; five sepals, united, 3–4 cm long; five petals; 10 stamens; one pistil.

Plant: Becoming woody with age, 70–200 cm tall; stems erect, hairless.

Fruit: Linear curved legume, 10–20 cm long, 3–4 mm wide; 30–40 seeds.

REASONS FOR CONCERN

Tall indigo contains saponin, which is toxic to livestock and humans. The seeds are the most toxic part of the plant and are consumed in the late summer, fall, or winter when other forage is scarce. Cattle are often affected when moved into new pastures containing the plant. It has been observed that cattle frequently develop a craving for the seeds. Cattle are often found dead in infested fields. Opened stomachs commonly reveal sprouted seeds and hemorrhagic inflammation of the intestines. Poisoned animals walk stiffly with an arched back, have shallow respiration, and have a weak rapid pulse. A coma generally precedes death.

Trifolium repens L.
WHITE CLOVER

Also known as: alsike clover, Dutch clover, wild white clover, landino clover

Scientific Synonym(s):
T. anomalum Schrank,
T. limonium Phil.,
T. nigrescens Schur,
T. nothum Stev.

QUICK ID: leaves alternate, compound with three leaflets; flowers white; stems often rooting at nodes

Origin: native to Europe; introduced as a forage crop; common in Canada prior to 1749

Life Cycle: perennial reproducing by seed and creeping stems

Weed Designation: none

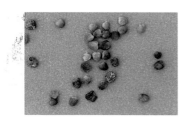

DESCRIPTION

Seed: Dull yellow to orangish-brown, round, 0.8–1.1 mm long and wide; surface smooth.
Viable for at least 25 years in dry storage; in field conditions less than 5 years.
Germination occurs from 10°C to 30°C (optimum 24°C).

Seedling: With oval cotyledons 2.4–4.1 mm long, 1–2 mm wide; first leaf simple, spatula-shaped; later leaves with three oval leaflets; stem does not elongate until the seventh leaf stage.

Leaves: Alternate, compound with three leaflets; leaflets oval, 1–3.5 cm long; stipules present, often forming a tube around the stem. Inflorescence a globe-shaped terminal raceme, 15–20 mm across, 20–40-flowered.

Flower: Pinkish-white to white, 7–11 mm long; five sepals, 3.5–7 mm long; five petals; 10–15 stamens; one pistil; seeds mature about 28 days after pollination but viable by day 12.

Plant: Prostrate, drought-intolerant; stems 10–50 cm long, often rooting at nodes; taproot shallow, lasting only 1 or 2 years; fibrous roots in top 10–25 cm of soil.

Fruit: Legume, 4–5 mm long, three- to six-seeded.

REASONS FOR CONCERN

White clover is a serious weed in lawns, waste areas, and abandoned fields. When white clover exceeds 20–40% of cattle feed, death from bloat may occur. It is host for alfalfa mosaic and pea mottle viruses.

Similar species: red clover (*T. pratense* L.), another introduced forage crop, is often found growing beside white clover. As the common name implies, it has reddish-pink flowers in a globe-shaped cluster. The flowers are 12–20 mm long. Each flower produces a pod containing one yellow or purple seed. The leaves of red clover are composed of three leaflets, each 2–5 cm long. Leaflets often have an inverted V on the upper surface.

red clover (*Trifolium pratense*)

Ulex europaeus L.
GORSE

Also known as: furze, whin, prickly broom

Scientific Synonym(s):
U. armoricanus Mabil.,
U. compositus Moench,
U. floridus Salisb.,
U. grandiflorus Pour.,
U. hibernicus G. Don,
U. major Thore., *U. mitis* hort., *U. ophistolepis* Webb, *U. strictus* Mackay, *U. vernalis* Thore

QUICK ID: spiny shrub; flowers yellow; pods 1–2 cm long

Origin: native to western Europe; introduced as a windbreak or hedgeplant; introduced into Oregon prior to 1894 from Ireland, widespread by 1950s

Life Cycle: perennial reproducing by layering and seeds; individual plants may produce up to 18,000 seeds per season

Weed Designation: USA: CA, HI, OR, WA; CAN: BC; MEX: federal

DESCRIPTION

Seed: Olive green to brown, bean-shaped; 2–3 mm long.
Viable in soil for 30–70 years.
Germination does not require light; optimal temperatures 15°–19°C; emergence from the top 5 cm of soil; heat from a brushfire stimulates germination of seeds 2–5 cm below the surface; soil temperatures 100°C are lethal.

Seedling: With oblong cotyledons, 5–7 mm long, leathery, stalkless; stem below cotyledons somewhat woody, hairless; first leaves simple, lance-shaped, 2–5 mm long, leathery, surface covered with stiff hairs; later leaves compound with two or three leaflets, 5–10 mm long; by day 25 rosette about 1.5 cm across; at about 5 cm tall (4–6 months of age) the first spines appear; no further leaves are produced, only spines.

Leaves: Alternate, reduced to flattened spines, 5–50 mm long.
Inflorescence a raceme or solitary flower.

Flower: Yellow, showy, fragrant, 15–20 mm long; five sepals, united and two-lipped, 10–15 mm long; five petals; 10 stamens; one pistil; flowering begins at about 18 months of age.

Plant: Spiny evergreen shrub, 1–6 m tall and 8 m in diameter, living to about 30 years of age, cannot withstand extremes of heat or cold; stems woody, green to brown, lower branches rooting when in contact with soil; roots deep.

Fruit: Black legume, hairy, 1–2 cm long, 6–8 mm wide; two- to 12-seeded (average four or five); explosive at maturity and ejecting seeds up to 5 m from parent plant; seeds require 2 months to mature after pollination.

REASONS FOR CONCERN

Burning off infested areas produces a large flush of seedlings from the seed bank in the soil. Fires with a temperature less than 100°C stimulate germination of seeds in the top 6 cm of the soil. Fires above 100°C produce a reduction of seed reserves by 33%. Repeated fires lead to the population decline but also attributed to soil erosion. Mature plants are highly flammable. Gorse is also a serious weed of open forests and pastures. Livestock avoid areas where it is growing due to its spiny nature. This often leads to a population increase and to dense impenetrable stands that exclude desirable vegetation in pastures.

Vicia cracca L.
TUFTED VETCH

Also known as: Canada pea, bird vetch, wild vetch, cow vetch, cat peas

Scientific Synonym(s):
V. tenuifolia Roth,
V. graccha Lepech.,
V. heteropus Freyn.,
V. scheuchzeri Brüg.,
V. semicincta Greene,
V. versicolor Salisb.

QUICK ID: plants climbing; tendrils present; flowers purple, borne in one-sided clusters

Origin: native to Europe and Asia; some native populations may exist in northeastern North America

Life Cycle: perennial reproducing by seeds and rhizomes; plants may produce up to 500 seeds

Weed Designation: none

DESCRIPTION

Seed: Olive green to reddish-brown or black, globe-shaped, 1.5–3 mm long, scar white or reddish-brown and extending one-quarter to one-third of the way around the seed.
Viability unknown; some reports indicate several years.
Germination requires open disturbed soil.

Seedling: With cotyledons remaining underground; first leaf scalelike and papery; first compound leaves composed of two to four leaflets, each oval to lance-shaped, 10–30 mm long, 3–5 mm wide, tendrils present; later leaves with 10–24 leaflets.

Leaves: Alternate, pinnately compound with 10–24 leaflets; leaflets oblong to lance-shaped, 1–3 cm long, bristle-tipped; tendril terminal; stipules with smooth margins.
Inflorescence a one-sided raceme; 15–40-flowered; long-stalked; borne in leaf axils.

Flower: Purple, turning blue with age, 9–13 mm long; five sepals, united, hairy, 2–3 mm long; five petals; 10 stamens; one pistil.

Plant: Climbing; stems weak 50–200 cm tall, angular; roots wiry and creeping.

Fruit: Light brown legume, 10–30 mm long, 4–7 mm wide, six- to 12-seeded; fruit walls becoming twisted on opening.

REASONS FOR CONCERN

Tufted vetch is a troublesome weed in cultivated crops and grains as its twining nature hampers harvest. During crop harvest, tufted vetch seeds are shed, aiding in its dispersal.

Similar species: narrow-leaved vetch (*V. sativa* ssp. *nigra* [L.] Ehrh.), an annual climbing vine, has stems up to 100 cm tall and pinnately compound leaves with two to five pairs of leaflets. The purple flowers produce pods, 4–7 cm long, containing 10–16 black seeds. Hairy vetch (*V. villosa* Roth), an annual or biennial, is distinguished by pods 20–40 mm long and 6–12 mm wide. The pods contain eight seeds with scars running one-twelfth to one-fifth of the way around the seed.

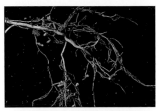

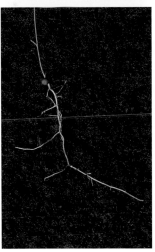

OTHER SPECIES OF CONCERN IN THE PEA FAMILY

ALHAGI MAURORUM MEDIK. — CAMELTHORN, CASPIAN MANNA, PERSIAN MANNA

Native to southwestern Asia and North Africa; shrub up to 1.2 m tall; roots creeping, up to 2 m deep and 8 m in all directions; stems branched, greenish, spiny; spines 1–2.5 cm long, yellow-tipped; leaves alternate, oval to lance-shaped, less than 2.5 cm long, simple, bluish-green, short-stalked; inflorescence a raceme, one- to eight-flowered; flowers brown to red or purple, pealike, 8–9 mm long; fruit a loment, reddish-brown, 1–3 cm long, one- to five-seeded; seeds dark brown to yellowish or greenish-brown, kidney-shaped. Weed designation: USA: AZ, CA, CO, NM, NV, OR, TX, WA.

CROTALARIA PALLIDA AIT. — SMOOTH RATTLEBOX

Probably native to tropical Africa; perennial, shrublike; stems 1–2 m tall; leaves alternate, stalks 2.2–6.7 mm long, compound with three leaflets; leaflets elliptical to oval, 2.8–13 cm long, 1.2–4.5 cm wide, prominently six- to 12-veined per side; stipules less than 1 mm long or absent; inflorescence a raceme, 15–40 cm long, 22–75-flowered; bracts 3–4 mm long; flowers yellow, 9–15 mm long; five sepals; five petals, standard 9–12 mm long, wings 6–10 mm long; 10 stamens; one pistil; fruit a legume, 3.5–3.8 cm long, brown, hairless to short-haired; 30–40 seeds, brown, up to 2.8 mm long, surface smooth. Weed designation: USA: AR.

Alhagi maurorum

GENISTA MONSPESSULANA (L.) L. JOHNSON — FRENCH BROOM

Native to Mediterranean region of southern Europe; shrub 3–5 m tall; stems covered in silver silky hairs; leaves alternate, stalks about 5 mm long, compound with three leaflets; leaflets oblong to lance-shaped, 10–15 mm long, lower surface hairy; stipules about 2 mm long; inflorescence a headlike raceme borne on stalks 1–3 mm long, four- to 10-flowered; flowers yellow to pale yellow, pealike; fruit a legume, 15–25 mm long, covered in silky hairs, brown to black at maturity; three to eight seeds, brown to black. Weed designation: USA: CA, HI, OR.

LATHYRUS TUBEROSUS L. — TUBEROUS VETCHLING

Native to Europe and western Asia; perennial with tubers up to 1 cm across; stems 30–100 cm tall, climbing and branches, often four-angled, hairless; leaves alternate, compound with two leaflets; leaflets narrowly elliptical, 2–5 cm long, short-stalked; tendril threadlike, three-branched; stipules pointed, about 1 cm long; inflorescence an axillary raceme 2–10 cm long, two- to five-flowered, stalk longer than the leaf; flowers bright red to pink or violet, resembling a sweet pea; five sepals; five petals; 10 stamens; one pistil; fruit a dull brown legume 25–35 mm long, 4–6 mm wide, hairless; three to six seeds, brown, 3–4 mm across. Weed designation: CAN: ON.

Lathyrus tuberosus

LESPEDEZA CUNEATA (DUM.-COURS.) G. DON — SERICEA LESPEDEZA

Native to eastern Asia; perennial; taproot to 1.5 m deep; crown woody; stems 1–2 m tall, branched; leaves alternate, compound with three leaflets; leaflets narrowly oblong to wedge-shaped, 6–25 mm long, 1–6 mm wide, spine-tipped, grayish-green due to densely haired surface; lower leaves stalked, upper stalkless; stipules thread or awl-like, 3–11 mm long; inflorescence a raceme, two- to four-flowered, containing flowers of two types: showy or inconspicuous and petal-less; showy flowers white to pale yellow with purple markings, 5–7 mm long; inconspicuous flowers not opening, self-fertilized; fruit a legume, 3–5 mm long, one-seeded. Weed designation: USA: CO, KS.

MIMOSA PIGRA L. — CATCLAW MIMOSA

Native to tropical Central and South America; shrub 3–6 m tall; stems greenish, becoming woody with age, recurved prickles 3–7 mm long; leaves alternate, 20–25 cm long, compound with five to 12 pairs of pinnae, prickly at junction; pinnae with 24–31 pairs of leaflets, sensitive to touch; leaflets about 8 mm long, margins hairy; inflorescence a globe-shaped head, 1–2 cm across, about 100-flowered; flowers pink; four sepals; four petals; eight stamens; one pistil; fruit a hairy loment, 6.5–7.5 cm long, 7–10 mm wide, 20–25-seeded; fruit often borne in clusters of seven; seeds brown or olive green, 4–6 mm long. Weed designation: USA: federal, AL, CA, FL, MA, MN, NC, OR, SC, VT.

Mimosa pigra

MIMOSA PUDICA L. — SENSITIVE PLANT, HUMBLE PLANT, TOUCH-ME-NOT, SLEEPING GRASS, SHAME PLANT, ACTION PLANT, LIVE-AND-DIE

Native to tropical areas of North and South America; annual, biennial, or short-lived perennial; stems erect to trailing, 20–100 cm long, branches glandular-hairy and prickly, somewhat woody at base, reddish-brown; nodes with reddish-brown to purple thorns and prickles; leaves alternate, bipinnately compound with two or four pinnae, each 2.5–5 cm long; leaflets 12–25 pairs, 9–12 mm long, collapsing when touched; inflorescence a head, 8–10 mm long, borne in leaf axils; flowers pinkish, 2–4 mm long; fruit a loment, 1–2 cm long, 3–5 mm wide, prickly, breaking into one to five segments; seeds flat, about 3 mm across, bristly. Weed designation: none.

MIMOSA STRIGILLOSA TORR. & GRAY — CREEPING SENSITIVE PLANT, POWDERPUFF

Native to tropical North America; perennial; rhizomes present; stems prostrate, 10–100 cm long; leaves alternate, bipinnately compound with three to six pairs of pinnae, each with 11–21 leaflets, collapsing when touched; inflorescence a cylindrical to globe-shaped head; flowers pink; five petals, free; 10 stamens, free; fruit a loment, hairy. Weed designation: none.

Mimosa strigillosa

PROSOPIS SSP. — MESQUITE

Native to North America (majority of species), Africa, Asia; tree or shrub, 3–15 m tall, taproot to 20 m depth; older bark rough and gray, young bark smooth, dark red to green; branches zigzagged, often thorny at each node; thorns 4–75 mm long; leaves alternate, bipinnately compound (fernlike), borne at zigzag nodes; inflorescence a spike, 5–8 cm long; flowers greenish-cream or yellow; five sepals; five petals; 10 stamens; one pistil; fruit a legume, up to 20 cm long, slightly constricted between the seeds; five to 20 seeds. Weed designation: USA: federal (over 20 ssp. listed), AL, CA, FL, HI, MA, MN, NC, OR, SC, VT.

PUERARIA MONTANA (LOUR.) MERR. — KUDZU

Native to Japan and China; twining or trailing semiwoody vine, 10–30 m long; roots tuberous, up to 5 m deep and 20 cm across, weighing up to 180 kg; stems ropelike, up to 10 cm in diameter, up to 30 per crown; leaves alternate, stalks 15–30 cm long, compound with three leaflets; leaflets 8–18 cm long, 6–20 cm wide, slightly lobed, hairy on both surfaces; inflorescence a raceme, 5–30 cm long; flowers fragrant (grape-scented), reddish-purple, pealike, 12–19 mm long; fruit a legume, 3–5 cm long, hairy; three to 10 seeds, oval. Weed designation: USA: CT, FL, IL, KS, KY, MA, MO, MS, OR, PA, TX, WA, WV.

ROBINIA PSEUDOACACIA L. — BLACK LOCUST

Native to eastern United States; tree up to 30 m tall; bark brown and furrowed; twigs with a pair of thorns (6–12 mm long) at nodes; leaves alternate, 20–35 cm long, compound with seven to 19 leaflets; leaflets elliptical to oval, 2–5 cm long, 12–19 mm wide; inflorescence a drooping raceme 10–20 cm long, borne on current year's growth; flowers fragrant, white, pealike, 1–2 cm long; fruit a legume, 5–10 cm long, poisonous; three to 14 seeds, dark brown to black. Weed designation: USA: CT, MA.

Robinia pseudoacacia

***SPHAEROPHYSA SALSULA* (PALLAS) DC. —**
ALKALI SWAINSONPEA, AUSTRIAN PEAWEED
Native to Asia; perennial with woody taproot and
rhizomes; stems 30–150 cm tall; leaves alternate,
compound with nine to 25 leaflets; stipules awl-
shaped, less than 5 mm long; leaflets oval to oblong,
5–20 mm long, covered in silvery hairs; inflorescence
a raceme, four- to eight-flowered; stalks 3–7 mm
long; flowers 6–25 mm long, brick red to purple;
fruit a legume, 15–35 mm long, inflated. Weed
designation: USA: CA, NV, OR, WA.

***SPARTIUM JUNCEUM* L. — SPANISH BROOM**
Native to southern Europe; shrub or small tree, 3–5
m tall; stems green; leaves linear to lance-shaped,
less than 1.5 cm long, simple, lower surface hairy;
stipules absent; inflorescence a raceme, five- to
20-flowered, borne on current year's growth; flowers
fragrant, yellow, pealike, 2.5–3 cm long; fruit a
legume, 5–10 cm long, 0.5–1 cm wide, 10–15-seeded.
Weed designation: USA: HI, OR, WA.

Geraniaceae

GERANIUM FAMILY

Herbs; leaves alternate or opposite; palmately
to pinnately lobed or compound; stipules
present; inflorescence a cyme or umbel;
flowers perfect, regular; five sepals, free to
partly fused; five petals; five, 10, or 15 stamens,
fused at the base; one pistil; three or five styles;
ovary superior; fruit a schizocarp, breaking into
mericarps at maturity. Worldwide distribution:
11 genera/780 species; predominantly found in
north temperate zone and South Africa.

Distinguishing characteristics: herbs; leaves
dissected; fruit splitting from the bottom
upward.

Members of this family include wild
and garden geraniums as well as storksbill.
The wild geraniums are represented by the
genus *Geranium* and garden geraniums by
Pelargonium.

Bicknell's geranium (*Geranium bicknellii*)

Also known as: pin clover, alfileria, red stem filaree, heron's-bill, pinweed, pingrass, wild musk, pink needle, thunder flower, cranesbill

Scientific Synonym(s):
E. aethiopicum (Lam.) Brumh. & Thellung

QUICK ID: leaves opposite, compound; flowers pink; fruit resembling a stork's bill

Origin: native to the Mediterranean region of southern Europe

Life Cycle: annual, winter annual, or biennial reproducing by seed; each plant producing up to 250 seeds

Weed Designation: USA: CO; CAN: MB

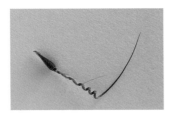

DESCRIPTION

Seed: Club-shaped, brown, about 5 mm long; tail 2–3 cm long, becoming coiled when dry; seed tail uncoils when moistened and coils when dry, working the seed into the soil.
Viability unknown; some reports indicate several years.
Germination occurs in the top 2.5 cm of soil; soil temperature range 4°–21°C.

Seedling: With three-lobed cotyledons, stalks hairy; first leaves densely haired, long-stalked, divided into toothed segments.

Leaves: Primarily basal, compound with irregular-shaped stalkless leaflets; leaflets 1–2.5 cm long, surface with short stiff hairs; stem leaves opposite when present.
Inflorescence an umbel; three- to 12-flowered.

Flower: Pink, 13–17 mm across; five sepals, hairy, bristle-tipped; five petals; 10 stamens (five fertile, five sterile); one pistil; style five-parted, 2–4 cm long; stalks 1–2 cm long; flowers appearing about 12 weeks after germination.

Plant: Low and spreading; stems 10–40 cm long, densely haired; taproot up to 8 cm long.

Fruit: Capsule, long-tailed 2–4 cm long, splitting from the bottom upward; five capsules per flower.

REASONS FOR CONCERN

Storksbill is often found in large populations and competes with crops for moisture and nutrients, adversely affecting the yield. Storksbill is reported to accumulate nitrates, which may cause livestock poisoning. It is host for the following viral diseases: aster yellows, beet curly top, peach yellow bud mosaic, strawberry green petal, tobacco yellow dwarf, and vaccinium false bottom.

OTHER SPECIES OF CONCERN IN THE GERANIUM FAMILY

Geranium bicknellii

GERANIUM BICKNELLII BRITT. — BICKNELL'S GERANIUM

Native to North America; annual or biennial; stems spreading, 10–60 cm long, branched, sticky-haired; leaves opposite, palmately three- to five-lobed, 2–7 cm across; lobes wedge-shaped, stalks sticky-haired; inflorescence a cyme, one- to two-flowered; flowers pale purple to pink; five sepals; five petals, slightly notched; 10 stamens, in two unequal sets; one pistil; stalks 1–3 cm long; fruit a schizocarp breaking into five mericarps at maturity; mericarps 2–3 mm long with a tail 15–20 mm long. Weed designation: none.

GERANIUM ROBERTIANUM L.—
ROBERT GERANIUM

Native to North America; annual or biennial; stems
spreading 10–60 cm long, branched, hairy, often
reddish-green; leaves opposite, 2–7.5 cm across,
divided Into three segments, each pinnately
lobed and cleft, dark green with a reddish tinge;
inflorescence a cyme, one- to two-flowered; flowers
pink, 10–15 mm wide; five sepals; five petals, 7–10
mm long; 10 stamens; one pistil; fruit a schizocarp
breaking into five mericarps at maturity; mericarps
about 2 mm long with tails 1–2 cm long. Weed
designation: USA: WA.

Geranium robertianum

Haloragaceae
WATER MILFOIL FAMILY

Aquatic herbs (occasionally shrubs or trees); leaves alternate, opposite, or whorled; simple to pinnately dissected; stipules absent; inflorescence a spike, raceme, panicle, or solitary flower; flowers perfect or imperfect, regular; three or four sepals, or absent (falling as the flower opens); petals, four or absent (falling as the flower opens); two to eight stamens; one pistil; two to four styles; ovary inferior; fruit a nut, drupe, or schizocarp. Worldwide distribution: eight or nine genera/100 species; predominantly found in the southern hemisphere.

Distinguishing characteristics: plants aquatic; leaves alternate, opposite, or whorled, dissected; four petals.

Species found in the northern hemisphere are aquatic plants.

northern water milfoil (*Myriophyllum sibiricum*)

Myriophyllum aquaticum (Vell.) Verdc.
PARROT FEATHER

Also known as: water milfoil; Brazilian water milfoil, water feather

Scientific Synonym(s): *Enydria aquatica* Vell., *M. brasiliense* Camb., *M. proserpinacoides* Gillies ex Hook. & Arn.

QUICK ID: aquatic plant; leaves in whorls of four to six; emergent leaves with six to 18 divisions

Origin: native to the Amazon River basin of South America; probably introduced into United States in 1890s by the aquaria trade, first observed in the wild in California in 1912; often grown as an ornamental aquatic

Life Cycle: perennial reproducing by stem and rhizome fragments, and winter buds (turions); stem fragments as small as 5 mm long may produce new plants

Weed Designation: USA: AL, CT, ME, MA, VT, WA

DESCRIPTION

Seed: Not produced by North American plants.
Viability unknown.
Germination unknown.

Seedling: Not produced by North American plants; reproduction by stem and rhizome fragments.

Leaves: In whorls of four to six; above water leaves bright green, oblong, 2–5 cm long, 7–8 mm wide, finely dissected with 12–36 divisions, each 4–8 mm long; submerged leaves reddish-green, oblong in outline, 1–3.5 cm long, 8–12 mm wide, featherlike with 20–30 divisions, each about 7 mm long; crowded near the tip of the stem. Inflorescence a solitary axillary flower; two bracts, about 1.5 mm long; stalks 0.2–0.4 mm long.

Flower: White to translucent, 1–3 mm across, female only; four sepals, triangular, 0.5–1.5 mm long, margins toothed; petals absent; stamens absent (four or eight in native range); one pistil; stigmas prominent with numerous fine white hairs.

Plant: Female in North America (male plants absent); stems brittle, 1–2 m long, yellowish-green, rising up to 20 cm above the water surface; lower nodes often rooted; roots threadlike; rhizomes creeping and tough.

Fruit: Nutlet; not produced by North American plants.

REASONS FOR CONCERN

Large populations of parrot feather can lead to reduction in light, dissolved oxygen, and changes in water chemistry, thereby affecting the biodiversity of the body of water. Infestations are known to impede water flow in drainage ditches and irrigation canals.

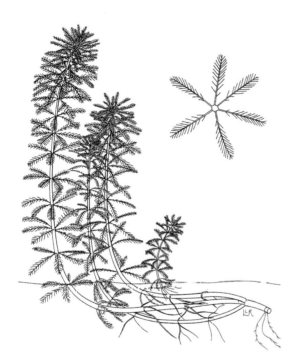

Myriophyllum spicatum L.
EURASIAN WATER MILFOIL

Also known as: spiked water milfoil

Scientific Synonym(s): none

QUICK ID: plant a submersed aquatic; leaves in whorls of four, leaflets with 28–48 threadlike segments; stems purplish-red with adventitious roots

Origin: native to Europe; introduced in the late 1800s at Chesapeake Bay in shipping ballasts; first Canadian report from Rondeau Provincial Park, Ontario, in 1961, and Okanagan Lake, British Columbia, in 1970

Life Cycle: perennial reproducing by seed, rhizomes, root and stem fragments, winter buds, and axillary buds; female spikes producing 12–40 seeds per stem; fragments 10–20 cm long forming new plants.

Weed Designation: USA: AL, CO, CT, FL, ID, ME, MA, MT, NV, NM, NC, OR, SC, SD, TX, VT, WA; CAN: AB, SK

DESCRIPTION

Seed: Round to slightly four-angled, 2.5–3 mm in diameter; surface rough with sharp projections. Viable for at least 7 years in dry storage. Germination often occurs at the time the seeds are shed; may lay dormant up to 7 years.

Seedling: Rarely seen in nature.

Leaves: In whorls of three to four, 1–3.5 cm long, featherlike with 28–48 threadlike segments; segments 6–12 mm long.

Inflorescence a terminal spike, 5–20 cm long, stalk usually pink and raised above the water, becoming thick and succulent to ensure spike is kept afloat; at maturity, the spike lays on the water surface, with shed seeds floating for up to 24 hours.

Flower: Borne in whorls of four; male or female; upper flowers male, four sepals, four wine-red petals, eight stamens; lower flowers female (often three to 10 whorls per spike), four sepals, four reddish petals, one pistil; petals of both flowers falling off soon after opening.

Plant: Aquatic, freshwater but tolerant of weak saline water (about one-third concentration of sea water); stems up to 12 m long, profusely branched; rhizome long and creeping; winter buds tight clusters of leaves, developing when water temperatures drop and day length shortens, breaking off and falling to the bottom, where they overwinter, rising to surface in spring to form new plants; fragmentation of stem and roots produces new plants, fragments less than 15 cm long produce over 80 new stems within 2 months; axillary buds produce in leaf axils, buds developing roots, then shed to form new plants by taking root in the soil; no fall dieback if water temperatures remain above 10°C.

Fruit: Globe-shaped and nutlike, breaking into four single-seeded sections; bracts surrounding the fruit.

REASONS FOR CONCERN

Eurasian water milfoil is known to clog waterways, irrigation ditches, and canals. It is a common weed of rice paddies. Large populations crowd out native vegetation, resulting in changes to fauna and flora of the body of water.

Similar species: the native species, northern water milfoil (*M. sibiricum* Komarov), is often confused with Eurasian water milfoil. It is distinguished by leaves with 12–22 threadlike divisions.

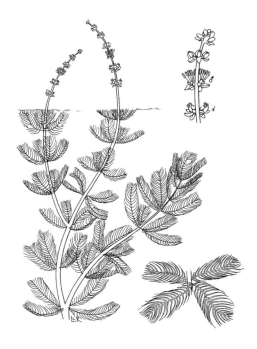

OTHER SPECIES OF CONCERN IN THE WATER MILFOIL FAMILY

Myriophyllum heterophyllum

MYRIOPHYLLUM HETEROPHYLLUM MICHX.— TWO-LEAF WATER MILFOIL

Native to eastern North America; aquatic perennial with rhizomes; winter buds produced at base of stem or rhizomes; stems reddish, 30–100 cm long, 2–8 mm diameter, branches numerous; leaves in whorls of four to six; submersed leaves 1.5–6 cm long, divided into six to 14 pairs of narrow segments (featherlike); emergent leaves green, bractlike, 4–12 mm long, margins with four to eight teeth; inflorescence a reddish-green spike, 5–30 cm long, rising above the water surface; flowers reddish-green, perfect or imperfect; four sepals, very small; four petals, 1–3 mm long; four stamens; one pistil; fruit a schizocarp breaking into four mericarps at maturity; mericarps 1–2 mm long, olive green. Weed designation: USA: CT, MA, ME, VT.

Hydrocharitaceae

WATERWEED FAMILY

Aquatic herbs, marine or freshwater; submersed leaves usually ribbonlike; leaves alternate, opposite, whorled, or basal; inflorescence an umbel or solitary; bracts spathelike; flowers imperfect or perfect; three sepals, green; three petals; one to many stamens; one pistil; two to 15 styles; ovary inferior; fruit a berry. Worldwide distribution: 15 genera/100 species; found in tropical to temperate regions.

Distinguishing characteristics: plants aquatic, submersed; leaves alternate, opposite, or whorled; flowers showy; three petals.

Several species are noxious weeds that have spread throughout the world via the aquaria industry.

Canada waterweed (*Elodea canadensis*)

male flower

female flower

flowering stem

Also known as: anacharis,
Brazilian or Argentine
elodea, common
waterweed, South
American waterweed,
dense-leaved elodea,
egeria, ditch moss, water
thyme

Scientific Synonym(s):
Philotra densa (Planch.)
Small & St. John., *Elodea
densa* (Planch.) Caspary,
Anacharis densa (Planch.)
Victorin

QUICK ID: plants aquatic,
submerged; leaves in
whorls of four to eight;
flowers white

Origin: perennial native to
Brazil and Argentina;
introduced by aquaria
trade; first observed in
the wild in New York state
in 1893

Life Cycle: perennial
reproducing by root
fragments

Weed Designation: USA: AL,
CT, ME, MA, OR, SC, VT, WA

DESCRIPTION

Seed: Spindle-shaped to elliptical, 5–8 mm long and
2 mm wide, beak 3–4 mm long, surface covered
in bumps; seeds not produced by plants in North
America.
Viability unknown.
Germination unknown, North American plants
not producing seed.

Seedling: Rarely produced in native habitat as male
plants outnumber female plants six to one;
seedlings not produced in North America

Leaves: In whorls of four to eight (lower leaves
occasionally opposite or in whorls of three),
linear to lance-shaped, 1–4 cm long, 2–5 mm
wide, tapering to a point, margins finely toothed;
upper leaves densely crowded.
Inflorescence a cyme, spathes two- to five-
flowered, 7.5–12 mm long; borne on the water
surface, originating in leaf axils on stalks 1–8 cm
long.

Flower: White and showy, 15–20 mm wide; male
flowers only in North America; three sepals,
green; three petals, 5–11 mm long, 6–9 mm wide;
nine stamens, yellow; female flowers absent in
North America, in the native range one flower per
spathe; three sepals, green; petals white, 5–8 mm
long; stamens sterile, yellow to reddish-orange;
one pistil.

Plant: Aquatic, preferring murky water and low
light; male or female, only male plants in North
America; stems up to 6 m long and 1–3 mm in
diameter, rooted in bottom substrate; roots
formed at double nodes, spaced every six to 12
nodes along the stem, distinguished by a very
short internode; stem fragments with a double
node producing new plants; spring growth starts
when water temperatures reach 10°C.

Fruit: Translucent capsule, 7–15 mm long, 3–6 mm wide; no fruit produced in North America.

REASONS FOR CONCERN

Brazilian waterweed forms large mats and crowds out native vegetation. It is also known to slow water movement and traps sediment. Plants do not tolerate freezing temperatures but can survive short periods under ice.

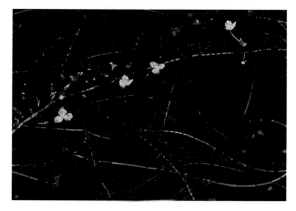

Elodea canadensis Michx.
CANADA WATERWEED

Also known as: water thyme, ditch moss, broad waterweed

Scientific Synonym(s): *Anacharis canadensis* (Michx.) Planch., *Philotra canadensis* (Michx.) Britt., *P. linearis* Rydb., *E. brandegeeae* St. John., *E. ioensis* Wylie, *E. linearis* (Rydb.) St. John, *E. planchonii* Caspary

QUICK ID: plants aquatic; leaves opposite or in whorls of three; plants male or female

Origin: native to western North America

Life Cycle: perennial reproducing by creeping roots, stem fragments, winter buds, and seed

Weed Designation: none

DESCRIPTION

Seed: Cylindrical, brown, about 4.5 mm long; surface smooth; rarely produced in nature. Viability unknown. Germination unknown.

Seedling: Rarely produced in nature.

Leaves: Opposite below and in whorls of three above, 6–20 mm long, 1–5 mm wide, dark green, translucent, stalkless, margins finely toothed. Inflorescence of solitary flowers borne in leaf axils; male spathes 8–14 mm long; female spathes 8–18 mm long.

Flower: White; male or female, appearing on separate plants; female flowers reaching the water surface on a threadlike stalk, 10–20 cm long, appearing before male flowers; three sepals, 3.5–5 mm long, 2–2.5 mm wide; three petals, about 5 mm long; one pistil; male flowers borne on short stalks, breaking from the plant and floating to the surface; three sepals, 3–5 mm long, 2–2.5 mm wide, dark-veined, falling off as the flower opens; three petals, about 5 mm long, less than 1 mm wide; nine stamens, outer six stalkless and spreading on opening of flower, inner three petal-like once pollen is shed; three staminodes, needle-like, less than 1 mm long; spathe 2–15 mm long.

Plant: Aquatic; male or female; stems 50–300 cm long; roots often appearing in leaf nodes; roots creeping, threadlike; winter buds (tight clusters of leaves) produced when water temperatures drop and day length shortens; buds breaking off and falling to the bottom, where they overwinter; in spring the buds anchor themselves to the bottom and produce new stems.

Fruit: Oval capsule, 6–9 mm long, 2–3 mm wide, stalkless, found in leaf axils; one- to two-seeded.

REASONS FOR CONCERN

Canada waterweed is a major weed problem in irrigation canals and drainage ditches. Large colonies often impede water flow.

Similar species: Nuttall's waterweed (*E. nuttallii* [Planchon] H. St. John), a native species, is distinguished by male spathes 2–4 mm long, leaves 1–1.7 mm wide, and male flowers with nine stamens.

Hydrilla verticillata (L. f.) Royle
HYDRILLA

Also known as: starvine, oxygen-plant, water thyme, Indian stargrass, Florida elodea

Scientific Synonym(s): *Serpicula verticillata* L. f., *Elodea verticillata* (L. f.) F. Muell, *Hydrilla lithuanica* (Andrz. ex Besser) Dandy

QUICK ID: plants aquatic; leaves in whorls of three to eight; leaf margins saw-toothed

Origin: native to Europe, Asia, and Africa; introduced into Florida in 1960 as an aquarium plant and naturalized into the environment by 1970

Life Cycle: perennial reproducing by seed, stem fragments, and turions; underground turions may produce up to 6,000 tubers per square meter.

Weed Designation: USA: federal, AL, AZ, CA, CO, CT, FL, ME, MA, MS, NV, NM, NC, OR, SC, TX, VT, WA

DESCRIPTION

Seed: Green to dark brown, elliptical, 2–2.5 mm long, surface shiny.
Viability of stem turions about 4 years; underground turions viable in sediment for 5 years, survive 3 days without water. Germination of stem turions occurs in the spring when water temperatures and light intensity increase.

Seedling: With two very small cotyledons; stem below the cotyledons up to 6 mm long; first leaves in a whorl of three; one to several branches occurring at the first node.

Leaves: In whorls of three to eight, submersed, linear; 6–20 mm long, 1–4 mm wide, stalkless, margins toothed, 11–39 teeth per cm; midrib often red; underside midvein with prickles. Inflorescence a spathe, one-flowered; male flowers released under water and floating to the surface, stalks about 0.5 mm long; female flowers 10–50 mm long, reaching water surface on long stalks.

Flower: Whitish or reddish; male flowers with three whitish-red or brown sepals, three whitish or reddish petals, and three stamens; female flowers have three whitish sepals, three translucent petals, and one pistil; wind-pollinated.

Plant: Of brackish (7% salinity of seawater) and freshwater; unisexual or bisexual; optimal growth temperatures 20°–27°C but survives at 10°C; rhizomes present; turions tuberlike, cream to brown, 5–12 mm long and wide; stems erect, up to 11 m long, branched near the water surface; stem turions dark olive green, 5–14 mm long, covered with short, stiff scales, consisting of

12–15 shortened internodes surrounded by fleshy leaves.

Fruit: Linear capsule, 5–15 mm long, 3–6 mm wide, borne inside the spathe; surface smooth or with a few short spines; zero- to six-seeded.

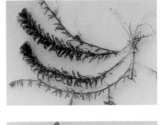

REASONS FOR CONCERN

Hydrilla can form thick mats and shade out native water plants. It is also known to slow water and clog irrigation and drainage canals. Hydrilla interferes with recreational and commercial boating, swimming, and fishing. As of December 2003, it was reported in 690 bodies of water in 190 drainage basins in 21 U.S. states.

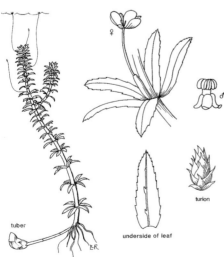

tuber

underside of leaf

turion

Hydrocharis morsus-ranae L.
EUROPEAN FROG-BIT

Also known as: floating frog-bit

Scientific Synonym(s): none

QUICK ID: plants floating; leaves heart-shaped; flowers white

Origin: native to Netherlands; introduced at Central Experimental Farm Arboretum, Ottawa, in 1932; spreading to Rideau Canal by 1939, and the Saint Lawrence River at Montreal by 1967

Life Cycle: annual reproducing by plantlets, turions, and seeds; single plants may produce up to 150 turions; seeds rarely produced

Weed Designation: USA: CA, ME, VT, WA

DESCRIPTION

Seed: Elliptical, 1–1.3 mm long; surface with fine network of veins; rarely produced in nature. Viability unknown.
Germination unknown.

Seedling: Information unavailable; rarely produced in nature.

Leaves: Basal, heart to kidney-shaped or round, leathery, emergent, or floating, 1.3–6.3 cm long, 1.2–6 cm wide, stalked, hairless, margins smooth; lower surface on floating leaves with spongy tissue confined to the midvein region. Inflorescence a cyme or solitary; male spathes with one- to five-flowered cyme, stalk up to 4 cm long; female flowers solitary, stalks up to 9 cm long.

Flower: White, about 1 cm across; male or female; male flowers with three sepals, three petals, and nine to 12 stamens; female flowers with three sepals, three petals, and one pistil; both flower types lasting a single day; after pollination, female flowers are pulled underwater, and reemerge as fruits 4–6 weeks later.

Plant: Free-floating; male or female; stems up to 20 cm tall; rhizomes absent; stolons present, cordlike; roots fibrous up to 30 cm long, becoming entangled in vegetation and helping to stabilize the colony; in summer numerous buds on the stolons give rise to new plants; in late fall, winter buds (turions) are produced on the stolons and released, then sink to the bottom; in spring the turions rise to the water surface to form small rosettes and later mature into recognizable plants.

Fruit: Spherical capsule, berry-like; seeds rarely produced.

REASONS FOR CONCERN

Dense, floating mats cause a reduction in available light, dissolved gases, and nutrients, which affects the growth of underwater flora and fauna. It also limits water flow in irrigation canals and movement of watercraft used for recreational activities.

Similar species: American spongeplant (*Limnobium spongia* [Bosc] Richard ex Steudel, p. 412), native to North America, resembles European frog-bit but has the lower surface of the leaf completely covered with spongy tissue.

OTHER SPECIES OF CONCERN IN THE WATERWEED FAMILY

LAGAROSIPHON MAJOR (RIDLEY) MOSS — OXYGEN WEED, AFRICAN ELODEA, CURLY WATERWEED

Native to South Africa; perennial submerged aquatic, male or female; rhizomes present; roots threadlike; stems up to 6.5 m long, brittle, 3–5 mm diameter; branches appearing every 10–12 nodes; leaves alternate, linear to lance-shaped, 6–20 mm long, 1–3 mm wide, three-veined, margins minutely toothed; flowers white to pinkish; female flowers reaching the water surface on threadlike stalks up to 25 cm long; male flowers produced in many-flowered spathes borne in leaf axils, breaking free and floating on the surface; fruit a capsule; seeds about 3 mm long. Weed designation: USA: AL, CA, FL, MA, NC, OR, SC, TX, VT, WA.

Limnobium spongia

LIMNOBIUM SPONGIA (BOSC) L. C. RICH. EX STEUD. — AMERICAN SPONGEPLANT

Native to North America; free-floating perennial plant; rhizomes absent; roots branched; stolons present; leaves basal, emergent, or floating, stalked, elliptical to round, 1–10 cm long, 1–7.8 cm wide; underside with spongy tissue from margin to margin; inflorescence a cyme; flowers greenish-white to yellow; male flowers with three sepals, three petals, and nine to 12 stamens; female flowers with three sepals, three petals, and one pistil; fruit an elliptical capsule, 4–12 mm diameter. Weed designation: USA: CA.

OTTELIA ALISMOIDES (L.) PERS. — DUCK LETTUCE

Native to Asia and Australia; plants annual or perennial; rhizomes and stolons absent; stems erect; leaves submerged, lance to oval, 7–22 cm long, 4.5–21 cm wide, stalks 8–50 cm long; margins smooth to crinkly; inflorescence a spathe, 2–6 cm long, five- to 10-winged; one-flowered; long-stalked; flowers white to pink or violet; three sepals; three petals; three or 15 stamens; one pistil; three to 15 styles;

fruit an oblong capsule, 15–40 mm long. Weed designation: USA: federal, AL, CA, MA, NC, OR, SC, VT.

STRATIOTES ALOIDES L.—WATER SOLDIERS, WATER PINEAPPLE

Native to northern Europe; aquatic perennial resembling the top of a pineapple or an *Aloe vera* plant, reproducing by vegetative offshoots; plants male or female; leaves alternate, straplike, 4–18 cm long and 4–8 mm wide, margins toothed; flowers white; three sepals; three petals; three stamens; one pistil; fruit a capsule; seeds rarely produced. Weed designation: AL, FL. Recently observed in the Trent-Severnway Waterway of Ontario.

Stratiotes aloides

VALLISNERIA AMERICANA MICHX.— AMERICAN EELGRASS, TAPE GRASS, WATER CELERY

Native to North America; aquatic perennial; rhizomes stout; roots up to 40 cm long, appearing at each rhizome node; stolons present; plants male or female; stems up to 9 m long; leaves basal, ribbon, or straplike, 5–550 cm long, 3–15 mm wide, thin, five to nine-veined with fine transverse connections; base sheathing; inflorescence a solitary axillary flower; flowers white, male or female; female flowers inconspicuous, borne on a very long spirally coiled stalk, reaching the water surface; three sepals; three petals; one pistil; one style; surrounded by a tubelike sheath 2–3 cm long; male flowers 1–1.5 mm across, enclosed by membranous sheath, breaking from the plant and floating to water surface; three sepals (two larger acting as floatation devices and the third a sail); two or three stamens; fruit a cylindrical capsule 5–10 cm long; 168–288 seeds, enclosed in a gelatinous material, shed when fruit wall decays. Weed designation: none; designated as the worst weed in the world.

Vallisneria americana

Hypericaceae
(Often Included in Clusiaceae)

SAINT-JOHN'S-WORT FAMILY

Herbs, trees, and shrubs; leaves opposite, simple, often dotted with translucent glands; inflorescence a cyme; flowers perfect, regular; four or five sepals; four or five petals; five to many stamens, borne in bundles opposite the petals; one pistil; two to five styles; ovary superior; fruit a capsule. Worldwide distribution: nine genera/560 species; predominantly temperate regions of the world.

Distinguishing characteristics: herbs; leaves opposite, often dotted with black glands; flowers yellow.

Several members of the Saint-John's-wort family are grown for ornamental and medicinal purposes. A few species are invasive weeds.

Saint-John's-wort (*Hypericum perforatum*)

Hypericum perforatum L.
SAINT-JOHN'S-WORT

Also known as: klamath weed, tipton weed, goatweed, eola weed, amber, rosin rose, herb john

Scientific Synonym(s): *H. officinale* Gater. ex Steud., *H. vulgare* Bauhin

QUICK ID: inflorescence flat-topped; flowers yellow; leaves and petals black-dotted

Origin: native to Europe, western Asia, and North Africa; first observed in Pennsylvania and Ontario in 1793

Life Cycle: perennial living up to 7 years, reproducing by seed and creeping rhizomes; plants produce up to 34,000 seeds

Weed Designation: USA: CA, CO, MT, NV, OR, SD, WA, WY; CAN: AB, MB, PQ

DESCRIPTION

Seed: Oblong, shiny black to light brown, 1–1.2 mm long, surface dull and roughened surface. Viable in soil for 10 years or more; 16 years in dry storage; 5 years submersed in fresh water. Germination occurs on the soil surface; seedlings emerging from 1 cm depth rarely survive; brief exposure to fire (100°–140°C) often increases germination.

Seedling: With oval to lance-shaped cotyledons, 1.5–3 mm long, 1–2 mm wide, upper surface with three prominent veins; stem below cotyledons purplish above the soil surface; first leaves opposite, oval to elliptical, black-dotted on the underside.

Leaves: Opposite, elliptical to oblong, 1–3 cm long, stalkless, surface with translucent dots, margins black-dotted; distinctly three- to five-veined. Inflorescence a flat-topped cyme, 25–100-flowered.

Flower: Showy, orange to yellow, 1.5–2.5 cm across; five sepals, 4–5 mm long, linear to lance-shaped; five petals, 8–12 mm long, black-dotted on edges; stamens yellow, numerous, borne in three groups; one pistil; three styles, 3–10 mm long.

Plant: Plants 30–120 cm tall; stems erect and branched, rust-colored, woody at base; short horizontal branches spreading from crown, sterile, 2–10 cm long, often reddish with black glands; taproot to 1.5 m deep; rhizomes extending to 0.5 m, forming new plants; root fragments resprout if located in top 5 cm of soil.

Fruit: Ovate capsule, 5–10 mm long, reddish-brown; three-celled; 61–502 seeds; surface sticky-haired; styles persistent.

REASONS FOR CONCERN

Saint-John's-wort contains hypericin, which causes photosensitization in fair-haired livestock. These animals should be given shelter from the sun's rays to prevent sun burn. Poisoned animals rarely die but suffer skin irritation and weight loss.

OTHER SPECIES OF CONCERN IN THE SAINT-JOHN'S-WORT FAMILY

HYPERICUM TETRAPTERUM FRIES. — SQUARE-STALKED SAINT-JOHN'S-WORT, SAINT-PETER'S-WORT

Native to Europe, western Asia, and North Africa; perennial up to 100 cm tall; rhizomes creeping; stems green to reddish-brown, four-sided, winged on the angles, branched; leaves opposite, oval, 1–3 cm long, light green, prominently veined, stalks clasping the stem, margins with black glands, surface with numerous transparent glands; inflorescence a cyme; flowers 7–12 mm across, yellow; five sepals; five petals, slightly longer than the sepals, scattered black dots on the margins; stamens numerous, borne in three groups; one pistil; fruit a capsule 5–10 cm long; seeds light brown, 1–1.5 mm long. Weed designation: none.

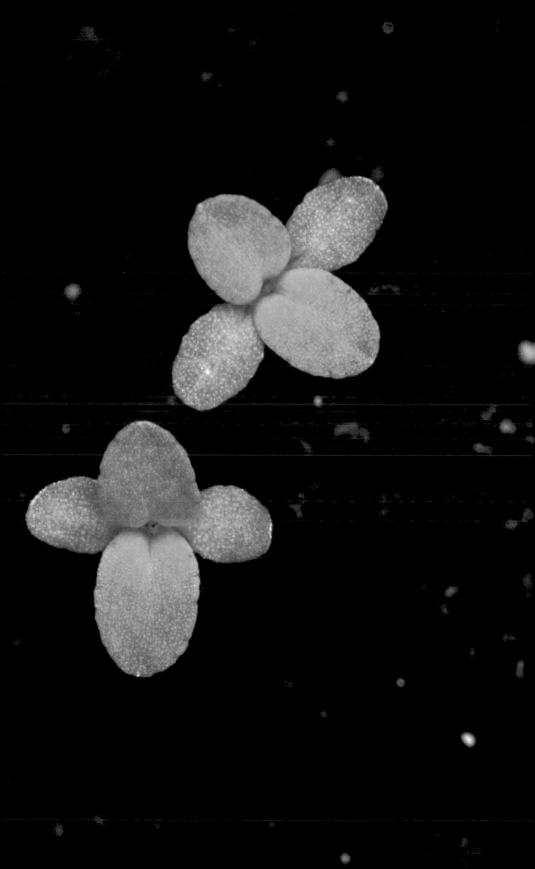

Iridaceae

IRIS FAMILY

Herbs with rhizomes, bulbs, or corms; leaves alternate or basal, sheathing and set edgewise to the stem; inflorescence a cyme, umbel, spike, or solitary; flowers perfect, regular or irregular; three sepals, petal-like; three petals; three stamens; one pistil; one style, three-branched; ovary inferior; fruit a capsule. Worldwide distribution: 70 genera/1,500 species; temperate and tropical regions of the world.

Distinguishing characteristics: herbs; leaves set edgewise to the stem; more than six petals.

Several species of iris, freesia, crocus, and gladiolus are grown for their ornamental value.

German iris (*Iris germanica*)

Iris pseudacorus L.
YELLOW FLAG

Also known as: yellow iris, pale yellow flag, water flag

Scientific Synonym(s):
I. acoriformis Boreau, *I. acoroides* Spach, *I. bastardii* Boreau, *I. curtopetala* F. Delaroche, *I. flava* Tornab., *I. lutea* Lam., *I. paludosa* Pers., *I. sativa* Mill.

QUICK ID: plants semiaquatic; leaves set edgewise to stem; flowers yellow

Origin: native to Europe, western Asia, and North Africa; introduced into North America about 1900 as an erosion control agent; first reported naturalized stands from Newfoundland in 1911, spreading to British Columbia by 1931

Life Cycle: perennial reproducing by seed, rhizomes, and fragments; average plants producing about 300 seeds

Weed Designation: USA: CT, MA, MT, NH, OR, WA;CAN: AB, BC

DESCRIPTION

Seed: D-shaped, pale to dark brown, flat, 6–7 mm across, corklike.
Viability unknown.
Germination does not require light; temperature range 15°–30°C (optimal 20°–30°C); submerged seeds not germinating; seeds may float for up to 7 months.

Seedling: Rare in nature; cotyledon grasslike; first leaf set edgewise to the stem; seedlings less than 2 months old are not drought-tolerant and cannot survive winter temperatures below −10°C.

Leaves: Primarily basal, about 10 per plant, set edgewise to the stem, green, 40–100 cm long, 1–3 cm wide, midribs prominent and raised; stem leaves equaling the flower cluster.
Inflorescence a cyme; four- to 12-flowered; bracts green with brown margins; outer bract keeled, 4.9–5.1 cm long, 7–10 mm wide; inner bract 6–9 cm long.

Flower: Showy, yellow, 8–10 cm across; three sepals, yellow or creamy-white streaked with purple or brown, 5–7.5 cm long, 3–4 cm wide; three petals, yellow, 2–3 cm long; three stamens; one pistil; three styles, 3–4 cm long, branches yellow.

Plant: Clump-forming, growing in moist areas; rhizomes pink, branched, 1–4 cm across, able to survive over 8 weeks submersed in water, also drought-tolerant; roots fleshy, up to 30 cm long; stems green, solid, 40–150 cm tall, fibrous remains of old leaves at base.

Fruit: Capsule, three-angled, 3.5–8 cm long, three-chambered; 32–47 seeds; average of 5.6 capsules per plant.

REASONS FOR CONCERN

Yellow flag is an aggressive ornamental plant that displaces native species by forming impenetrable stands. It is poisonous to livestock. Symptoms include moderate to severe bouts of abdominal pain, gastroenteritis, nausea, vomiting, diarrhea, spasms, staggering, and paralysis.

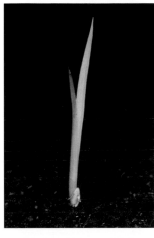

OTHER SPECIES OF CONCERN IN THE IRIS FAMILY

Iris missouriensis

IRIS MISSOURIENSIS NUTT. — ROCKY MOUNTAIN IRIS

Native to North America; perennial, rhizomes branched, 2–3 cm diameter; roots fleshy; stems 25–60 cm tall; leaves pale green, whitish at base, 45–60 cm long, 3–12 mm wide; two or three stem leaves; inflorescence a cyme, one- to three-flowered; bracts subopposite, oval to lance-shaped, keeled; outer bract 3.5–4 cm long, inner bract 5.5–7 cm long; flowers blue to white, the veins deep violet; three sepals, the veins purplish, claw yellowish-white and purple-veined and dotted, 3.7–7 cm long, 1.2–3.2 cm wide; petals oblong to lance or spatula-shaped, 3.6–7 cm long, 5–12 mm wide; three stamens; one pistil; style 2.4–4 cm long; fruit a capsule 4–5 cm long, and 1.5 cm wide; seeds light brown. Weed designation: USA: CA.

Juncaginaceae
ARROWGRASS FAMILY

Herbs of freshwater and saline marshes; leaves grasslike, sheathing at base; inflorescence a spike or raceme; flowers perfect or imperfect, regular; three sepals, green or red; three petals, similar to the sepals; six stamens, short-stalked; one pistil; three to six styles; ovary superior; fruit a cluster of follicles or achenes. Worldwide distribution: four genera/26 species; tropical to subarctic regions.

Distinguishing characteristics: plants grasslike, preferring moist habitats; flowers nonshowy, borne on racemes or spikes.

Several species of this family contain triglochinin, a toxin responsible for several human and livestock deaths.

Seaside arrowgrass (*Triglochin maritimum*)

flower

Triglochin maritimum L.
SEASIDE ARROWGRASS

Also known as: arrowgrass, spikegrass, saltmarsh arrowgrass, shore arrowgrass

Scientific Synonym(s): *T. concinnum* var. *debile* (M.E. Jones) J.T. Howell, *T. debile* (M.E. Jones) A. & D. Love, *T. elatum* Nutt.

QUICK ID: leaves grasslike, basal and somewhat fleshy; flowers greenish-white; plants of moist habitats

Origin: native to North America

Life Cycle: perennial reproducing by seed; each plant capable of producing up to 600 seeds

Weed Designation: none

DESCRIPTION

Seed: Elliptical, yellowish-brown, 1.5–3.5 mm long, 0.7–1 mm wide.
Viability unknown.
Germination occurs with soil temperatures between 20°C and 30°C.

Seedling: With one dark green cotyledon, 3–6 mm long, almost round in cross-section, and grasslike; tip of cotyledon often bent or hooked.

Leaves: Basal, grasslike, 2.2–15 cm long, 1–6 mm wide, somewhat fleshy; semicircular in cross-section; four to 10 emerging from sheaths of the previous year's growth; sheaths 7–25 mm long, 1–2 mm wide; ligule hoodlike.
Inflorescence a terminal raceme, 10–50 cm long, 1.5–7 mm wide.

Flower: Greenish-white, 1–4 mm wide; three sepals, green, 1.3–1.7 mm long; three petals, green, about the same size as the sepals; six stamens; one pistil; six stigmas, feathery, collecting pollen from the wind.

Plant: Of moist alkaline areas; stems 20–100 cm tall; rootstalk short; roots fibrous; leaf bases from previous season present at base.

Fruit: Oblong capsule, 5–7 mm long; single-seeded; three to six fruits per flower.

REASONS FOR CONCERN

Seaside arrowgrass contains triglochinin and taxiphillin, toxic substances that release hydrogen cyanide when ingested causing respiratory failure. Plants are most dangerous in the early spring when livestock are feeding on new growth. Livestock deaths have occurred when 0.5% of body weight in plant material had been ingested. The toxins must be ingested over a short period of time as the poisonous substances are not cumulative. Hay containing large amount of seaside arrowgrass is as poisonous as young emerging plants.

OTHER SPECIES OF CONCERN IN
THE ARROWGRASS FAMILY

Triglochin palustre

TRIGLOCHIN PALUSTRE L. —
MARSH ARROWGRASS

Native to North America; perennial; stems 9–43 cm tall; leaves grasslike, 6–25 cm long, 1–3 mm wide, sheath 3–5 cm long and 1.5–5 mm wide; ligule not hoodlike; inflorescence a raceme, 5–21 cm long and 2–5 mm wide; flower stalks 0.4–4.5 mm long; flowers greenish-purple; three sepals; three petals; six stamens; six pistils; fruit a schizocarp breaking into six mericarps at maturity. Poisonous. Weed designation: none.

Lamiaceae
(Formerly Labiatae)

MINT FAMILY

Herbs, occasionally shrubs and trees; stems often four-sided; leaves opposite or whorled, simple, often hairy with glands secreting volatile oils; stipules absent; inflorescence an axillary cyme or verticil; flowers perfect, irregular; five sepals, united; five petals, united and two-lipped (upper lip with two petals, the lower with three); two or four stamens, borne on the petals; one pistil; ovary superior; fruit usually four nutlets. Worldwide distribution: 180 genera/3,500 species; Mediterranean region the center of greatest diversity.

Distinguishing characteristics: herbs; stems square; leaves opposite; flowers irregular.

Several spices used in flavoring food are members of the mint family. They include lavender, mint, basil, sage, oregano, and thyme. Species such as bugleweed, coleus, and salvia are grown for their ornamental value.

henbit (*Lamium amplexicaule*)

Dracocephalum parviflorum Nutt.
DRAGONHEAD

Also known as: small-flowered dragonhead, false dragonhead

Scientific Synonym(s): *Moldavica parviflora* (Nutt.) Britt.

QUICK ID: stems square; leaves opposite; petals light blue or violet

Origin: native to North America

Life Cycle: annual or biennial reproducing by seed; plants capable of producing up to 500 seeds

Weed Designation: CAN: MB

DESCRIPTION

Seed: Oval, brown to black, 2–3 mm long. Viable in soil for up to 3 years. Germination enhanced by the presence of light; optimal soil temperature 25°C; germinate readily after a fire or disturbance.

Seedling: With oval cotyledons, 10–22 mm long, 5–9 mm wide, base with two backward-pointing lobes; stem below cotyledons with a few short hairs at or near the soil surface; first leaves opposite, oval, margins with rounded teeth.

Leaves: Opposite, oval to oblong, 2–6 cm long; stalked; margins coarsely toothed, underside hairy below Inflorescence a terminal spike, 2–5 cm long, 2–3 cm across, borne in leaf axils.

Flower: Blue to purple; five sepals, united, 10–14 mm long, 15-nerved; five petals; four stamens; one pistil; bracts leaflike and spine-tipped.

Plant: Plants 20–90 cm tall, becoming stiff and bristly by late summer; stems square, branched, smooth to hairy; taproot thick.

Fruit: Nutlet; four nutlets per flower.

REASONS FOR CONCERN

American dragonhead is a serious competitor in cultivated fields and row crops. It is a common weed of grains in western Canada.

Similar species: thyme-leaved dragonhead (*D. thymiflorum* L.), another introduced species, is distinguished from American dragonhead by spineless bracts and by flowers in interrupted whorled clusters along the upper stem.

Galeopsis tetrahit L.
HEMP NETTLE

Also known as: dog nettle, bee nettle, wild hemp, flowering nettle, ironweed, brittle-stem hemp nettle, ironwort, simon's weed

Scientific Synonym(s): none

QUICK ID: stem square and bristly; leaves opposite; flowers pink

Origin: native to Europe and Asia; introduced into Canada prior to 1884

Life Cycle: annual reproducing by seed; plants capable of producing up to 2,800 seeds

Weed Designation: USA: AK; CAN: AB, MB, PQ; MEX: federal

DESCRIPTION

Seed: Egg-shaped, mottled grayish-brown, 3–4 mm long.
Viable in soil for up to 10 years.
Germination occurs at 1–4 cm depth.

Seedling: With oval cotyledons, 5–10 mm long, 2–5 mm wide, base with two pointed lobes, tip slightly notched; stem below cotyledons often purplish at soil level; first leaves opposite, oval, coarsely toothed, prominently veined, surface rough due to fine wiry hairs.

Leaves: Opposite, oval to lance-shaped, 3–12 cm long, surface bristly haired, margins with 5–10 coarse teeth per side; stalks 1–3 cm long. Inflorescence a terminal raceme.

Flower: White, pink, or variegated, with two yellow spots, about 1 cm across; five sepals, united, 7–11 mm long, prominently 10-ribbed and spine-tipped; five petals, united into a tube, 15–22 mm long; four stamens, hairy; one pistil.

Plant: Plants 30–100 cm tall; stems square, bristly haired, branched; nodes swollen, numerous stiff downward-pointing hairs present; taproot slender.

Fruit: Nutlet; four nutlets per flower.

REASONS FOR CONCERN

Hemp nettle is a serious competitor with crops for moisture and soil nutrients. Yield losses of 24% in wheat and 25% in canola have been reported. Hemp nettle often forms dense stands in pastures, roadsides, and waste ground. Seeds are usually shed before crops are harvested.

Glechoma hederacea L.
GROUND IVY

Also known as: gill-over-the-ground, creeping charlie, catsfoot, field balm, run-away-robin, alehoof, gillale

Scientific Synonym(s): *Nepeta glechoma* Benth., *N. hederacea* (L.) Trevison, *G. hirsuta* W. & K.

QUICK ID: plants prostrate; stems rooting at nodes; flowers purple

Origin: native to Europe and Asia; first North America record from 1829

Life Cycle: perennial reproducing by creeping stems and seeds

Weed Designation: USA: CT

DESCRIPTION

Seed: Oval, dull brown, quarter circle in cross-section, 1.6–1.7 mm long, 1–1.1 mm wide; surface smooth with a slight sheen.
Viable in soil for less than 2 years; some seeds may persist for at least 5 years.
Germination requires light.

Seedling: With round to oval cotyledons, slightly hairy; first leaves opposite, heart-shaped, margins with rounded teeth.

Leaves: Opposite, heart- to kidney-shaped, 1–4 cm long, purplish-green, stalks 1–10 cm long, margins with rounded teeth; long hairs found at base of the leaf stalk; mintlike odor when crushed.
Inflorescence an axillary cluster of two to seven flowers.

Flower: Blue to purple flowers, tube-shaped, 10–23 mm across, less than 15 mm long; often male or female; large flowers often perfect, smaller flower female and lacking stamens; stalks less than 10 mm long; five sepals, 5–7 mm long, spine-tipped, prominently 15-nerved; five petals; four stamens; one pistil.

Plant: Weak and trailing, often forming large mats, somewhat shade-tolerant; stems 10–90 cm long, square, purplish-green, often rooting at nodes; roots fibrous; stem and leaves with a strong rancid mintlike odor when crushed.

Fruit: Nutlet; four nutlets per flower; surrounded by the bristly calyx.

REASONS FOR CONCERN

Some sources report that ground ivy may be poisonous to horses when eaten in large quantities. There is no evidence of poisoning to other livestock. Ground ivy is often a nuisance weed in turf.

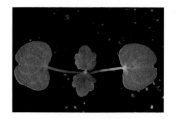

Lamium amplexicaule L.
HENBIT

Also known as: dead nettle, blind nettle, bee nettle, giraffe head

Scientific Synonym(s): *Galeobdolon amplexicaule* Moench, *L. stepposum* Kossko ex Klok., *Lamiopsis amplexicaulis* (L.) Opiz

QUICK ID: stems square; leaves opposite, stalkless; flowers pink to purple

Origin: native to Europe and western Asia

Life Cycle: annual or biennial reproducing by seeds; plants producing up to 200 seeds

Weed Designation: CAN: AB, MB

DESCRIPTION

Seed: Oval, speckled grayish-brown, 1.6–2 mm long, 0.9–1 mm wide, surface smooth, dull. Viable in soil for up to 25 years. Germination is shallow, usually in the top 2 cm of soil.

Seedling: With oval cotyledons, 3–12 mm long, 1–4 mm wide, base with two lobes, tip usually notched; stem below cotyledons usually purplish near the soil surface; first leaves opposite, oval, margins with rounded teeth, surface covered with soft hairs.

Leaves: Opposite, oval to round, 5–30 mm long, 10–30 mm across, prominently veined, surface with a few hairs; lower leaves long-stalked, margins with two to four rounded teeth per side; upper leaves stalkless, appearing to surround the stem. Inflorescence a whorl of six to 10 axillary flowers.

Flower: Dark purple, helmet-shaped, stalkless; five sepals, 5–8 mm long, five-nerved, hairy; five petals, 12–18 mm long, hairy on the outside; four stamens; one pistil.

Plant: Plant 10–40 cm tall; stems weak, square, branched from the base; roots shallow and fibrous.

Fruit: Nutlet; four nutlets per flower.

REASONS FOR CONCERN

Henbit is not a common weed of cultivated crops due to its shade intolerance. It is a common weed in row crops and gardens. Henbit is poisonous to livestock, especially sheep, causing the animal to stagger. Henbit is host for aster yellows, tobacco etch, and tobacco mosaic viruses.

Prunella vulgaris L.
SELF-HEAL

Also known as: healall, carpenter weed, thimbleflower, heart-of-the-earth

Scientific Synonym(s): *Brunella officinalis* Crantz, *B. vulgaris* var. *indivisa* Neilreich, *P. reptans* Dumort.

QUICK ID: stems square; flowers purple; bracts round to kidney-shaped, bristly

Origin: native to Europe; some native North American populations may exist

Life Cycle: perennial reproducing by seed

Weed Designation: CAN: federal

DESCRIPTION

Seed: Pear-shaped, brown to black, 1.8–1.9 mm long, surface smooth.
Viable in soil for at least 5 years.
Germination requires exposure to light; emergence from the top 1 cm of soil, although some seedlings emerge from 7.5 cm.

Seedling: With oval cotyledons, tip notched; first leaves opposite, egg-shaped, tips pointed, margins smooth to wavy-toothed.

Leaves: Opposite, oval, 2.5–10 cm long, 7–40 mm wide, often purplish-green, stalked, margins smooth to wavy or irregularly toothed; upper leaves short-stalked to stalkless.
Inflorescence a dense terminal spike 2–5 cm long; three flowers per bract; bracts round to kidney-shaped, bristly.

Flower: Blue to purple, 11–14 mm long, stalkless; five sepals, 7–10 mm long, purplish-green, united and two-lipped, irregularly 10-nerved, spine-tipped; five petals, united and two-lipped, the upper lip hoodlike; four stamens, upper two short, the lower longer than the petals; one pistil.

Plant: Plants 5–60 cm tall; stems prostrate, erect to ascending, square, rough-haired.

Fruit: Nutlet; four nutlets per flower.

REASONS FOR CONCERN

Self-heal is a common weed of turf, often growing below the mower's blade zone.

Stachys palustris L.
MARSH HEDGENETTLE

Also known as: woundwort, hedgenettle

Scientific Synonym(s):
S. maeotica Postrig., *S. paluster* L., *S. wolgensis* Vilens.

QUICK ID: stems square with opposite leaves; flowers mottled pink and purple; leaves hairy on both surfaces

Origin: native to North America

Life Cycle: perennial reproducing by seed and creeping rhizomes

Weed Designation: CAN: NS

DESCRIPTION

Seed: Oval, dark brown, 1.8–2.2 mm long, 1.2–1.8 mm wide.

Viability unknown.

Germination occurs in the top 2 cm of soil.

Seedling: With oval cotyledons, 3.5–10 mm long, 2–5 mm wide; stem below cotyledons with several minute hairs; first leaves opposite, oval, margins with rounded teeth; by third pair of leaves creeping rhizomes have started to develop.

Leaves: Opposite, oblong to lance-shaped, 3–15 cm long, 1–4 cm wide, stalkless, surface hairy, margins with rounded teeth.

Inflorescence a terminal raceme, 2.5–25 cm long; flowers in whorls of upper leaf axils.

Flower: Pinkish-purple; five sepals, united into a funnel 6–9 mm long, five-lobed, sharp-tipped, prominently five- to 10-nerved; five petals, 11–16 mm long, hairy; four stamens (upper two short, the lower two longer); one pistil.

Plant: Plants 30–100 cm tall; stems square, covered in short stiff hairs; rhizomes white, branched, succulent, tuber-bearing.

Fruit: Nutlet; four nutlets per flower.

REASONS FOR CONCERN

Marsh hedgenettle is a serious weed in moist low-lying areas of cultivated fields and row crops. Once established it is difficult to eradicate.

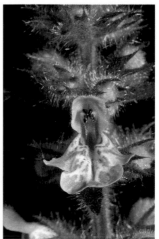

OTHER SPECIES OF CONCERN IN
THE MINT FAMILY

Leonurus cardiaca

Nepeta cataria

LEONURUS CARDIACA L. — MOTHERWORT
Native to Europe and Asia; aromatic perennial with short rhizomes; stems stiff, 40–150 cm tall, square, hollow; leaves opposite, 5–12 cm long, 3–15 cm wide, dull green, palmately three-lobed, stalks 3–14 cm long and four-angled, margins toothed or lobed, underside with bristly hairs; upper leaves less lobed, short-stalked; inflorescence a verticel with six to 15 flowers per node; flowers pink to pale purple, 10–15 mm long, two-lipped; five sepals, prickly; five petals; four stamens; one pistil; fruit a nutlet; four nutlets per flower, partially surrounded by the persistent prickly calyx. Weed designation: none.

NEPETA CATARIA L. — CATNIP, CATMINT
Native to Europe and Asia; aromatic perennial spreading by short rhizomes and seeds; stems pale green, erect 30–100 cm tall, square, branched; leaves opposite, 3–8 cm long, triangular to heart-shaped, margins toothed, surface with soft hairs; inflorescence a compact spike, 2–6 cm long; flowers borne in whorls, white with pink or purple dots on lower lip, irregular, 10–15 mm long; five sepals, fused into a tube, 15-nerved; five petals; four stamens; one pistil; fruit a reddish-brown nutlet; four nutlets per flower. Weed designation: none.

SALVIA AETHIOPIS L. — MEDITERRANEAN SAGE

Native to southern Europe; aromatic annual, biennial, or short-lived perennial; stems 30–100 cm tall, square, covered in woolly hairs; leaves opposite, grayish-green, oval to triangular, 5–30 cm long, margin irregularly lobed to toothed, lower surface covered in white woolly hairs; basal leaves oval, margins with rounded teeth, surface with woolly grayish-white hairs; inflorescence a verticel with four to 10 flowers per node; 20–90 cm long; bracts about 1 cm long, spine-tipped; flowers pale yellow to white, 1.5–2 cm long, two-lipped; five sepals, two-lipped, spine-tipped; five petals; two stamens; one pistil; fruit a smooth brown nutlet, about 4 mm long; four nutlets per flower. Weed designation: USA: CA, CO, NV, OR, WA.

SALVIA SCLAREA L. — EUROPEAN SAGE

Native to southern and eastern Europe; aromatic biennial or perennial; stems 20–100 cm tall, square, bristly haired near the top; leaves opposite, oval to oblong, 10–20 cm long, margins coarsely and irregularly toothed, long-stalked; inflorescence a verticel with 4–8 flowers per node; lower bracts up to 4 cm long; flowers blue to white with yellow markings, two-lipped, 2–2.5 cm long; five sepals, two-lipped, spine-tipped; five petals; two stamens; one pistil; fruit a nutlet; four nutlets per flower. Weed designation: USA: WA.

Salvia sclarea

Liliaceae

LILY FAMILY

Herbs, shrubs, and woody vines with bulbs, rhizomes, and corms; leaves alternate, whorled, or basal, simple, parallel-veined; inflorescence a raceme or solitary flower; flowers perfect (rarely imperfect), regular; three sepals, often petal-like; three petals; six stamens; one pistil; ovary superior, occasionally inferior; fruit a berry or capsule. Worldwide distribution: 40 genera/6,500 species; found throughout the world.

Distinguishing characteristic: herbs with bulbs, rhizomes, or corms; leaves parallel-veined; flowers with more than six petals.

The lily family is an important family in the horticultural industry. It includes lilies, tulips, hostas, daffodils, agave, yucca, and lily-of-the-valley. Other species—such as onion, garlic, and asparagus—are grown for their food value.

white camas (*Zigadenus elegans*)

Allium vineale L.
WILD GARLIC

Also known as: wild onion, crow garlic, field garlic; stag's garlic, scallions

Scientific Synonym(s): none

QUICK ID: plants with garlic odor; flowers pink; leaves hollow near base

Origin: native to Europe, western Asia and North Africa

Life Cycle: perennial reproducing by seed, bulbils, and bulb off-shoots; umbels may have up to 300 bulbils

Weed Designation: USA: AR, CA, HI

DESCRIPTION

Seed: Black, 3–4 mm; rarely produced.
Viability for bulbils is usually less than 2 months; some hard-shelled bulbs remain dormant up to 6 years.
Germination occurs in late summer to midwinter; new plants from bulbils and seed do not flower the first growing season but produce bulbs at base; new plants from bulbs may flower depending on soil moisture and may produce bulbs at base.

Seedling: Resembles young grass; one cotyledon; garlic odor present in crushed leaves.

Leaves: Alternate, two-ranked; linear or grasslike; blades hollow below middle, circular, 15–60 cm long, 2–10 mm wide; bases sheathing; all leaves with a garlic or onion scent; two- to four-leaved at flowering.
Inflorescence a headlike umbel, 2–5 cm across, zero- to 50-flowered; spathe dry and thin, beaked, two- to several-veined; flowers often replaced by bulbils.

Flower: Greenish-white, pink to purplish, bell-shaped, 3–4 mm long; three sepals; three petals; six stamens, anthers purple; one pistil; flower stalks 10–20 mm long; bulbils ovoid, 4–6 mm long, 2–3 mm wide, stalkless, often tipped by a slender green leaf.

Plant: With a garlic odor, tolerant to drought, cold, and waterlogged soils; flowering stems 30–120 cm tall, 1.5–4 mm wide; stem leafy to middle; five to 20 bulbs in a cluster, hard-shelled, 1–2 cm long and wide, outer coats brownish-yellow, inner coats white to pale brown; bulbs surrounded by round to egg-shaped bulblets, outer coats papery; roots fibrous.

Fruit: Capsule, three-celled.

REASONS FOR CONCERN

Wild garlic is a troublesome pest of grains, pastures, and hayfields. Grains may become tainted by the garlic odor or taste at the time of harvest. Cattle feeding on hay contaminated with wild garlic produce off-tasting milk and meat.

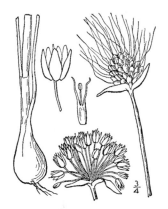

Asparagus densiflorus Kunth.
ASPARAGUS FERN

Also known as: Sprenger's asparagus fern, emerald fern

Scientific Synonym(s):
A. aethiopicus L.,
A. sprengeri Regel

QUICK ID: plants with sprawling spiny stems; flowers white; fruit a red berry

Origin: native to South Africa; widely planted as a ground cover or ornamental

Life Cycle: perennial reproducing by seed and rhizomes

Weed Designation: none

DESCRIPTION

Seed: Black, globe-shaped to angular. Viability unknown. Germination occurs when soil temperatures 22°–35°C.

Seedling: With a linear cotyledon, papery, similar to later leaves; young stem often purplish.

Leaves: Alternate, scalelike, and membranous, diamond-shaped, 1–2 mm long, attached in the middle and borne at the base of branchlets; three to 20 cladophylls per node, linear to needle-like, flattened, 8–22 mm long, 1.5–2 mm wide, one-veined.
Inflorescence an axillary raceme; five- to nine-flowered.

Flower: White or pinkish-white, fragrant, wheel- to bell-shaped; sepals and petals, three each, 3–4 mm long, 1.5–2 mm wide; six stamens; one pistil; style three-branched at tip.

Plant: Shrublike or herbaceous, 20–60 cm tall, hardy to –1°C; stems stiff or wiry, arched to sprawling, up to 200 cm long, often spiny; branches numerous; branchlets flat, needle-like, 1–2.5 cm long; roots fibrous, tuber-bearing.

Fruit: Bright red berry, 5–8 mm across; three-seeded.

REASONS FOR CONCERN

Asparagus fern has escaped cultivation and spread into the natural environment. In areas where it is a problem, it has displaced native vegetation and formed large impenetrable colonies.

Similar species: another species, garden or common asparagus (*A. officinalis* L.) has escaped cultivation and spread into the natural environment in temperate regions. Mature plants and berries are suspected to cause poisoning in cattle. The erect stem, 1–2.5 m tall, has scalelike leaves 3–4 mm long and cladophylls 1–3 cm long, borne in clusters of four to 15 per node. The bell-shaped, yellow to yellowish-green flowers 3–8 mm long and 1–2 mm wide, produce hard red berries 6–10 mm long with two to four seeds.

tuberous root

Zigadenus venenosus S. Wats.
DEATH CAMAS

Also known as: soap plant, alkaligrass, meadow death camas, hog potatoes

Scientific Synonym(s): *Zygadenus gramineus* Rydb., *Toxicoscordion gramineum* (Rydb.) Rydb., *T. venenosum* (S. Wats.) Rydb.

QUICK ID: leaves grasslike, V-shaped, pale green; flowers white

Origin: native to western North America

Life Cycle: perennial reproducing by seed and bulb offsets

Weed Designation: CAN: MB

DESCRIPTION

Seed: Oval, light brown, 5–6 mm long and 1.5 mm wide; surface wrinkled.
Viability unknown.
Germination information unavailable.

Seedling: With a single grasslike cotyledon; lower part of the leaf becoming bulbous with age.

Leaves: Alternate, but appearing basal, grasslike, 10–50 cm long, 2–10 mm wide, whitish-green, channeled, veins parallel.
Inflorescence a terminal raceme, 2–20 cm long, 2–5 cm wide, often with one or two branches near the base; 10–50-flowered; bracts green to whitish, papery, 5–25 mm long.

Flower: Creamy white, 9–12 mm across; three sepals, 4–5 mm long; three petals, 4–5 mm long; six stamens; one pistil; flower stalks 3–25 mm long.

Plant: Grasslike, preferring dry prairies and hillsides, shade-intolerant; stems 20–70 cm tall, hairless; roots fibrous; bulb onionlike, 6–20 cm below the ground surface; bulb 10–30 mm long, 8–20 mm wide.

Fruit: Papery capsule, 8–20 mm long, 4–7 mm wide; three-celled; several-seeded.

REASONS FOR CONCERN

Death camas, as the name implies, is very poisonous to livestock. The bulb is the most poisonous part of the plant. It is easily pulled from the soft ground in spring when most poisonings occur. Due to its early spring growth, livestock are at risk until better grazing conditions develop. The lethal dose in sheep is estimated to be 0.6–2% of body weight in plant material. Death usually occurs within 30 hours. Research has shown that 20-year-old dried plant material contains significant amounts of the toxin zygacine. There have been several reports of honeybee poisonings. It has been determined that the nectar and pollen contain the toxic constituents.

Similar species: white camas (*Z. elegans* Pursh) is closely related to death camas. It is distinguished by greenish to creamy white flowers 15–20 mm across and bracts tinged with pink or purple. A native plant of western North America, white camas is not as poisonous as death camas, although poisonings have occurred. A lethal dose in sheep is estimated to be 2–6% of body weight.

white camas (*Zigadenus elegans*)

OTHER SPECIES OF CONCERN IN THE LILY FAMILY

Allium canadense

***ALLIUM CANADENSE* L. — WILD ONION**
Native to North America; perennial; one to four bulbs, ovoid, 1–2.5 cm long, 6–30 mm wide; outer coats brown to gray, fibrous; inner coats whitish; two to six leaves, solid, flat, 20–50 cm long, 1–7 mm wide; sheaths less than 15 cm long; flowering stem 10–60 cm long, 1–5 mm wide; inflorescence an umbel, zero- to 60-flowered, some flowers often replaced by bulbils; three to four bracts, oval to lance-shaped, prominently three- to seven-veined; flowers white to pink or lavender, urn- to bell-shaped, 4–8 mm long, stalks 8–70 mm long; three sepals; three petals; six stamens, anthers yellow; one pistil; fruit a capsule. Weed designation: none.

***ASPHODELUS FISTULOSUS* L. — ONIONWEED**
Native to Europe and Asia; perennial with swollen rhizomes; roots fibrous; flowering stems 20–70 cm tall; leaves basal, grasslike, 5–35 cm long, 2–4 mm wide, hollow; sheaths membranous; margins smooth; inflorescence a raceme, 15–70 cm long, many-flowered; bracts numerous and papery, 4–7 mm long; flowers white to pale pink, closing in evening and during cloudy days; sepals and petals, three each, 5–12 mm long, vein dark pink or brown; six stamens; one pistil; fruit a globe-shaped capsule, 5–7 mm long; three or six seeds, black, 3–4 mm long. Weed designation: USA: federal, AL, CA, FL, MA, MN, NM, NC, OR, SC, VT.

HEMEROCALLIS FULVA (L.) L.—
ORANGE DAYLILY, TAWNY DAYLILY

Native to Asia; perennial, clump-forming, spreading by fleshy root and rhizome fragments; flowering stems 70–150 cm tall, often branched; leaves numerous, primarily basal, yellowish-green and straplike, 70–100 cm long and 1–3 cm wide; inflorescence a cyme, 10–20-flowered; flowers orange, often yellow at base, funnel-shaped, not fragrant and lasting a single day; six tepals, 7–8.5 cm long and 1.8–3.5 cm wide; stamens 4.5–6.5 cm long; pistil 9–11 cm long; fruit a leathery capsule, rarely produced; seeds black, rarely produced. The common cultivated daylily has been distinguished as the cultivar 'Europa', which was introduced into North America in the 1600s. Weed designation: none.

Hemerocallis fulva

ORNITHOGALUM UMBELLATUM L.—
SLEEPY DICK, NAP-AT-NOON

Native to Europe and Asia; perennial; bulbs ovoid 1–2 cm long, 1–2.5 cm wide, outer coats white to pale brown, papery; bulblets numerous; flowering stem 20–30 cm tall; four to nine leaves, basal, linear to lance-shaped, 20–30 cm long, 3–5 mm wide; inflorescence a flat-topped corymb, eight- to 20-flowered; bracts white and papery, 1–4 cm long; flowers white, opening at noon and closing at sunset (not opening on cloudy days); sepals and petals, three each, green-striped, 15–22 mm long, 7–8 mm wide; six stamens, the outer often shorter than the inner; one pistil; one style; fruit an oblong to ovoid capsule, six-angled; seeds numerous. Plants contain digitalis-like toxins that are poisonous to humans and livestock. Weed designation: USA: AL, CT.

Ornithogalum umbellatum

Lythraceae
LOOSESTRIFE FAMILY

Herbs, shrubs, or trees; leaves opposite or whorled, simple; stipules small or absent; inflorescence a cyme, panicle, raceme, or solitary flower; flowers perfect, regular; four to eight sepals (occasionally three to 16); four to eight petals (occasionally zero to 16); eight to 16 stamens, in two rows, the inner shorter; one pistil; ovary superior; fruit a capsule. Worldwide distribution: 25 genera/550 species; found throughout the world.

Distinguishing characteristics: plants semiaquatic; flowers showy; four to eight petals; eight to 16 stamens.

One species, purple loosestrife (*Lythrum salicaria*), is a serious weed of wetlands and marshes. Once planted widely for its ornamental value, many jurisdictions have designated it as a noxious weed.

purple loosestrife (*Lythrum salicaria*)

Lythrum salicaria L.
PURPLE LOOSESTRIFE

Also known as: purple
lythrum, spiked
loosestrife, rainbow weed

Scientific Synonym(s):
L. palustre Salis.,
L. spicatum S.F. Gray,
L. spiciforme Dulac

QUICK ID: leaves opposite;
flowers pink in dense
terminal spikes; plant of
moist habitats

Origin: native to Europe;
introduced as an
ornamental plant in the
late 1800s; common in
natural environment by
1814

Life Cycle: perennial
(living up to 22 years)
reproducing by seed,
stem, and root fragments;
once believed to be
sterile and not capable of
producing seeds, purple
loosestrife was widely
planted in gardens; large
plants producing up to
2.7 million seeds in 900–
1,000 capsules

Weed Designation: USA: AL,
AZ, AR, CA, CO, CT, FL, ID,
IA, IN, MA, MN, MO, MT,
ND, NE, NM, NV, OH, OR,
PA, SC, SD, TN, TX, UT, VT,
VA, WA, WY; CAN: federal,
AB, BC, MB, PEI, SK

DESCRIPTION

Seed: Brown to black, 0.2–0.4 mm long.
Viable in soil for up to 20 years.
Germination occurs at 20°C with no germination
below 14°C; emergence of seedlings from top
2 cm of soil; seeds very light, wind-dispersed
and may float great distances on water; research
shows that over 90% of seedlings with leaves
removed then resprout and continue to grow.

Seedling: With oval cotyledons, 3–6 mm long,
2–3 mm wide, hairless; first pair of leaves
opposite, similar to the cotyledons; research has
demonstrated germination on water surface with
seedlings floating to new locations.

Leaves: Opposite or in whorls of three, lance-
shaped, 3–10 cm long, stalkless, base wide and
heart-shaped, surface slightly hairy; leaves in
inflorescence may be alternate.
Inflorescence a raceme, 10–100 cm long; 5–8
cm of stem flowering at any given time; flowers
borne in whorls.

Flower: Showy, pinkish-purple, 15–20 mm across;
four to eight sepals, prominently eight-, 10-, or
12-nerved, green; four to eight petals, 7–10 mm
long; eight to 16 stamens; one pistil.

Plant: Of marshes; stems square, 50–240 cm tall,
branched, soft-haired; root system extensive;
spongy tissue forming on root tissue allows
oxygen flow to submersed roots.

Fruit: Capsule, 4–6 mm long and 1.5 mm wide;
83–130 seeds.

REASONS FOR CONCERN

Purple loosestrife spreads rapidly and replaces all native vegetation in wetland areas. It is a concern along canals and rivers where it slows the flow of water. Purple loosestrife is host for cucumber mosaic virus.

Similar species: swamp loosestrife (*Decodon verticillatus* [L.] Ell.) is often mistaken for purple loosestrife. This native species is distinguished by five to seven short sepals, five pink petals 10–15 mm long, and 10 stamens (five short and five long). The leaves, 5–15 cm long and 1–3 cm wide, are opposite or in whorls of three or four.

OTHER SPECIES OF CONCERN IN THE LOOSESTRIFE FAMILY

***LYTHRUM VIRGATUM* L. (INCLUDES GARDEN CULTIVARS 'MORDEN'S PINK' AND 'MORDEN'S GLEAM')—EUROPEAN WAND LOOSESTRIFE**
Native to Europe and Asia; perennial; stems 50–240 cm tall; leaves opposite or in whorls of three, lance-shaped, 3–10 cm long, base narrow, surfaces hairless; inflorescence a raceme, 10–100 cm long; flowers borne in whorls; flowers showy, pinkish-purple, 15–20 mm across; four to eight sepals, prominently eight-, 10-, or 12-nerved, green; four to eight petals, 7–10 mm long; eight to 16 stamens; one pistil; fruit a capsule. Distinguished from *Lythrum salicaria* by hairless leaves with a narrow base. Weed designation: USA: IA, MI, MN, MT, NE, NV, NC, ND, SD, TN, VA, WA, WI; CAN: MB.

ROTALA INDICA (WILLD.) KOEHNE — INDIAN TOOTHCUP

Native to Asia; aquatic, annual or perennial; stems 10–30 cm long, ascending, four-angled, hairless; leaves opposite, four-ranked, 4–20 mm long, 3–7 mm wide, oblong to oval, short-stalked; inflorescence a spike, bract leaflike; flowers pinkish or white; four sepals, triangular, 2–3 mm long; four petals; two to six stamens (commonly four); one pistil; fruit an oblong capsule, about 1.5 mm long; seeds numerous, pear-shaped. Weed designation: none.

Rotala indica

Malvaceae
MALLOW FAMILY

Herbs or shrubs; leaves alternate, simple and often palmately lobed; stipules present; inflorescence a cyme or solitary flower; flowers perfect, regular; five sepals, united; five petals; stamens numerous, fused and forming a tube around the style; one pistil; ovary superior; fruit a capsule or a schizocarp that breaks into several mericarps at maturity. Worldwide distribution: 85 genera/1,500 species; found throughout the world.

Distinguishing characteristics: herbs; flowers with stamens fused into a tube surrounding the style; fruit a schizocarp.

The mallow family is very important economically. Cotton and okra are important crop species, while hollyhock and hibiscus are grown for ornamental value. The family also contains weedy species, such as mallow and velvetleaf.

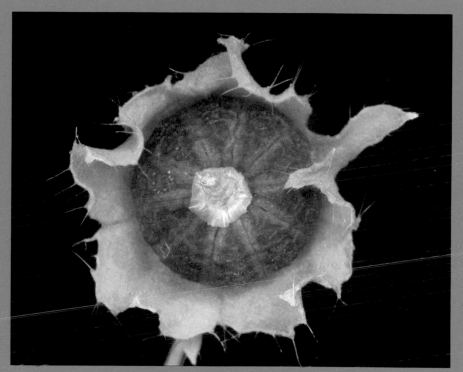

round-leaved mallow (*Malva rotundifolia*)

Abutilon theophrasti Medik.
VELVETLEAF

Also known as: Indian mallow, butter print, velvetweed, Indian hemp, cottonweed, buttonweed, piemaker, elephant ear

Scientific Synonym(s): *A. abutilon* (L.) Rusby, *A. avicennae* Gaertn., *Sida abutilon* L.

QUICK ID: flowers orangish-yellow; leaves alternate, heart-shaped, covered in soft hairs

Origin: native to India; introduced in the United States as a potential fiber crop about 1750

Life Cycle: annual reproducing by seed; large plants capable of producing over 17,000 seeds

Weed Designation: USA: CO, IA, OR, WA; CAN: federal, BC, NS, PQ

DESCRIPTION

Seed: Kidney-shaped, flattened, dull grayish-brown, 2.9–3.4 mm long, 2.6–2.9 mm wide, less than 1.6 mm thick.
Viable in soil for at least 50 years.
Germination occurs between 24°C and 30°C (no germination above 35°C); optimal depth 1.9 cm; emergence from top 7.6 cm of soil.

Seedling: With round cotyledons, 6–10 mm across, stalks up to 10 mm long, surface covered in short hairs; stem below cotyledons covered in short hairs, often purplish at soil level; first leaves prominently veined below, surface covered in short velvetlike hairs.

Leaves: Alternate, heart-shaped, 5–20 cm wide, surface covered in soft velvety hairs, margins irregularly toothed and hairy; leaves drooping during the nighttime.
Inflorescence a solitary axillary flower.

Flower: Orangish-yellow, 15–20 mm across; five sepals; five petals; stamens numerous, united to form a column around the style; one pistil.

Plant: Robust, 30–230 cm tall; stems hairy; taproot white.

Fruit: Green to black schizocarp, 2–2.5 cm across; breaking into 12–15 mericarps at maturity, each five- to 45-seeded.

REASONS FOR CONCERN

Velvetleaf is a serious weed in corn and soybeans. It is commonly found in cultivated fields, gardens, fencerows, and waste areas. Herbicide applications must be conducted during midday when the leaves are fully erect. Leaves fold at night, thereby reducing leaf surface and adsorptive area. It is also suspected to produce chemicals that affect water uptake and food production in crops, especially soybeans. It is host for tobacco streak and turnip mosaic viruses.

Hibiscus trionum L.
FLOWER-OF-AN-HOUR

Also known as: Venice
mallow, bladder ketmia,
shoofly, rose mallow
Scientific Synonym(s):
Trionum trionum (L.)
Woot. & Standl.

QUICK ID: calyx inflated
and surrounding the
capsule; leaves alternate,
three-lobed; flowers pale
yellow with a purple
center
Origin: native to southern
Europe; introduced
ornamental plant; first
Canadian report from
Prince Edward Island in
1952
Life Cycle: annual
reproducing by seed
Weed Designation: USA: CO,
WA

DESCRIPTION
Seed: Round to triangular, purplish-brown to black,
2–2.2 mm long and wide; surface rough and dull;
under high magnification surface covered in
small rounded bumps.
Viability unknown; some reports indicate several
years in soil.
Germination occurs with soil temperatures
between 18°C and 22°C.
Seedling: With heart- to kidney-shaped cotyledons,
4–17 mm long, 4–9 mm wide, yellowish-green
when they first emerge; stem below cotyledons
covered in gland-tipped hairs; first leaf round; by
third leaf stage leaves are three-lobed; stipules
also appear at this stage.
Leaves: Alternate, three-lobed, stalks 1–3 cm long,
margins coarsely toothed; upper leaves deeply
three-lobed; lobes up to 5 cm long and 2.5–3 cm
wide; stipules 3–7 mm long and 1 mm wide.
Inflorescence a solitary flower borne in leaf axils;
stalks about 2.5 cm long, densely haired.
Flower: Pale yellow with a purplish-brown eye,
about 4 cm wide; five sepals, papery, prominently
purple-veined, surface hairy; five petals; stamens
numerous, anthers orange; one pistil; 10 bracts,
narrow and papery, 7–10 mm long; flowers
opening only for a few hours.
Plant: Plants 30–50 cm tall; stems erect, branched at
the base, hairs bristly, star-shaped to forked.
Fruit: Globe-shaped capsule, about 14 mm long,
bristly haired, opening at the top; calyx inflated
and surrounding the capsule, papery, purple-
veined.

REASONS FOR CONCERN

Flower-of-an-hour is a common weed of cultivated fields, row crops, orchards, and rangeland. It competes with crops for available light, moisture, and soil nutrients.

Malva rotundifolia L.
ROUND-LEAVED MALLOW

Also known as: cheeses, low mallow, running mallow, blue mallow

Scientific Synonym(s): *M. pusilla* Sm.

QUICK ID: leaves round to kidney-shaped; flowers pale blue to white; plants with low spreading stems

Origin: native to Europe; common in New York state by 1870s; first Canadian report from Quebec in 1878

Life Cycle: annual, winter annual, or biennial reproducing by seed; plants capable of producing up to 5,200 seeds

Weed Designation: CAN: MB, SK

DESCRIPTION

Seed: Kidney-shaped, dark brown, 1.7–2 mm across, less than 1 mm thick, surface wrinkled.
Viable in soil for up to 100 years.
Germination occurs from 5°C to 30°C; optimal temperature 20°C.

Seedling: With heart-shaped cotyledons, 4.5–7 mm long, 3–3.5 mm wide, prominently three-veined, upper surface shiny; stems below cotyledons with a few short soft hairs; first leaves kidney-shaped, margins irregularly lobed with rounded teeth.

Leaves: Alternate, round to kidney-shaped, 2–5 cm across, shallowly five-lobed, stalks 10–15 cm long, margins with rounded teeth.
Inflorescence an axillary cyme; two- to five-flowered.

Flower: Pale blue to white, 4–6 mm across; five sepals, united; five petals, about 6 mm long; stamens many; one pistil; three bracts, small, borne below each flower.

Plant: Prostrate to ascending; stems 20–125 cm long, branched and forming large mats.

Fruit: Schizocarp; breaking into eight to 15 mericarps at maturity; each segment one-seeded; calyx becoming enlarged and veiny with maturity.

REASONS FOR CONCERN

Round-leaved mallow is a strong competitor with cultivated crops. It is also a weed of gardens, waste areas, and farmyards. Round-leaved mallow is host for the following viral diseases: aster yellows, beet curly top, tobacco streak, tomato spotted wilt, and hollyhock mosaic.

Similar species: high mallow (*M. sylvestris* L., p. 477), an introduced annual or biennial weed, has dark bluish-purple flowers and leaves with three to seven lobes. Another introduced species, whorled mallow (*M. verticillata* L., p. 477), has white to pale purple flowers and stems up to 2 m tall. The leaves are round to kidney-shaped with five to nine lobes.

Sida rhombifolia L.
ARROWLEAF SIDA

Also known as: rhombic-leaved sida, Cuban jute, big Jack, bloom weed, coffee bush, Paddy's lucerne, Queensland hemp, teaweed

Scientific Synonym(s):
S. retusa L.

QUICK ID: plants shrubby; leaves alternate, diamond-shaped; flowers yellow to orange

Origin: uncertain, but probably Europe; introduced into the United States as a potential fiber crop in late 1800s

Life Cycle: perennial (living up to 3 years) or annual (cool climates) reproducing by seed; average plants producing about 11,600 seeds per season

Weed Designation: none

DESCRIPTION

Seed: Dark brown to black, 2–3 mm long, surface smooth.
Viability unknown.
Germination occurs at 5–20 mm depth; no emergence from 5 cm depth; optimal temperature is 25°–35°C with no germination above 40°C; fresh seed is dormant.

Seedling: With heart-shaped to round or oval cotyledons, distinctly veined; first leaf round to oval or diamond-shaped, toothed.

Leaves: Alternate, diamond-shaped, 1.5–5 cm long, 7–22 mm wide; stalks 2–6 mm long; margins coarsely toothed; stipules linear, 6–9 mm long, somewhat spiny-tipped.
Inflorescence a solitary axillary flower.

Flower: Yellow to yellowish-orange, 6–10 mm wide; five sepals; five petals, 4–8 mm long; stamens numerous; one pistil; stalks 1–4 cm long.

Plant: Shrublike, little to no growth when temperatures below 20°C; stems 20–120 cm tall, woody and flexible, covered in star-shaped hairs; taproot with numerous fine roots.

Fruit: Schizocarp, about 1 cm across sharp-pointed, prominently beaked; eight to 12 mericarps, about 3–4.5 mm long, 10–14-seeded.

REASONS FOR CONCERN

Arrowleaf sida is a principal weed of pasture, beans, cassava, sugarcane, and maize in South America. It is also a weed of coffee in the Dominican Republic, as well as sorghum in Mexico.

Similar species: prickly or spiny sida (*S. spinosa* L., p. 479), native to tropical North America, may be mistaken for arrowleaf sida. It is distinguished by two or more pale yellow flowers per axil, stalks less than 12 mm long, and schizocarps with five one-seeded segments.

Urena lobata L.
CAESAR WEED

Also known as: African jute, aramina, hibiscus bur, pink Chinese bur, bur mallow, grand cousin

Scientific Synonym(s): *U. trilobata* Vell.

QUICK ID: plants shrubby; flowers pink; leaves shallowly five-lobed

Origin: uncertain, probably native to southern Asia

Life Cycle: perennial (living 2–3 years) or annual reproducing by seed

Weed Designation: USA: HI

DESCRIPTION

Seed: Brown, kidney-shaped, 2–3 mm across; surface with fine hairs.
Viability unknown.
Germination occurs when soil temperatures reach 35°C; dormant period required after being shed.

Seedling: Round to oval cotyledons, long-stalked, prominently veined, tip shallowly notched; first leaves round to oval.

Leaves: Alternate, grayish-green due to star-shaped hairs, oval to angular or heart-shaped, shallowly five-lobed, 1–12 cm long and wide, stalks 1–9 cm long, margins toothed; stipules linear, 2–3 mm long.
Inflorescence an axillary cyme, two- or three-flowered, or a solitary flower.

Flower: Pink, 1–2 cm across; five sepals; five petals; stamens numerous; one pistil; stalks 2–3 mm long (up to 5 mm long in fruit), covered in star-shaped hairs.

Plant: Shrubby, shade-intolerant; bark tough and fibrous; stems 60–300 cm tall, branched, covered in star-shaped hairs; taproot tan-colored.

Fruit: Schizocarp, 6–7 mm across, breaking into five mericarps at maturity, each 4–5 mm long, awn-tipped.

REASONS FOR CONCERN

Caesar weed tends to increase on overgrazed pastures. The burs become entangled in livestock fur, which aids in its spread.

OTHER SPECIES OF CONCERN IN THE MALLOW FAMILY

ANODA CRISTATA (L.) SCHLECHT. — CRESTED OR SPURRED ANODA

Native to tropical North America; annual; stems branched, 30–120 cm tall, covered in bristly hairs; leaves alternate, triangular, 2–10 cm long, distinctly three-lobed, margins toothed, stalked, both surfaces hairy; stipules present; inflorescence a solitary axillary flower; flowers pale blue to pink or lavender, 7–12 mm across; five sepals; five petals; stamens numerous; one pistil; fruit a schizocarp; 10–20 mericarps, 2.8–3.2 mm long, beaked, covered in bristly hairs. Weed designation: USA: CO.

MALACHRA ALCEIFOLIA JACQ. — YELLOW LEAFBRACT

Native to tropical North and South America; plants shrubby; stems 50–250 cm tall, branched, covered in star-shaped or stiff simple hairs; leaves alternate, oval to elliptical, 2–15 cm long, margin toothed to shallowly three- to five-lobed; stipules up to 2 cm long; inflorescence a head, bracts whitish with green veins, 1–2.5 cm long; flowers yellow; five sepals, white with red veins; petals 1–2 cm long; stamens numerous; one pistil; fruit a schizocarp; five mericarps, papery, 3–3.5 mm long, prominently red-veined; seeds black, about 2.5 mm long. Weed designation: USA: HI.

MALVA SYLVESTRIS L. — HIGH MALLOW

Native to Europe, Asia, and North Africa; biennial; stems up to 1.25 m tall; leaves alternate, 2–4 cm long and 2–5 cm wide, three-lobed; stalks 2–6 cm long, hairy; stipules papery, about 5 mm long; flowers borne in axillary clusters of one to four; five sepals, partly fused, 3–6 mm long; five petals, pinkish-purple, about 1 cm wide, notched at the tip; stamens numerous and forming a column around the pistil; one pistil; fruit a schizocarp with 10–12 veiny mericarps; seeds brown, about 2.5 mm long. Weed designation: none.

MALVA VERTICILLATA L. — WHORLED MALLOW

Native to China; annual or biennial; stems up to 2.5 m tall, green to purple with star-shaped hairs; leaves alternate, round, 2–23 cm long and broad, heart-shaped at base; stalks 4–24 cm long; margins toothed to five-lobed; stipules about 5 mm long; flowers borne in axillary clusters of two or more flowers; five sepals, partly fused; petals purplish, 7–9 mm long; stamens numerous, forming a column; one pistil; fruit a schizocarp enclosed by the persistent veiny calyx; schizocarp breaking into 10–12 mericarps; seeds dark brown to black, 1–1.5 mm long. Weed designation: none.

Malva sylvestris

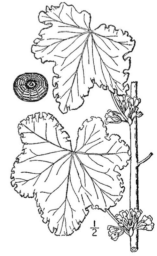

Malva verticillata

$\frac{3}{5}$

Malvella leprosa

MALVELLA LEPROSA (ORTEGA) KRAPOV. — ALKALI MALLOW

Native to western North America; perennial with creeping roots 15–20 cm below surface; stems 20–40 cm tall, covered with white bristly star-shaped hairs; leaves alternate, fan- to kidney-shaped or triangular, 1.5–4.5 cm across, stalks 10–30 mm long, surfaces covered with white star-shaped hairs, margins toothed to wavy; inflorescence a cyme, two- to five-flowered, or solitary, stalks 1–3 cm long, one to three bracts, about 3 mm long; flowers creamy white to yellow, 2.5–3.5 cm across; five sepals; five petals, 10–15 mm long; stamens numerous; one pistil; fruit a schizocarp, 5–8 mm across; six to 10 mericarps, about 3 mm long, one-seeded. Weed designation: USA: CA.

SIDA CORDIFOLIA L. — HEART-LEAVED SIDA

Shrublike herb, 20–120 cm tall, covered in simple and star-shaped hairs; leaves alternate, 1–4.5 cm long, 7–35 mm wide, round to oval or lance-shaped; stalks 5–30 mm long, woolly haired; margins wavy to toothed; stipules 2–6 mm long, threadlike; one to two flowers, borne in leaf axils; five sepals, 5–10 mm long, hairy; five petals, yellow; stamens numerous, forming a column around the pistil; one pistil; fruit a disc-shaped schizocarp, breaking into nine to 10 veiny mericarps; seeds brown to dark brown, flattened, about 2 mm long. Weed designation: none.

Sida cordifolia

SIDA SPINOSA L. — SPINY SIDA

Shrublike annual or perennial; stems 30–100 cm tall, covered in star-shaped hairs; leaves alternate, lance-shaped to oval, 5–40 mm long, 3–25 mm wide; margins toothed; one to three spiny structures at leaf base; stalks 2–20 mm long; stipules threadlike, 2–5 mm long; flowers borne in axillary clusters of one to five; five sepals, 4–5 mm long; five petals, white; stamens numerous, forming a column around the pistil; one pistil; fruit a globe-shaped schizocarp breaking into five mericarps; seeds brown to black, about 1.5 mm long. Weed designation: none.

Sida spinosa

Molluginaceae

(Often Included in Aizoaceae)

CARPETWEED FAMILY

Herbs or shrubs; leaves alternate, opposite, or whorled, simple; stipules absent; inflorescence a cyme or solitary flowers; flowers perfect, regular; four or five sepals; petals absent; five to 10 stamens; one pistil; one to five styles; ovary superior; fruit a capsule. Worldwide distribution: 14 genera/95 species; native to tropical and subtropical regions of the world.

Distinguishing characteristics: herbs; four or five sepals; five to 10 stamens; fruit a capsule.

carpetweed (*Mollugo verticillata*)

Mollugo verticillata L.
CARPETWEED

Also known as: Indian chickweed, whorled chickweed, devil's grip, green carpetweed

Scientific Synonym(s):
M. berteriana Ser.

QUICK ID: plants with prostrate stems; leaves in whorls of three to eight; flowers white

Origin: native to tropical North America and South America

Life Cycle: annual reproducing by seed

Weed Designation: none

DESCRIPTION

Seed: Kidney-shaped, dark orange to brown, 0.5–0.6 mm long and wide; under magnification surface shiny and ridged.
Viability unknown.
Germination in late spring/early summer but grows quickly.

Seedling: With oblong cotyledons, 1.5–3.5 mm long, less than 1.3 mm wide, thick, surface hairless; stem below cotyledon often brown; first leaf spatula-shaped, dull green above, pale below; cotyledons often present throughout the growing season.

Leaves: In whorls of three to eight, linear to elliptical or spatula-shaped, 5–40 mm long, 0.5–15 mm wide; stalks 1–4 mm long; dull green above and pale below, lower part of margin with a few hairs. Inflorescence an umbel; two- to six-flowered; borne in leaf axil.

Flower: White, stalks 3–20 mm long; five sepals, 1.5–2.5 mm long, green on outside, white inside; petals absent; three to four stamens; one pistil.

Plant: Prostrate; stems yellowish-green, 3–45 cm long, profusely branched, and forming mats, hairless.

Fruit: Oval capsule, 2.5–3.3 mm long, thin-walled, opening by three teeth; 15–35-seeded.

REASONS FOR CONCERN

Carpetweed is a serious weed in gardens and row crops. It is also found on roadsides, railways, and waste areas. It is reported to be host for cucumber mosaic, tobacco etch, and tobacco mosaic viruses.

Similar species: Indian or whorled chickweed, other common names for carpetweed, imply that it resembles common chickweed (*Stellaria media* [L.] Cyrill., p. 250). Common chickweed is distinguished from carpetweed by its oval-shaped opposite leaves and five deeply lobed petals.

OTHER SPECIES OF CONCERN IN THE CARPETWEED FAMILY

GLINUS LOTOIDES L. — DAMASCISA, LOTUS SWEETJUICE

Native to Europe, Asia, and Africa; annual; stems 5–35 cm long, branched at the base, covered in star-shaped hairs; leaves alternate or whorled, oblong to oval, round or spatula-shaped, 5–25 mm long, 0.4–17 mm wide; stalks 1–7 mm long; inflorescence a cyme, three- to 15-flowered; flowers whitish-green; five sepals, 3.5–4 mm long; five to 20 petals or absent; five stamens; one pistil; fruit a capsule, 3.6–4.5 mm long, 30–125-seeded; seeds orangish-brown. Weed designation: none.

Moraceae

MULBERRY FAMILY

Shrubs and trees with milky juice; leaves alternate; two stipules; inflorescence a spikelike head; flowers male or female; male flowers with four sepals, four stamens; female flowers with zero or four sepals, one pistil; fruit an achene, drupe, or multiple. Worldwide distribution 53 genera/1,400 species; found throughout subtropical regions of the world, with a few temperate species.

Distinguishing characteristics: tree; leaves alternate; two stipules; flowers male or female; fruit resembling a raspberry.

Several species in this family are grown food or for their ornamental value. Economically important species include Indian rubber tree, breadfruit, and figs.

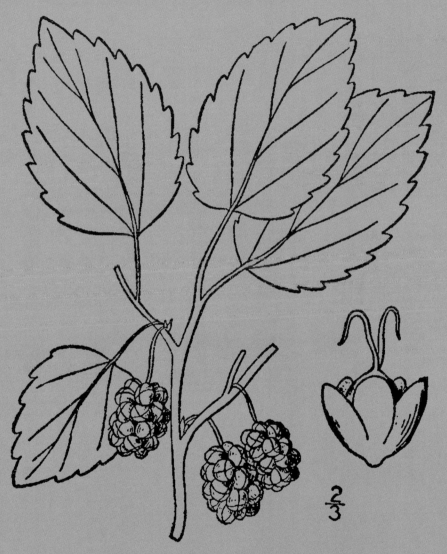

$\frac{2}{3}$

white mulberry (*Morus alba*)

Morus alba L.
WHITE MULBERRY

Also known as: Russian
  mulberry, silkworm
  mulberry, Chinese
  white mulberry
Scientific Synonym(s):
  M. indica L., *M. multicaulis*
  Perr.

QUICK ID: tree up to 15
  m tall; leaves irregularly
  two- to five-lobed; fruit
  red to purple or black,
  raspberry-like
Origin: native to China;
  introduced into North
  America about 1750 in
  an attempt to develop a
  silkworm industry
Life Cycle: perennial
  reproducing by seed
Weed Designation: none

DESCRIPTION
Seed: Light brown, oval, 2–3 mm long.
  Viability unknown.
  Germination requires 1–3-month cold period;
  seedlings not tolerant of direct sunlight for the
  first few weeks.
Seedling: With oval to oblong cotyledons; first leaf
  oval.
Leaves: Alternate, oval, deeply to irregularly two-
  to five-lobed, bright green, 6–10 cm long, 3–6
  cm wide; stalks 25–50 mm long, short-haired;
  margins toothed; stipules oval to lance-shaped,
  5–9 mm long; upper surface shiny, lower surface
  pale.
  Inflorescence a catkin; male catkins 25–40 mm
  long; female catkins 5–8 mm long.
Flower: Green; male or female; male flowers, four
  sepals, green with red tips; four stamens; female
  flowers, four sepals, green, of two sizes.
Plant: Shrub or tree with milky sap, 4–15 m tall;
  intolerant of shade, tolerant to drought, poor soil,
  salt, and pollution; bark gray, turning yellowish-
  brown, shallowly furrowed; branches orangish-
  brown to dark green; lenticels reddish-brown and
  prominent; sap milky.
Fruit: Achene, surrounded by enlarged fleshy calyx,
  red turning black to purple or pink to white at
  maturity; borne in clusters about 2.5 cm long.

REASONS FOR CONCERN

White mulberry may hybridize with the native red mulberry (*M. rubra* L.), raising concerns about genetic contamination of native stands. There have also been reports of appearance of white mulberry in low-till cropland. All parts of the plant, except the fruit, are poisonous to humans.

Similar species: red mulberry (*M. rubra* L.), native to eastern North America, is distinguished by larger, thicker leaves with a downy-haired lower surface, red to black fruit, and fruiting clusters 2–3 cm long.

red mulberry (*Morus rubra*)

OTHER SPECIES OF CONCERN IN
THE MULBERRY FAMILY

FICUS THONNINGII BLUME — LAUREL FIG

Native to Asia (India to Malaysia); evergreen tree up
to 30 m tall, aerial roots present, bark gray, milky
sap; branches brown; leaves elliptical to oblong or
oval, 3–13 cm long, 1.5–6 cm wide, leathery; stalks
5–10 mm long; stipules 7–9 mm long; margins
smooth; surfaces hairless; two or four basal veins,
10–18 lateral veins; male flowers with two to six
sepals and one to two stamens; female flowers with
one pistil; fruit a syconium, borne in pairs, purple or
black, oblong to oval or pear-shaped, 9–11 mm long,
5–6 mm wide; lower bracts oval to lance-shaped
1.5–3.5 mm long; apical bracts 1–3 mm wide. Weed
designation: none.

Onagraceae

EVENING PRIMROSE FAMILY

Herbs; leaves alternate or opposite, simple; stipules when present falling off early; inflorescence a spike, raceme, panicle, or solitary flower; flowers perfect, regular; four sepals (occasionally two, three, or five); four petals (occasionally two); two, four, or eight stamens; one pistil; ovary inferior; fruit a capsule, berry, or nutlet. Worldwide distribution: 20 genera/650 species; primarily temperate regions.

Distinguishing characteristics: herbs; flower parts in multiples of four; ovary inferior; fruit a capsule.

The evening primrose family has limited economic importance. Most species are grown for ornamental value and include evening primrose, clarkia, and fuchsia.

yellow evening primrose (*Oenothera biennis*)

Epilobium hirsutum L.
HAIRY WILLOW HERB

Also known as: fiddle grass, codlins and cream, apple-pie, cherry-pie, blood vine, purple rocket

Scientific Synonym(s): none

QUICK ID: leaves opposite; flowers with four pink petals; fruit a capsule

Origin: native to Europe and North Africa; possibly introduced as a garden ornamental in 1850s

Life Cycle: perennial reproducing by seed and rhizomes

Weed Designation: USA: MA, WA

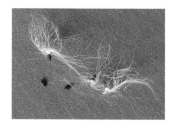

DESCRIPTION

Seed: Oblong to oval, 0.8–1.5 mm long, 0.4–0.5 mm wide; hairs dull white, 5–7 mm long, falling off easily.
Viability unknown.
Germination information unavailable.

Seedling: With oval to round cotyledons, long-stalked; first leaves opposite, oblong to oval, margins smooth, midrib prominent; second set of leaves similar, margins minutely toothed; rhizomes up to 60 cm long at the time of flowering stem development.

Leaves: Opposite, oblong to lance-shaped, 2–12 cm long, 0.5–4.5 cm wide, surfaces with woolly hairs and scattered glandular hairs, stalkless and clasping the stem, margins sharply toothed, prominently veined.
Inflorescence a raceme, borne in leaf axils.

Flower: Rose purple, erect, stalks 5–18 mm long; four sepals, 6–11 mm long, densely haired; four petals, 6–20 mm long, notched at tip; eight stamens; one pistil; styles 5–11 mm long; seeds ripe 4–6 weeks after flowering.

Plant: Semiaquatic, often with corklike tissue at the base when growing in water, intolerant of shade; stems erect, 20–250 cm tall, branched, covered in long, soft hairs; rhizomes creeping; stolons ropelike, thick, and fleshy.

Fruit: Capsule, 3–10 cm long, stalks 5–20 mm long, woolly haired; seeds covered with white hairs.

REASONS FOR CONCERN

Hairy willowherb is a problem weed along irrigation canals and drainage ditches. It is commonly found growing alongside purple loosestrife (*Lythrum salicaria* L., p. **XXX**) in wetlands, where both species are responsible for crowding out native vegetation. Research has shown that it outcompetes purple loosestrife when temperatures are below 18°C.

Similar species: a native species, fireweed (*Chamerion angustifolium* ssp. *angustifolium* [L.] Holub), may be mistaken for hairy willowherb. Fireweed is distinguished by long terminal racemes with 15–80 pink flowers. The flowers, 15–25 mm across, have petals that are not notched at the tip.

fireweed (*Chamerion angustifolium* ssp. *angustifolium*)

Also known as: evening primrose, hoary evening primrose, king's cure-all

QUICK ID: four petals, yellow; eight stamens; stems reddish-green

Scientific Synonym(s): *O. pycnocarpa* Atkinson & Bartlett, *O. muricata* L.

Origin: native to North America; often cultivated in gardens for use as a vegetable or medicinal herb

Life Cycle: biennial reproducing by seed; large plants producing up to 118,500 seeds

Weed Designation: none

DESCRIPTION

Seed: Oblong to triangular or angular with three or four faces, dark orange to reddish-brown, 1–1.7 mm long, 0.6–1.4 mm wide, surface dull, ridged lengthwise.
Viable for up to 80 years in soil.
Germination occurs in the top 5 mm of soil.

Seedling: With spatula-shaped cotyledons, 6–11 mm long, 4–6 mm long, surface dull green above, veins shiny, base with a few short hairs; stem below cotyledons bright red; first leaves alternate but appearing opposite.

Leaves: Alternate, oblong to lance-shaped, 1–12 cm long, 2–3 cm wide, upper leaves stalkless, margins wavy to toothed; reduced in size upward; basal leaves elliptical to oblong, stalked, midrib pinkish, margins wavy.
Inflorescence a leafy terminal spike up to 40 cm long.

Flower: Bright yellow, 2–5 cm across; opening in the evening; four sepals, yellowish-green, 20–22 mm long, bent backward; four petals, 12–25 mm long; eight stamens, 3–11 mm long; one pistil.

Plant: Producing a basal rosette of leaves in the first season; flowering stem erect, up to 2 m tall, grayish- to reddish-green; taproot thick and deep.

Fruit: Erect capsule, 1–3.5 cm long and 6 mm wide, four-valved and splitting from the top.

REASONS FOR CONCERN

Yellow evening primrose is a weed of roadsides and waste areas. It is serious problem in cultivated fields of winter wheat or fall rye.

OTHER SPECIES OF CONCERN IN THE EVENING PRIMROSE FAMILY

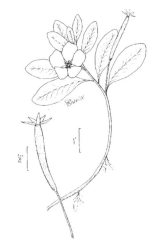

Ludwigia peploides

LUDWIGIA PEPLOIDES (KUNTH) P. H. RAVEN — FLOATING PRIMROSE-WILLOW, YELLOW WATER PRIMROSE

Native to South America; perennial of wetlands and marshes; stems green to reddish-brown, floating to creeping, 10–300 cm long, branched, rooting at nodes; leaves alternate, oblong to round, 1–10 cm long; stalks 3–60 mm long; margins smooth; upper surface hairless to hairy, prominently veined; inflorescence a spike; bracts triangular; flowers yellow; five sepals, green, 3–12 mm long; five petals, 7–24 mm long; 10 stamens in two sets, one pistil; fruit a capsule, five-angled, stalk 6–90 mm long; seeds 1–1.5 mm long. Weed designation: none.

LUDWIGIA URUGUAYENSIS (CAMB.) HARA — LARGE-FLOWERED WATER PRIMROSE, URUGUAY PRIMROSE

Native to South America; perennial of wetlands and marshes; roots feathery, appearing at nodes; stems erect to ascending, green to reddish-brown; leaves alternate, elliptical to lance-shaped, willowlike, 3–13 cm long, 3–25 mm wide, stalks 3–23 mm long; inflorescence a spike; stalks 2–3 cm long; bracts oblong to oval; flowers yellow; five or six sepals, 13–20 mm long; five or six petals, 20–30 mm long; 10 stamens; one pistil; fruit a capsule, 13–25 mm long, prominently 10-ribbed, stalk 10–60 mm long; seeds yellowish, 1.2–1.5 mm long. Weed designation: NC, SC, WA.

OENOTHERA LACINIATA HILL. —
CUTLEAF EVENING PRIMROSE

Native to subtropical North America; annual or short-lived perennial; stems erect to spreading, 5–100 cm long, simple or branched, often soft-haired; leaves alternate, 2–10 cm long, 5–35 mm wide, oblong to lance-shaped, short-stalked, margins deeply lobed to toothed; inflorescence a spike; flowers opening in evening and lasting 1 day, yellow to pale yellow and fading to orange; four sepals, 5–15 mm long; four petals, 5–22 mm long; eight stamens; one pistil; fruit a capsule, 2–5 cm long; seeds brown, 1–2 mm wide. Weed designation: none.

Oenothera laciniata

Orobanchaceae

BROOMRAPE FAMILY

The broomrapes are a family of parasitic and semiparasitic plants. Members of the parasitic genera lack chlorophyll, roots, and well-developed leaves, while the semiparasitic genera are partly photosynthetic with green stems and leaves. Flowers are produced in spikes or racemes with one bract at the base and two bracts below each flower. Flowers have four to six sepals, a two-lobed tubular corolla, four stamens, and one pistil. The ovary is superior. The fruit is a capsule with numerous dustlike seeds. Worldwide distribution: 99 genera/2,060 species; found throughout the world.

Broomrapes attach themselves to the roots of other plants with swollen structures called haustoria. Haustoria are formed when the rudimentary seedling root penetrates the root tissue of the host, causing an enlargement at the point of attachment.

clustered broomrape (*Orobanche fasciculata*)

clustered broomrape
(*Orobanche fasciculata*)

Orobanche ramosa L.
HEMP BROOMRAPE

Also known as: branched broomrape

Scientific Synonyms: none

QUICK ID: stem and leaves yellow; flowers whitish-blue to purple, tube-shaped; parasitic on roots of other plants

Origin: native to Eurasia and North Africa

Life Cycle: annual or short-lived perennial reproducing by seed; larger plants producing up to 250,000 seeds

Weed Designation: USA: AL, AZ, CA, FL, MA, MN, NC, OR, SC, TX, VT

DESCRIPTION

Seeds: Dustlike, 0.2–0.3 mm across, surface with a netted vein pattern.
Viability is at least 10 years in field soil. Germination triggered when certain root chemicals of an adequate host are nearby. Optimal soil temperatures of 18°–23°C are required for germination.

Seedlings: With reduced or absent cotyledons; seedling root has 2–4 days to attach itself to the host plant, otherwise it withers and dies.

Leaves: Alternate, scalelike, and yellowish, 3–10 mm long.

Flower: Cluster a spike 2–25 cm long with numerous flowers, each spike with a bract 6–10 mm long; below each flower, two small bracts. Flowers whitish-blue to purple with four or five yellowish sepals, each 6–8 mm long, a glandular, hairy two-lipped tube-shaped corolla, 10–22 mm long, four stamens, and one pistil.

Plant: Parasitic on the roots of other plants; chlorophyll and roots absent, deriving all nutrients from the host; slender branched yellow stems, 10–60 cm tall covered in short sticky hairs; stem base swollen and tuberous and attached to the roots of the host plant.

Fruit: Capsule 8–11 mm long containing up to 1,200 dustlike seeds.

REASON FOR CONCERN

Hemp broomrape is parasitic on plant species in at least 20 plant families. It is especially parasitic on members of the nightshade family (Solanaceae), which includes tomato, potato, eggplant, and tobacco. Crop losses of 100% have been reported in tomato fields. The seed viability limits use of that cropland for several years. The exceptionally light seeds are easily distributed by farm machinery, wildlife, and wind. Seeds are often spread in manure, as they pass through an animal's digestive tract unaffected.

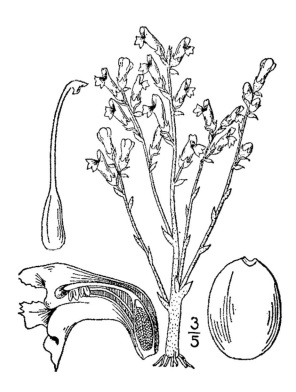

OTHER SPECIES OF CONCERN IN THE BROOMRAPE FAMILY

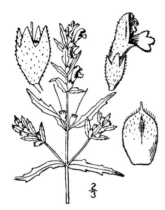

Odontites serotina

ODONTITES SEROTINA (LAM.) DUM.— RED BARTSIA

Native to Europe; annual parasitic on the roots of other plants; stems 10–40 cm tall; leaves opposite, lance-shaped, 1–3 cm long, stalkless, margin with two or three pairs of blunt teeth; inflorescence spikelike, composed of solitary axillary flower; flowers light red, tube-shaped, about 1 cm long; four sepals, fused, bell-shaped; four petals; four stamens; one pistil; fruit a capsule, about 7 mm long; average plants producing about 1,400 seeds. Weed designation: CAN: federal, AB, MB, SK.

STRIGA SSP. (TREATMENT FOR S. ASIATICA [L.] KUNTZE) — WITCHWEED

Native to tropical Asia and Africa; annual parasitic on the roots of grasses (including corn and sorghum); roots reduced, succulent; underground leaves white and scalelike, turning blue when exposed to air; stems 10–30 cm tall, square, bristly haired; leaves opposite, green, linear, 10–40 mm long, 1–4 mm wide, stalkless, bristly haired; inflorescence a spike, 10–15 cm long; two bracts below each flower, about 5 mm long; flowers red to yellow or white, 10–15 mm long, 6–9 mm wide, two-lipped; five sepals; five petals; four stamens; one pistil; fruit an ovoid capsule, black; 550–1,400 seeds, brown, dustlike. Weed designation: USA: federal, AL, AZ, AR, CA, HI, MA, MN, NC, OR, SC, VT; MEX: federal.

Oxalidaceae

WOOD SORREL FAMILY

Herbs; rhizomes fleshy; leaves alternate, pinnately or palmately compound; stipules absent; inflorescence an umbel; flowers perfect; regular; five sepals; five petals; 10 stamens, fused at the base; one pistil; five styles; ovary superior; fruit a capsule, often explosive. Worldwide distribution: eight genera/950 species; primarily tropical and subtropical regions.

Distinguishing characteristics: herbs; leaves compound; stamens fused at base; five styles.

A few species are cultivated for their ornamental value.

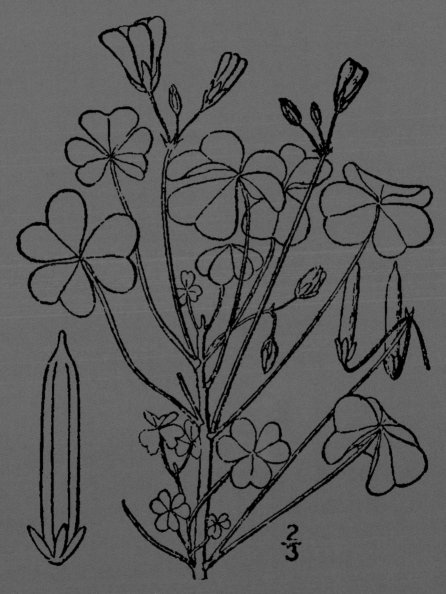

yellow wood sorrel (*Oxalis stricta*)

Oxalis corniculata L.
CREEPING WOOD SORREL

Also known as: yellow
wood sorrel, sour clover,
creeping lady's sorrel
Scientific Synonym(s):
O. pusilla Salisb., *O. repens*
Thunb., *O. villosa* Bieb.,
O. langloisii (Small) Fedde,
Xanthoxalis corniculata
(L.) Small., *X. langloisii*
Small, *X. repens* (Thunb.)
Moldenke, *Acetosella*
corniculata (L.) Kuntze

QUICK ID: leaves alternate,
compound with three
heart-shaped leaflets;
flowers yellow; stems
often rooting at nodes
Origin: native to Asia,
Malaysia, or Australia;
introduced into North
America about 1732
Life Cycle: perennial
reproducing by seed only
Weed Designation: none

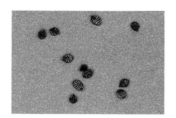

DESCRIPTION
Seed: Elliptical, brown, 1–1.4 mm long, about 1 mm
wide, surface dull with nine cross ridges and two
or three faint cross ridges.
Viable for up to 4 years in soil.
Germination occurs in light and temperatures
about 20°C
Seedling: With oval to oblong cotyledons, about 2.3
mm long and 1 mm wide; stem below cotyledons
very short; first leaves alternate, compound with
three heart-shaped leaflets; leaves after fourth
leaf stage with hairy margins and petioles.
Leaves: Alternate, often purplish-green, long-
stalked, compound with three leaflets, sour-
tasting; leaflets heart-shaped, 1–2 cm across;
stipules present, but small.
Inflorescence a cyme, one- to four-flowered,
borne in leaf axils.
Flower: Yellow, 4–10 mm across; five sepals, 3.5–
7 mm long; five petals, 7–11 mm long; 10–15
stamens; one pistil.
Plant: Plants 10–50 cm tall; stems erect to prostrate,
grayish-green, often root at the nodes.
Fruit: Cylindrical capsule, 1.2–2.5 cm long, hairy;
many-seeded; exploding at maturity, ejecting
seeds up to 2 m from parent.

REASONS FOR CONCERN

Creeping wood sorrel is a common weed in lawns, gardens, and greenhouses. It has been reported to cause oxalate poisoning in sheep. In Mexico, it is a common weed of beans, corn, coffee, and potatoes. It is reported to be a weed in 17 crops in 44 countries.

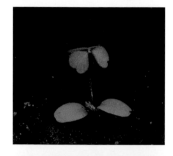

Similar species: when not in flower, creeping wood sorrel is often confused with black medic (*Medicago lupulina* L., p. 366) and white clover (*Trifolium repens* L., p. 376). The distinguishing characteristic between these three species is the shape of the leaflet. The leaflets of black medic and white clover are not heart-shaped.

OTHER SPECIES OF CONCERN IN
THE WOOD SORREL FAMILY

OXALIS STRICTA L. — YELLOW WOOD SORREL

Native to North America; plants producing rhizomes; stems erect to ascending, 5–50 cm long, green to pinkish, hairy; leaves alternate, 5–30 mm wide, compound with three leaflets; leaflets 1–2 cm wide, heart-shaped, margins hairy; inflorescence an umbel, one- to five-flowered; flowers yellow, 5–13 mm across; five sepals; five petals; 10–15 stamens; one pistil; fruit a capsule 1–3 cm long, surface densely hairy, borne on bent stalks. Weed designation: none.

Oxalis stricta

Papaveraceae
POPPY FAMILY

Herbs, occasionally shrubs; sap milky or colored; leaves alternate, opposite, or whorled, simple; inflorescence an umbel or solitary flowers; flowers perfect, regular; two or three sepals, falling as the flower opens; four to eight petals (occasionally eight to 15); 12 or more stamens; one pistil; ovary superior; fruit a capsule. Worldwide distribution: 26 genera/200 species; subtropical and temperate North America.

Distinguishing characteristics: herbs with milky juice; flowers showy; fruit a capsule.

Several members of the poppy family are grown for their ornamental or medicinal value. A few species are weedy.

Icelandic poppy (*Papaver nudicaule*)

Argemone mexicana L.
PRICKLY POPPY

Also known as: Mexican prickly poppy, yellow prickly poppy, Mexican thistle

Scientific Synonym(s):
A. leiocarpa Greene

QUICK ID: leaves prickly lobed, bluish-green; stems prickly; flowers yellow; sap yellow

Origin: native to tropical America; introduced garden ornamental throughout the world, becoming weedy where introduced

Life Cycle: annual reproducing by seed; large plants producing up to 3,600 seeds

Weed Designation: none

DESCRIPTION

Seed: Black to brown, globe-shaped, 1.5–2 mm across, surface with a network of fine lines; aril present.
Viability unknown.
Germination occurs after 24 days of dry storage at 4°–32°C.

Seedling: With linear cotyledons, 15–35 mm long, 1–3 mm wide; first leaf wedge-shaped, margin with two to three pairs of teeth.

Leaves: Alternate, 6–20 cm long, 3–8 cm wide, bluish-green, stalkless and somewhat clasping, margins unevenly toothed; prickles on margin and underside; basal leaves stalked.
Inflorescence a solitary flower; one or two bracts, leaflike; flower buds globe-shaped, 10–15 mm long, 9–13 mm wide, sparingly prickled.

Flower: Bright to pale yellow, 4–7 cm wide; two or three sepals, prickly; six petals, in two whorls of three, oblong to oval; 30–50 stamens; one pistil; stigma dark red, three- to six-lobed.

Plant: Thistle-like, bluish-green; stems 30–100 cm tall, prickly; stem and foliage with yellow sap.

Fruit: Capsule, 2–4.5 cm long, 12–20 mm across, 300–400-seeded; surface prickles 6–10 mm long; 60–90 fruits per plant.

REASONS FOR CONCERN

Prickly poppy is toxic to poultry, sheep, cattle, horses, and humans. The toxic substances—berberine and protopine—are present in the sap. It is a serious weed of corn, cereals, cotton, potatoes, and sugarcane. It is reported as a serious weed in 30 countries.

Chelidonium majus L.
CELANDINE

Also known as: greater celandine, rock poppy, swallowwort

Scientific Synonym(s): none

QUICK ID: plants with yellowish-orange sap; flowers yellow; fruit podlike

Origin: native to Europe and Asia; believed to be introduced as a wart treatment prior to 1672

Life Cycle: biennial or short-lived perennial reproducing by seed

Weed Designation: none

DESCRIPTION

Seed: Black, 1–3 mm long; surface pitted. Viability unknown.
Germination requires a dormant period; optimal temperatures 20°–30°C.

Seedling: With oval to lance-shaped cotyledons; first leaf weakly three-lobed, pale green; second leaf weakly five-lobed.

Leaves: Alternate, 5–35 cm long, deeply five- to nine-lobed, margins wavy to irregularly toothed; stalks 2–10 cm long, bristly haired.
Inflorescence an umbel; stalk 2–10 cm long; few-flowered; bracts present.

Flower: Yellow, up to 2 cm wide; stalks 5–35 mm long; two sepals, up to 1 cm long, falling off as the flower opens; four petals; 12 to many stamens; one pistil.

Plant: Shade-tolerant; sap yellowish-orange; stems 20–100 cm tall, somewhat succulent and bristly, branched; bristly haired at the base.

Fruit: Linear to oblong capsule, erect, 2–5 cm long, hairless, opening on two sides from the bottom upward; seeds few.

REASONS FOR CONCERN

Celandine's ability to grow in shade has allowed it to spread throughout forests, displacing native vegetation. The roots are also reported to be poisonous to humans and livestock.

Similar species: a native species, celandine poppy (*Stylophorum diphyllum* [Michx.] Nutt.), closely resembles the celandine. Celandine poppy has bright yellow flowers 2–3 cm across, hairless nodding capsules 20–35 mm long, and five- to seven-lobed leaves up to 50 cm long.

celandine poppy (*Stylophorum diphyllum*)

Also known as: field poppy, Flander's poppy, redweed, common poppy

Scientific Synonym(s): *P. commutatum* Fisch. & C. A. Mey., *P. insignitum* Jord., *P. intermedium* Beck, *P. roubiaei* Vig., *P. tenuissimum* Fedde, *P. trilobum* Wallr., *P. tumidulum* Klokov, *P. strigosum* (Boenn.) Schur.

QUICK ID: plants with milky sap; leaves basal; petals with a dark spot at base

Origin: native to eastern Mediterranean region of Europe, southwest Asia and North Africa

Life Cycle: annual reproducing by seed; average plants producing up to 20,000 seeds; large plants may release up to 800,000 seeds

Weed Designation: none

DESCRIPTION

Seed: Kidney-shaped, dull purplish-gray to dark brown, 0.7–1 mm long and 0.5 mm wide, surface net-veined.

Viability in soil after 26 years is about 13%. Germination may occur at 2°C, but optimal is 7°–13°C; light enhances germination.

Seedling: With lance-shaped cotyledons; first leaves oval to elliptical, margins smooth, surface hairy.

Leaves: Primarily basal, up to 15 cm long, light green, deeply lobed, stalked; stem leaves few, alternate, stalkless.

Inflorescence a solitary flower borne on long stalks; stalks covered with stiff spreading hairs; flowers nodding when in bud.

Flower: Deep scarlet, red to pink or white, purple or bluish, often dark spotted at base of petal, 3–10 cm across; two sepals, hairy, falling as the flower opens; four petals, 2–6 cm long; stamens numerous, anthers bluish; one pistil; stigmatic disc flat with five to 18 stigmas.

Plant: Plants 10–90 cm tall; stem erect, covered with stiff hairs; stem and foliage with milky juice.

Fruit: Globe-shaped capsule, 1–2 cm wide, tan to brown, hairless; large plants with up to 400 capsules; capsules may contain up to 2,000 seeds.

REASONS FOR CONCERN

Corn poppy is often a weed of annual crops, winter cereals, gardens, meadows, and disturbed sites. Crop losses of 43% have been reported. It is easily controlled by herbicides, but seed bank and viability are a concern.

Similar species: long-headed poppy (*P. dubium* L.), native to Europe and Asia, forms hybrids with corn poppy. It is distinguished by red to orange petals up to 3 cm long (dark basal spot present), capsules 2–6 cm long, and flower stalks with appressed bristly hairs.

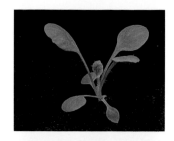

OTHER SPECIES OF CONCERN IN THE POPPY FAMILY

Glaucium flavum

GLAUCIUM FLAVUM CRANTZ — YELLOW HORNPOPPY

Native to the Black Sea and Caucasus region of eastern Europe and southwestern Asia; biennial or perennial with yellow sap; stems up to 80 cm tall, leafy, branched; basal leaves numerous, up to 30 cm long, stalked, hairy; stem leaves lyre-shaped with seven to nine lobes, stalkless and clasping the stem; inflorescence a solitary axillary flower; stalks up to 4 cm long; flowers yellow or orangish with a reddish or purple basal spot; two sepals, 2–3 cm long; four petals, 2.5–4 cm long; stamens many; one pistil; fruit a linear capsule up to 30 cm long, curved; seeds dark brown. Weed designation: USA: MA.

Phytolaccaceae
POKEWEED FAMILY

Herbs, shrubs, or tropical trees; leaves alternate, simple; stipules absent; inflorescence a spike or raceme; flowers perfect, regular; four or five sepals; petals absent; three to 30 stamens; one to six pistils, partially fused; ovary superior; fruit a berry, drupe, or achene. Worldwide distribution 17 genera/120 species; predominantly tropical and subtropical North and South America.

Distinguishing characteristics: herbs; leaves alternate; sepals petal-like; fruit a fleshy berry.

Members of this family are of limited economic importance. One species, pokeweed (*Phytolacca americana*), is occasionally grown as a vegetable The seeds, fruit, and roots are extremely poisonous. Blanching the young leaves in three changes of water makes them edible. The toxins are not easily removed from mature leaves. Extreme caution should be exercised when using this plant as a vegetable.

pokeweed (*Phytolacca americana*)

Phytolacca americana L.
POKEWEED

Also known as: Virginia poke, scoke, pigeonberry, garget, inkberry, red ink plant, coakrum, American cancer, cancer jalap, inkweed, pokeberry, common pokeweed, American pokeweed, pocan bush

Scientific Synonym(s): *P. decandra* L.

QUICK ID: stems red to pinkish-green; flowers white; fruit a dark purplish-black berry with crimson juice

Origin: perennial native to eastern North America; now introduced into Europe and western Asia; widely planted as a vegetable; juice of berries used for dye

Life Cycle: perennial reproducing by seed; often grown as a spring vegetable, although poisonous

Weed Designation: none

DESCRIPTION

Seed: Lens-shaped, 2.8–2.9 mm long, 2.5–2.8 mm wide, black to purple, surface smooth and glossy. Viable in soil for up to 40 years.
Germination occurs in the top 3 cm of soil; optimal depth is top 1 cm or at the soil surface.

Seedling: With elliptical to egg-shaped cotyledons, 6–11 mm wide, 17–33 mm long, unequal, leaflike, very pale beneath; stem below cotyledons bright magenta, succulent; first and later leaves resembling cotyledons.

Leaves: Alternate, oblong to lance- or egg-shaped, 10–50 cm long, 3–18 cm wide, dark to light green, veins prominent and pinkish; stalks 1–6 cm long, hairless; margins smooth to wavy; ill-scented when crushed.
Inflorescence a raceme, 10–25 cm long, red-stemmed; stalk up to 15 cm long

Flower: Greenish-white and tinged with pink, 5–7 mm across; stalks 3–13 mm long; five sepals, 2.5–3.3 mm long, greenish-white to pinkish; petals absent; 10 stamens, one pistil.

Plant: Plants 2–3 m tall with a spread of 1–2 m, often falling over due to the size; stems hollow, branched above, pink to reddish-green; taproot fleshy, white, up to 15 cm across; dying back to the ground each fall and growing quickly in the spring.

Fruit: Globe-shaped berry, 5–10 mm across, juice crimson; 10-seeded; green and turning dark purple-black at maturity.

REASONS FOR CONCERN

The taproot and fruit of pokeweed are poisonous. There have been numerous reports of cattle and pig fatalities. Symptoms range from stomach cramps to convulsions and respiratory failure. The showy berries are often spread by birds that feed on the fruit. It is also reported to becoming a weed of no-till cropland. The large plants tend to be a problem in corn and soybean fields.

OTHER SPECIES OF CONCERN IN THE POKEWEED FAMILY

***PHYTOLACCA BOGOTENSIS* KUNTH. —**
SOUTHERN POKEBERRY

Native to South America; plants up to 2 m tall; leaves elliptical to lance-shaped, up to 18 cm long and 7 cm wide; stalks 5–30 mm long; inflorescence a dense raceme up to 20 cm long; stalks 1–5 cm long; flowers white to red; five sepals; seven to 12 stamens in one whorl; one pistil; stalks 2–4 mm long; fruit a greenish-brown to dark brown berry, 5–8 mm across; seeds black 2.5–3.3 mm long, shiny. Weed designation: none.

***PHYTOLACCA ISOCANDRA* L. —**
TROPICAL POKEWEED, JABONCILLO

Native to Central and South America; plants up to 3 m tall; leaves elliptical to oblong or lance-shaped, up to 30 cm long and 15 cm wide; stalks 5–60 mm long; inflorescence a spike up to 30 cm long, stalk 1–10 cm long; flowers white or pink to pinkish-red; five sepals; 10–22 stamens, borne in two whorls; one pistil; stalks 0.5–2 mm long; fruit a purplish-black berry, 7–8 mm across; seeds black, 2.5–3 mm long, shiny. Weed designation: none.

PHYTOLACCA OCTANDRA L. — RED INKPLANT

Native to tropical regions of the world; plants up to 2 m tall; leaves elliptical to lance-shaped, up to 22 cm long and 7.5 cm wide; stalks 5–30 mm long; Inflorescence a spike up to 14 cm long, stalk 5–30 mm long; flowers white to pinkish or reddish; five sepals; eight to 10 stamens in one whorl; one pistil; stalks less than 2 mm long; fruit a greenish berry, 4.5–6 mm across; seeds black, about 2 mm long, shiny. Weed designation: none.

Plantaginaceae
PLANTAIN FAMILY

Herbs, occasionally aquatic, leaves alternate (appearing basal), opposite, or whorled; stipules absent; inflorescence a spike, raceme, or head; flowers perfect; four or five sepals, united; four or five petals, united and often two-lipped; four stamens, a staminode may be present; one pistil; ovary superior; fruit a capsule or nutlet. Worldwide distribution: 90 genera/1,900 species; found throughout the world.

Distinguishing characteristics: herbs; leaves basal or opposite; four or five petals united.

A few members of the plantain family are grown for their ornamental value. One species, foxglove (*Digitalis purpurea*), was the original source of the cardiac medicine, digoxin. Several species are common or noxious weeds.

hoary plantain (*Plantago media*)

toadflax (*Linaria vulgaris*)

Chaenorrhinum minus (L.) Lange
DWARF SNAPDRAGON

Also known as: small snapdragon, small toadflax, railroad weed
Scientific Synonym(s): *Linaria minor* (L.) Desf.

QUICK ID: lower leaves opposite, upper leaves alternate; flowers light blue with a yellow throat; plants less than 40 cm tall
Origin: native to the Mediterranean region of Europe; first U.S. report from New Jersey in 1874; first Canadian report from New Brunswick in 1881
Life Cycle: annual reproducing by seed; large plants producing up to 7,500 seeds
Weed Designation: USA: WA

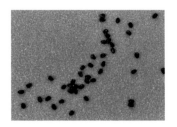

DESCRIPTION
Seed: Elliptical to oval, brown, 0.8–1.2 mm long, surface rough with several ridges.
Viability less than 10% after 2 years in soil. Germination enhanced by light; emergence from 1 cm soil depth; optimal temperature 20°C.
Seedling: With lance-shaped cotyledons, 2.5–10 mm long, less than 4 mm long, withering soon after the first leaves appear; first pair of leaves spatula-shaped, slightly hairy.
Leaves: Opposite below and alternate above, lance- to club-shaped, 1–2.5 cm long, 1–5 mm wide, stalkless, margins smooth.
Inflorescence a solitary flower, borne in leaf axils.
Flower: Light blue to purplish with a yellow throat, 6–8 mm long, two-lipped and irregular; five sepals, united; five petals, united, spurred; four stamens; one pistil; spur purplish, 1.5–2 mm long; stalks 10–15 mm long.
Plant: Bushy, prefers gravelly or sandy soils (often found on railway grades), breaking off at the ground at maturity and becoming a tumbleweed; stems 5–40 cm tall, slightly zigzagged in appearance, surface with short sticky hairs.
Fruit: Round capsule, about 5 mm across; seeds released through two small pores.

REASONS FOR CONCERN

Small snapdragon is a weed in cultivated fields, gardens, waste areas, roadsides, and railway grades. Recently, it has also been recognized as a serious weed problem in strawberry fields.

Linaria dalmatica (L.) Mill.
DALMATIAN TOADFLAX

Also known as: smooth toadflax

Scientific Synonym(s):
L. macedonica Griseb.

QUICK ID: flowers yellow; leaves alternate, bluish-green and hairless

Origin: native to southeastern Europe; introduced into North America around 1900 as an ornamental plant

Life Cycle: perennial (living 3–15 years) reproducing by rhizomes and seed; each plant capable of producing up to 500,000 seeds

Weed Designation: USA: AZ, CA, CO, ID, MT, ND, NM, NV, OR, SD, WA, WY; CAN: federal, AB, BC, MB

DESCRIPTION

Seed: Oval to somewhat angular, black, 1.3–1.9 mm long.
Viable in soil for up to 10 years.
Germination occurs at 1–3 cm depth (optimal depth 5 mm); emergence occurs when soil temperatures reach 8°C at 2.5 cm depth.

Seedling: With oval cotyledons, 3–9 mm long, 1–3 mm wide, tip distinctly pointed, hairless; first pair of leaves opposite; later leaves alternate.

Leaves: Alternate (may appear opposite), oval to lance-shaped, 3–8 cm long, 1–2 cm wide, bluish-green, hairless; stalkless and clasping the stem with heart-shaped bases.
Inflorescence a terminal raceme.

Flower: Yellow (buds often purplish-red), 14–35 mm long, stalks 2–4 mm long; five sepals, united, 6–8 mm long; five petals, united, spurred; four stamens, short; one pistil; spur 13–20 mm long.

Plant: Stout, 30–120 cm tall, older plants with up to 10 stems; stems branched, bluish-green, hairless; rootstalk woody.

Fruit: Cylindrical capsule, 7–10 mm long, 4–8 mm wide; 140–250-seeded; average plants produce 250 capsules.

REASONS FOR CONCERN

Dalmatian toadflax is an aggressive weed of pastures, rangeland, and roadsides. It is a deep-rooted perennial that, once established, is difficult to eradicate.

Similar species: toadflax (*L. vulgaris* Hill., p. 534) is closely related to Dalmatian toadflax. It has numerous narrow leaves, 2–10 cm long, and smaller yellow flowers. These yellow flowers, 2–3 cm long, have an orange throat.

Linaria vulgaris Hill.
TOADFLAX

Also known as: butter-and-eggs, wild snapdragon, yellow toadflax, flaxweed, ramsted, eggs-and-bacon, perennial snapdragon, Jacob's ladder, rabbit flower, impudent lawyer

Scientific Synonym(s): *L. linaria* (L.) Karst.

QUICK ID: flowers yellow with a spur; leaves alternate; stems with numerous narrow leaves

Origin: native to Europe; introduced in mid-1800s as a garden plant; reported populations as far north as Fort Smith, Northwest Territories, and Fairbanks, Alaska

Life Cycle: perennial reproducing by seed and rhizomes; plants capable of producing up to 8,700 seeds; root fragments about 10 cm long can colonize an area up to 2 m in diameter

Weed Designation: USA: ID, MT, NV, NM, OR, SD, WA, WY; CAN: federal, AB, BC, MB, SK; MEX: federal

DESCRIPTION

Seed: Flattened oval, dull black, 1.5–2 mm across, margin with a papery wing.
Viability is about 67% at 13 years of dry storage. Germination occurs in the top 3 cm of soil; some surface germination occurs.

Seedling: With oval cotyledons, 1.3–10 mm long, 1–4 mm wide, tip with a distinct point, hairless; first leaves opposite, oval; later leaves alternate, lance-shaped; shoots from root buds appearing about 3 weeks after emergence, with 90–100 buds by the end of the first season, and occupying 1 m in diameter of soil.

Leaves: Alternate (crowded and may appear opposite), lance-shaped, 2–10 cm long, 1–5 mm wide, pale green, hairless, short-stalked, margins smooth.
Inflorescence a dense terminal raceme.

Flower: Yellow with an orange throat (resembling snapdragons), 2–3.5 cm long, stalks about 5 cm long; five sepals, united; five petals, united, spurred; four stamens; one pistil; spur 2–3 cm long.

Plant: Plants 30–130 cm tall, often forming large colonies; stems hairless; rhizomes extensive and creeping, buds numerous.

Fruit: Oval to egg-shaped capsule, 8–12 mm long; three- to 110-seeded.

REASONS FOR CONCERN

Toadflax is a potential weed problem in zero- and minimal-till areas because of its prolific seed production and creeping rhizome. It is an aggressive weed in rangeland, where it quickly replaces grasses and herbs. Some sources report that toadflax is poisonous to cattle. Toadflax is host for tobacco mosaic virus.

Similar species: Dalmatian toadflax (*I dalmatica* [L.] Mill., p. 532), a closely related species, is more robust than toadflax. Stems, up to 1 m tall, have numerous oval to lance-shaped leaves that clasp the stem. Flowers are yellow and 3–4 cm long.

Plantago lanceolata L.
RIBGRASS

Also known as: English plantain, black plantain, ribwort, narrowleaved plantain, buckhorn, buck plantain, black jacks, ripple plantain, snake plantain, dog's-ribs, cat's-cradles, rat-tail, hen-plant, lanceleaf Indianwheat

Scientific Synonym(s):
P. altissima auct. non L.

QUICK ID: leaves basal, much longer than broad; leaves with prominent ribs; flowers borne in dense spikes

Origin: native to eastern Europe and north central Asia; first reported in North America at Nova Scotia in 1829

Life Cycle: perennial reproducing by seed; each plant capable of producing over 10,000 seeds

Weed Designation: USA: AR, IA; CAN: federal, PQ

DESCRIPTION

Seed: Elliptical, orange-brown to black, 2–3 mm long, 1–1.2 mm wide; under low magnification surface appears rough.
Viable in soil for up to 16 years.
Germination takes place in the top 0.5 cm of soil; seeds sticky when wet, aiding in their dispersal.

Seedling: With narrow linear cotyledons 17–22 mm long and 1 mm wide; first leaves lance-shaped, hairless, underside three- to five-ribbed.

Leaves: Basal, lance-shaped, grasslike, 8–30 cm long, 1–4 cm wide, prominently three- to seven-veined, margins smooth, tufts of brown hairs at leaf base.
Inflorescence a cylindrical to globe-shaped spike, up to 12 cm long; large plants with up to 30 spikes.

Flower: Brownish-greenish, 4–6 mm across; four sepals, united, 3–3.5 mm long; four petals, united, 2–3 mm long; four stamens, yellow, longer than the petals; one pistil; bracts papery, brownish; wind-pollinated.

Plant: Reaching maturity in 51 days, living to 12 years; flowering stem 10–80 cm tall, leafless; rootstalk short, roots to 1 m deep but most confined to top 10 cm of soil.

Fruit: Elliptical capsule, 3–4 mm long, one- to two-seeded.

REASONS FOR CONCERN

The seeds of ribgrass are similar in size to red clover and alfalfa, making separation of the seed difficult. Ribgrass in host for rosy apple aphid and the following viral diseases: beet yellows, tobacco mosaic, and tobacco ring spot. Ribgrass has been reported as a weed in over 50 countries.

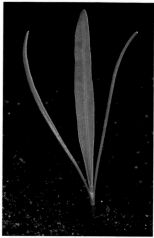

Plantago major L.
COMMON PLANTAIN

Also known as: white man's foot, ground plantain, greater plantain, major plantain, dooryard plantain, English plantain, birdseed plantain, ripple-seed plantain, waybread

Scientific Synonym(s): *P. asiatica* auct. non L., *P. halophila* Bickn.

QUICK ID: leaves basal, prominently three- to five-ribbed; flowers greenish-white, borne in dense spikes

Origin: native to Europe; first observed in North America in 1748; first Canadian report from Montreal in 1821 and spreading to British Columbia by 1899

Life Cycle: annual, biennial, or perennial reproducing by seed or ramets (side shoots); large plants producing up to 14,000 seeds

Weed Designation: CAN: PQ

DESCRIPTION

Seed: Triangular to four-sided, brownish-black, 0.9–1.2 mm long, 0.5–0.7 mm wide; surface rough and dull; sticky when wet.
Viable in soil for up to 60 years.
Germination requires the presence of light.

Seedling: With spatula-shaped cotyledons, 1.5–7 mm long, less than 1 mm wide, faintly three-veined, withering soon after the appearance of the first leaves; first leaves elliptical, prominently three- to five-ribbed, hairless.

Leaves: Basal, oval, 5–30 cm long, up to 12 cm wide, stalks with winged margins, margins smooth; underside prominently three- to five-ribbed, sparsely haired.
Inflorescence a spike, 7–30 cm long

Flower: Greenish-white, 2–3 mm across, short-stalked; four sepals, united, 1.5–2 mm long; four petals, united, papery; four stamens, longer than sepals or petals; one pistil; wind-pollinated.

Plant: Found as far north as 200 km south of the arctic treeline, able to withstand temperatures of −40°C; stems 10–60 cm tall, leafless, hairless; rootstalk short and thick, roots fibrous, up to 80 cm deep and wide; ramets remain attached to parent plant for 2–3 years.

Fruit: Egg-shaped capsule, 2–4 mm long, splitting around the middle; five- to 30-seeded.

REASONS FOR CONCERN

Common plantain is a common weed in cultivated fields, lawns, roadsides, and waste areas. It is host for tobacco mosaic, aster yellows, beet curly top, beet yellows, tobacco streak, and tomato spotted wilt viruses. It is reported to be a weed in over 50 countries.

Similar species: blackseed plantain (*P. rugelii* Decne.) is often mistaken for common plantain. The hairless leaves of blackseed plantain are prominently three-veined and have toothed to wavy margins. Another characteristics is the base of the leaf stalk that is reddish-purple.

OTHER SPECIES OF CONCERN IN
THE PLANTAIN FAMILY

Digitalis lanata

Limnophila sessiliflora

DIGITALIS LANATA EHRH.—
GRECIAN FOXGLOVE
Native to southern Europe; perennial; stems 50–100 cm tall; leaves alternate, oblong to lance-shaped, stalkless, margins smooth; inflorescence a spike, densely haired; flowers white to pale yellow with brown veins, 25–30 mm long, irregular; five sepals; five petals; four stamens; one pistil; fruit a capsule. Weed designation: none.

LIMNOPHILA SESSILIFLORA (VAHL) BLUME—
ASIAN MARSHWEED, DWARF AMBULIA
Native to India and Sri Lanka; aquatic or semiaquatic perennial; stems up to 4 m long, often rooting at nodes; submersed leaves in whorls of six to 10, finely divided and featherlike, 5–40 mm long, oval to elliptical or lance-shaped; aerial leaves lance-shaped, 5–18 mm long, 3–4 mm wide, borne in whorls of five to eight, margins irregularly torn; inflorescence a solitary axillary flower; flowers blue, violet, or pink; sepals 4–7 mm long, hairy; petals 5–10 mm long, upper lip with two blue dots; four stamens; one pistil; fruit a capsule, 3.5–5.5 mm long, green to dark brown; 200–300 seeds. Weed designation: USA: federal, AL, CA, FL, MA, NC, OR, SC, VT.

PLANTAGO ARISTATA MICHX. —
LARGEBRACTED PLANTAIN

Native to North America; plants annual, hairless; flowering stems 10–40 cm tall; leaves basal, linear, 5–15 cm long and less than 1 cm wide, margins smooth, surface with cobweb to woolly hairs; inflorescence a dense spike, 1–13 cm long, 7–8 mm across; bracts linear, about 3 cm long and 2 mm wide, densely haired; four sepals; four petals, 1.5–2.3 mm long, often three spreading and one erect; four stamens, anthers yellow; one pistil; fruit a capsule; two seeds. Weed designation: USA: AR; CAN: PQ; MEX: federal.

Plantago aristata

PLANTAGO CORONOPUS L. —
BUCKHORN PLANTAIN

Native to Europe; annual or perennial with a stout taproot; stems 10–45 cm tall, often hairy throughout; leaves basal, oblong to lance-shaped in outline, 1–15 cm long, 5–20 mm wide, margins deeply toothed to lobed; inflorescence a compact spike, 8–50 mm long; bracts 1.5–2.8 mm long; flowers greenish-white; four sepals, 2.3–3 mm long; four petals, 0.8–1 mm long; four stamens; one pistil; fruit a capsule, 2–2.5 mm long; three to five seeds. Weed designation: none.

PLANTAGO MEDIA L. — HOARY PLANTAIN

Native to Europe and Asia; perennial; taproot thick and cylindrical; leaves basal, elliptical to oval, 4.5–13 cm long, 1.5–5 cm wide, covered in white hairs, seven or nine veins; stalks 5–80 cm long, winged; margins smooth; spikes cylindric, 3–8 cm long, borne on stalks 15–45 cm long; bracts 2–3 mm long; four sepals, about as long as the bract; four petals, white, 1.7–2.3 mm long; stamens pinkish, long-stalked; one pistil; fruit a capsule, 2.5–4 mm long; seeds yellowish-brown, 1.5–2 mm long. Weed designation: none.

PLANTAGO PSYLLIUM L. — FLAXSEED PLANTAIN

Native to Europe and Asia; annual; stems erect, 10–50 cm tall, surface with soft hairs; leaves opposite or in whorls of three, linear, 3–10 cm long; inflorescences an elliptical to globe-shaped spike, 1–2 cm long, borne on stiff stalks 3–8 cm long originating in leaf axils; lower bracts with green tips, 2–5 mm long; flowers greenish-white; four sepals; four petals; four stamens; one pistil; fruit a capsule; seeds black and shiny, about 2.5 mm long and 0.8 mm wide. Weed designation: CAN: PQ; MEX: federal.

Plantago media

Plantago psyllium

VERONICA OFFICINALIS L.— COMMON SPEEDWELL

Native to North America; perennial, hairy throughout; stems prostrate to ascending, 10–30 cm long; leaves opposite, elliptical, 2.5–5 cm long, 1–2.5 cm wide, short-stalked, margins toothed; inflorescence a spikelike raceme, 3–6 cm long, bracts linear to lance-shaped; flowers pale blue to violet with dark veins, 5–7 mm wide; four sepals; four petals; two stamens; one pistil; fruit a heart-shaped capsule, 4.5–5 mm wide. Weed designation: none.

Veronica officinalis

Poaceae
(Formerly Gramineae)

GRASS FAMILY

Herbs (tree or shrublike in bamboo); rhizomes, stolons, and fibrous roots often present; stems (culms) round with swollen nodes and hollow internodes; leaves alternate, two-ranked, often differentiated into blade, sheath, ligule, and auricles; sheath usually open; ligule when present consisting of membrane or hairs, found at the junction of the blade and sheath; auricles when present, small projections at the top of the sheath; inflorescence a spikelet, often forming spikes, racemes, or panicles; spikelets with two bracts at base (technically called first and second glume); flowers perfect or imperfect; flowers with two bractlets, the lemma, and palea; sepals and petals reduced to very small structures called lodicules; three stamens, occasionally two or six; one pistil; two stigmas, often feathery; ovary superior; fruit a caryopsis (grain). Worldwide distribution: 600 genera/10,000 species; found throughout the world.

Distinguishing characteristics: herbs; stems hollow; leaves two-ranked; sepals and petals absent.

Economically, the grasses are the most important plant family in the world. Important species include wheat, rice, corn, sorghum, sugarcane, and bamboo. Several species are noxious weeds.

panicle

spikelet

palea

lemma

glumes

flower

Aegilops cylindrica Host
JOINTED GOATGRASS

Also known as: goatgrass

Scientific Synonym(s):
  Cylindropyrum cylindricum (Host) A. Löve, *Triticum cylindricum* (Host) Ces., Pass. & Gib., *A. tauschii* auct. non Coss.

QUICK ID: stems jointed; ligule and auricles present; sheath open with overlapping margins

Origin: native to Russia; introduced into North America in the early 1900s; greatest concentration occurs in wheat-growing regions

Life Cycle: annual or winter annual reproducing by seed; plants may produce up to 3,000 seeds; large plants producing up to 100 spikes and 1,500 spikelets

Weed Designation: USA: AZ, CA, CO, ID, NM, OR, WA; CAN: federal, AB, BC; MEX: federal

DESCRIPTION

Seed: Oblong, pale yellow to green, 4–5 mm long and 3 mm in diameter; awn variable, 1.5–60 mm long depending on position in the spike; resembling long grains of winter wheat. Viable in soil for 3–5 years. Germination occurs in soil temperatures between 3°C and 35°C (optimum 15°–20°C); emergence from 0–5 cm depth (2–3 cm optimal), although studies have shown emergence from 12.5 cm depth, favoring compacted soil such as wheel tracks; lowest seed in spikelet dormant until following spring, upper seeds not dormant and germinating when conditions allow; seeds grown in high moisture conditions have higher germination rate than those produced in drier sites.

Seedling: With one reddish- to brownish-green cotyledon; ligule membranous, less than 0.5 mm long.

Leaves: Leaves 2–12 cm long, 2–5 mm wide, hairy; sheath open with overlapping hairy margins; ligule membranous, less than 2 mm long, upper margins fringed; auricles sickle-shaped, about 0.5 mm long, with hairs 1–3 mm long. Inflorescence a cylindrical panicle, 5–10 cm long with five to 10 spikelets; spikelets one per node, two- to five-flowered (lower two flowers fertile), set edgewise to the stem; upper spikelets 8–10 mm long; glumes 7–9 mm long with awns 9–18 mm long; lemmas of terminal spikelets 9–11 mm long, five-veined, awn 4–5 cm long; lemmas of lower spikelets sharp-pointed or one-awned, awns 1–9 cm long.

Plant: Tufted, up to 135 tillers from the base, maturing before most cereal crops, susceptible to wheat rust; stems jointed, 20–90 cm tall; roots fibrous; typical plants growing in wheat fields produce about 130 seeds.

Fruit: Caryopsis (grain).

REASONS FOR CONCERN

Jointed goatgrass is host for wheat streak mosaic virus. It is also an overwintering host for winter wheat pests, such as the Russian wheat aphid. Densities of five to eight plants per square foot have been reported to reduce winter wheat yields by 25–32%. In North America, it infests 5 million acres of winter wheat, and 2.5 million acres of fallow land. It is spreading at the rate of 50,000 acres per year. Jointed goatgrass seed harvested with wheat increases dockage and reduces grain value by as much as $1 per bushel.

Similar species: stout goatgrass (*A. crassa* Boiss.), a native of the Middle East, is distinguished by thicker hairy spikes and awnless glumes. Low goatgrass (*A. geniculata* Roth), a native to the eastern Mediterranean region, is distinguished by an oval-shaped spike with two to five spikelets.

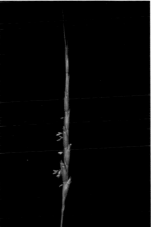

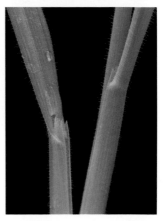

Avena fatua L.
WILD OAT

Also known as: oat grass, poor oats, wheat oats, flaxgrass, drake, haver-corn, hever, black oats

Scientific Synonym(s): *A. hybrida* Peterm. ex Reichenb. p.p.

QUICK ID: ligule present; auricles absent; sheath open, margins transparent

Origin: native to Europe and Asia; introduced as a contaminant in seed; cultivated by settlers in Newfoundland in 1600s

Life Cycle: annual reproducing by seed; plants capable of producing up to 500 seeds (100–150 common)

Weed Designation: CAN: federal, AB, BC, MB, PQ, SK; MEX: federal

DESCRIPTION

Seed: Elliptical, light yellow to black, about 1 cm long; base with numerous brown hairs, the scar often referred to as a suckermouth; awn black to dark brown, 2–5 cm long, twisted and bent at a 90° angle.
Viable in soil for 75 years, although 4–10 years is common.
Germination requires a dormant period; no germination when exposed to light; optimal temperature 15°–21°C; no germination at temperatures below 5°C.

Seedling: With cotyledon twisted counterclockwise; slightly hairy; seed retained on the root of the seedling; growth is slow for the first 2 weeks but soon surpasses the cultivated crop in height.

Leaves: Flat, 20–60 cm long, 4–18 mm wide, twisted counterclockwise; sheath open with transparent edges, slightly hairy; ligule papery, irregularly torn, 2–5 mm long; auricles absent.
Inflorescence an open panicle, pyramidal, and nodding, 10–40 cm long; spikelets 2–2.5 cm long, two- to three-flowered, borne singly at the end of drooping branches; glumes smooth, 18–26 mm long, seven- to 11-veined, fine-lined; lemmas 14–20 mm long, seven- to nine-veined, awns twisted and slightly bent, 25–40 mm long.

Plant: Tufted, three to five tillers from the base; seeds often shed before the cultivated crop is harvested, maturity reached in 60–90 days; stems erect 40–150 cm tall; three to five nodes, dark-colored; roots fibrous, extensive; straw may be allelopathic and may cause damage to sensitive plants.

Fruit: Caryopsis (grain).

REASONS FOR CONCERN

Wild oat, one of the most serious weeds in North America, has the potential to reduce crop yields by 50%. Dockage losses, lower crop grade and quality, and the ability to produce chemicals that inhibit germination of other plants are the main concerns for wild oat populations. It is also host for barley yellow dwarf, lucerne dwarf, ryegrass mosaic, and wheat streak mosaic viruses. Wild oat is a weed in more than 20 crops in 55 countries.

Similar species: cultivated oat (*A. sativa* L.) is distinguished from wild oat by a short straight awn. The seeds of cultivated oat are not shed at maturity.

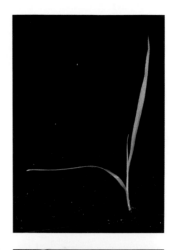

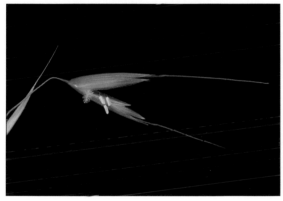

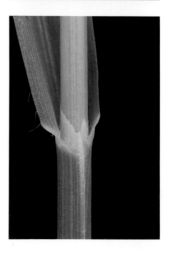

Bromus inermis Leyss.
SMOOTH BROME

Also known as: awnless brome, Hungarian brome, Russian brome, Austrian brome

Scientific Synonym(s): *B. pumpellianus* (Scribn.) Wagon.

QUICK ID: ligule present; auricles absent; sheath closed

Origin: native to Europe; introduced as a forage crop before 1875; native subspecies occur

Life Cycle: perennial (living to 60 years) reproducing by seed and creeping rhizomes; plants capable of producing up to 200 seeds

Weed Designation: none

DESCRIPTION

Seed: Elliptical, pale yellow to dark brown, 9.5–10.6 mm long, 1.9–2.7 mm wide; awn short, less than 3 mm long may be present.
Viable up to 14 years in soil; 19 years in dry storage.
Germination is enhanced by light but light is not required; optimal temperature 20°–30°C, although germination can occur at 1°C.

Seedling: With one cotyledon; first leaves covered in fine silky hairs; base of young stem often pink to pale red; rhizomes begin to form about 3 weeks after germination; at 8 weeks (about fifth leaf stage) tillers appear and rhizomes are about 15 cm long.

Leaves: Flat, 15–40 cm long, 5–15 mm wide, nearly hairless; sheaths closed with a small V-shaped notch, a few hairs present; ligule 1–2 mm long, brownish at the base; auricles absent.
Inflorescence an open panicle, nodding, 5–20 cm long, 4–10 cm wide, one to four branches per node, 1–10 cm long; branches with several spikelets; spikelets 1.5–3 cm long, 3–5 mm wide, purplish-brown, seven- to 13-flowered; glumes 6–11 mm long, prominently one- to five-nerved; lemmas 7–16 mm long, five- to nine-nerved, awns less than 3 mm long.

Plant: Plants 50–150 cm tall; stems erect, hairless; rhizome creeping and extensive, dark-colored, jointed, the internodes covered in large brown to black scaly sheaths; roots and shoots produced at each node.

Fruit: Caryopsis (grain); seven to 10 per spikelet.

REASONS FOR CONCERN

Smooth brome, often planted as a forage crop, persists after cultivation and infests later crops. Smooth brome is host for the following viral diseases: barley stripe mosaic, barley yellow dwarf, brome mosaic, and oat pseudo-rosette.

Similar species: nodding brome (*B. anomalus* Rupr. ex Fourn.), another native brome grass, is distinguished from smooth brome by a drooping panicle and fibrous roots. Rhizomes are not present. It is found in moist grassland areas of western North America.

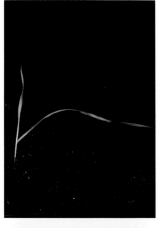

Also known as: downy chess, early chess, cheatgrass, cheatgrass brome, slender brome, drooping brome grass, thatch grass, military grass

Scientific Synonym(s): *Anisantha tectorum* (L.) Nevski

QUICK ID: ligule present; auricles absent; leaves covered in soft hairs

Origin: native to the Mediterranean region of southern Europe; first observed near Denver, Colorado, in the 1890s

Life Cycle: annual or winter annual reproducing by seed; large plants producing up to 5,000 seeds

Weed Designation: USA: CO, CT; CAN: AB, MB, SK

DESCRIPTION

Seed: Elliptical, pale brown with a red tinge, 10.5–12.1 mm long, 1.1–1.5 mm wide; awn straight, 12–17 mm long.
Viable for 4 years in soil; 12 years in dry storage. Germination occurs in the top 2.5 cm of soil, with no emerging from 10 cm depth; optimal temperature 10°–20°C, inhibited by temperatures above 30°C.

Seedling: With one cotyledon; first leaf tall and narrow; early leaves soft-haired, twisted, midrib prominent.

Leaves: Flat, 5–12 cm long, 2–4 mm wide, surface with long soft white hairs; sheaths closed, soft-haired; ligule thin and transparent, 1–5 mm long, irregularly toothed; auricles absent.
Inflorescence a nodding panicle, 7–20 cm long, purplish-green, surfaces with soft white hairs, branches drooping; spikelets 2–4 cm long (including awn), three- to 10-flowered; glumes 6–13 mm long, woolly, one-nerved; lemmas 9–16 mm long, lance-shaped, woolly, awns 10–25 mm long.

Plant: Tufted, flowering in early spring and maturing by midsummer; stems 2.5–60 cm tall, two to five nodes, often rooting at lower nodes, surface with silky hairs; roots fibrous, to 1.5 m deep.

Fruit: Caryopsis (grain); five to 10 per spikelet.

REASONS FOR CONCERN

Downy brome has been reported to reduce wheat yields by 92%. It is an aggressive species that invades pastures and rangeland. Early spring growth is readily eaten by livestock. By midsummer, the sharp spikelets and rough awns may injure eyes and mouths of livestock. Downy brome is host for barley stripe mosaic, barley yellow dwarf, ryegrass mosaic, and wheat streak mosaic viruses.

Similar species: Japanese chess (*B. japonicus* Thunb.), another introduced brome grass, is distinguished from downy brome by the number of nerves on the first glume. Japanese chess has three nerves while downy brome has one. Japanese chess grows from 20 to 60 cm tall and is covered in soft white hairs. Weed designation: CAN: AB, SK.

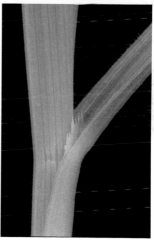

Cenchrus longispinus (Hack.) Fern.
LONGSPINE SANDBUR

Also known as: bur grass, sandbur, sandbur grass, bear grass, hedgehog grass, innocent weed, mat sandbur

Scientific Synonym(s): *C. carolinianus* Walt.

QUICK ID: fruit burlike; ligule of fine hairs; auricles absent

Origin: native to Europe, South Africa, and Australia

Life Cycle: annual reproducing by seed; spikelets may produce 10–30 seeds; early germinating plants may produce up to 1,120 seeds in one season

Weed Designation: USA: CA, WA

DESCRIPTION

Seed: Oval, reddish-brown, 2.2–3.8 mm long, 1.5–2.6 mm wide; one to four per bur.
Viable in soil for up to 4 years.
Germination is delayed by the presence of light; optimal depth is 1–3 cm, with some seedlings emerging from 25 cm.

Seedling: With one cotyledon; leaves folded in bud; blade rough, sheath smooth.

Leaves: Flat, 63–187 mm long, 3–8 mm wide, dark green; sheaths inflated (loose), flat, and keeled, margins hairy; tuft of short and long hairs at the collar; ligule a ring of fine hairs, 0.7–1.7 mm long; auricles absent.
Inflorescence a compact panicle with six to 15 spikelets, 4.1–10.2 cm long, 1.2–21.2 cm wide; rachis internodes 2–5 mm long; one spikelet, stalkless, two- to three-flowered; glumes shorter than lemmas, first glume 1.5–3.8 mm long, one-nerved, second glume 4–16 mm long, three- to five-nerved; lemma 5–7.6 mm long, five- to seven-nerved.

Plant: Tufted, 10–90 cm tall, often rooting at lower nodes; tillers appear in a radial or pinwheel arrangement from the crown.

Fruit: Caryopsis (grain); borne in globe-shaped burs, 8.3–11.9 mm long, 3.5–6 mm wide, surface with spines 3.5–7 mm long and 1 mm wide, usually 40 or more per bur; mature burs bronze or purple.

REASONS FOR CONCERN

The burs are spread by attaching themselves to tires, clothing, and livestock fur. The burs may also cause injury to sheep by becoming entangled in wool. Large populations of this species also hinders fruit and vegetable picking.

Similar species: few-flowered sandbur (*C. spinifex* Cav.), native to Mexico, has upright stems and leaves 1.5–4 mm wide. The spike, 1–5 cm long and 1–1.5 cm wide, produces burs 8–12 mm wide with 12 or more spines.

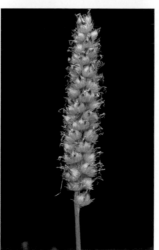

few-flowered sandbur
(*Cenchrus spinifex*)

Cynodon dactylon (L.) Pers.
BERMUDA GRASS

Also known as: devil grass, quick grass, Bahama grass, couch grass, dog's tooth grass, Indian doob, scotch grass

Scientific Synonym(s): *Capriola dactylon* (L.) Kuntze, *Panicum dactylon* L.

QUICK ID: stems prostrate; inflorescence a whorl of fingerlike spikes; ligule a ring of white hairs

Origin: native to tropical Africa, India, and Malaysia; introduced as a pasture or lawn grass; once used in prevention of soil erosion

Life Cycle: Perennial reproducing by seed, rhizomes and stolons;

Weed Designation: USA: AR, CA, UT; MEX: federal

DESCRIPTION

Seed: Elliptical (one side nearly straight, the other somewhat rounded), straw-colored to orange-red, 1.0–2.2 mm long, 0.8–1 mm wide.
Viable in soil for up to 4 years.
Germination is best at soil temperatures about 38°C.

Seedling: With one cotyledon; leaves rolled in bud; ligule present; blades smooth, short, and narrow; sheaths green, smooth, and hairless.

Leaves: Flat, 2–16 cm long, 3–5 mm wide, grayish-green, slightly hairy to smooth; fringe of hair just above the collar; sheath two opposite per node, up to 15 mm long, flattened, slightly hairy to smooth; ligule a ring of white hairs, less than 0.5 mm long.
Inflorescence a panicle 2–10 cm long, with 3–7 fingerlike spikes, 1.5–10 cm long; spikelets purplish, borne in two rows, one-sided; glumes 0.5–1 mm long, lemma 2–2.5 mm long, three- to five-veined.

Plant: Prostrate, often forming dense mats; stems 5–45 cm long, wiry, often reddish; stolons flat, hairless, bearing a bladeless sheath at each joint, readily rooting; rhizomes hard, scaly, and pointed, 1.5–3.3 mm in diameter.

Fruit: Caryopsis (grain).

REASONS FOR CONCERN

Bermuda grass is often referred to as one of the most serious weeds in the grass family. It is reported to be a weed in corn, cotton, sugarcane, vineyards, and plantation crops in 80 countries.

Similar species: African bermudagrass (*C. transvaalensis* Burtt Davy), native to South Africa, is distinguished by an inflorescence with one to three branches each 1–2 cm long. The leaves, 10–30 cm long and 1–2 mm wide, are prominently three-veined. Another introduced species, stargrass (*C. plectostachyus* [Schumann] Englem.), has an inflorescence with two to seven whorls, and stiff leaves, 10–30 cm long. The transparent ligule is about 1 mm long.

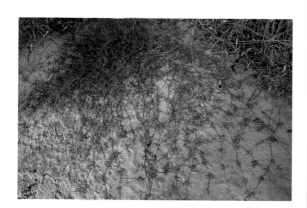

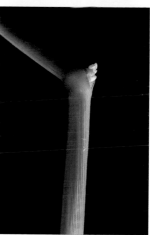

Dactylis glomerata L.
ORCHARD GRASS

Also known as: cocksfoot, barnyard grass, cockspur

Scientific Synonym(s):
D. hispanica Roth

QUICK ID: ligule membranous, 5–7 mm long; auricles absent; spikelets borne on one-sided clusters

Origin: perennial introduced from Europe and North Africa

Life Cycle: perennial reproducing by seed

Weed Designation: PQ

DESCRIPTION

Seed: Elliptical to lance-shaped, tan to light brown, 3–8 mm long.
Viable in soil for up to 14 years.
Germination occurs in the top 7.5 cm of soil; optimal temperature 20°–30°C.

Seedling: With one cotyledon; first blade perpendicular to ground, blade 7–12 cm long, 13–23 mm wide, light green; sheaths flattened, strongly keeled, hairless, whitish at base; ligule membranous, 3–5 mm long; auricles absent.

Leaves: Flat to strongly keeled (V-shaped), 7–30 cm long, 2–14 mm wide, bluish-green, surface and margins rough; sheath hairless, flattened, partly opened; ligule 5–7 mm long, membranous; auricles absent.
Inflorescence a panicle 5–25 cm long; spikelets 5–7 mm long, borne on one-sided clusters, three- to six-flowered, flat; first glume 5–6 mm long; second glume 6–7 mm long, sharp-pointed and keeled, bristly haired; lemmas 5–8 mm long, five-nerved, awns 0.5–1.5 mm long.

Plant: Tufted (clump-forming), requires 120 days frost-free to reach maturity; stems coarse, 60–150 cm tall; roots primarily in to 8 cm of soil, but may extend to 46 cm depth.

Fruit: Caryopsis (grain).

REASONS FOR CONCERN

Orchard grass has been associated with barley yellow dwarf virus, cocksfoot streak, and ryegrass mosaic virus.

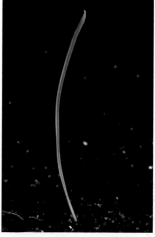

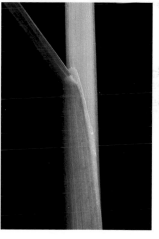

Dactyloctenium aegyptium (L.) Willd.
CROWSFOOT GRASS

Also known as: Durban crowfoot grass, Egyptian grass

Scientific Synonym(s): *Cynosurus aegyptius* L.

QUICK ID: flower cluster of two to seven fingerlike spikes; ligule membranous; seeds orangish-brown

Origin: annual or perennial native to Old World tropics

Life Cycle: annual or perennial reproducing by seed; large plants producing up to 66,000 seeds

Weed Designation: none

DESCRIPTION

Seed: Orangish-brown, 0.8–1.1 mm long. Viability up to 19 years in laboratory storage. Germination greatest at 20°C; no germination at 10°C and less than 10% at temperatures above 30°C.

Seedling: With one cotyledon; leaves and sheaths hairless, margins with long stiff hairs; ligule membranous and fringed.

Leaves: Flat, 3–25 cm long, 2.5–7.5 mm wide, surface hairless or sparsely haired, the hairs with swollen bases; sheaths compressed, hairless; ligule membranous, margin jagged and ending in hairs; auricles absent.
Inflorescence a panicle with two to nine fingerlike spikes; spikes 1–3 cm long, 6–8 mm wide; spikelets 2.5–3 mm long; glumes 1.5–2.2 mm long, one-veined, awns 0.8–4 mm long; lemmas 2.6–4 mm long, three-veined.

Plant: Spreading to ascending, often rooting at nodes and forming mats, prefers light soils with low moisture; stems flattened, branched, 20–40 cm tall; stolons present.

Fruit: Caryopsis (grain).

REASONS FOR CONCERN

Crowsfoot grass is a principal weed of cotton and peanuts. It is reported to be a serious weed in 45 countries. It is found in sugarcane, corn, coffee, sweet potatoes, and millet.

Digitaria sanguinalis (L.) Scop.
LARGE CRABGRASS

Also known as: purple
crabgrass, fingergrass,
Polish millet, crowfoot
grass, pigeongrass, hairy
fingergrass, redhair grass
Scientific Synonym(s):
Syntherisma sanguinalis
(L.) Dulac, *Panicum
sanguinale* L.

QUICK ID: panicle of three
to 13 fingerlike terminal
spikes; auricles absent;
ligules present
Origin: native to Europe;
introduced as a forage
crop in 1849
Life Cycle: annual
reproducing by seed;
large plants producing up
to 150,000 seeds
Weed Designation: CAN:
federal, PQ

DESCRIPTION
Seed: Elliptical to lance-shaped, dull olive to brown,
2.7–3.0 mm long, 0.8–0.9 mm wide.
Viable for at least 3 years in soil.
Germination requires light; optimal
temperatures of 20°–35°C.
Seedling: With one cotyledon; first leaf short and
wide, tip blunt.
Leaves: Flat, 5–20 cm long, 4–10 mm wide; sheath
with overlapping hairy margins; numerous long
white hairs found at the junction of the blade
and sheath; ligule membranous, less than 2 mm
long; auricles absent.
Inflorescence a panicle of three to 13 fingerlike
spikes, each 5–18 cm long, whorled at the top
of the stem; spikelets about 3 mm long, one-
flowered, borne on one side of the stem; first
glume minute or absent; second glume 0.8–1.5
mm long, three-veined; lemmas 2.5–3.3 mm long,
five-nerved, pale brown.
Plant: Continuing to grow until killed by frost; may
produce up to 700 tillers from the base; stems
ascending to prostrate, 30–120 cm long, three to
eight nodes, often rooting at lower nodes; roots
fibrous, extending to a depth of 2 m.
Fruit: Caryopsis (grain).

REASONS FOR CONCERN

Large crabgrass is a serious weed in row crops, cultivated fields, and lawns. It is host for barley stripe, wheat streak mosaic, sugarcane mosaic, and lucerne dwarf viruses. Large crabgrass is found in 33 crops in 56 countries.

Similar species: small crabgrass (*D. ischaemum* Schreb. ex Muhl.) is closely related to large crabgrass. Small crabgrass has hairless leaves with a few long hairs at the junction of the blade and the sheath. The membranous ligule is 2–3 mm long. Small crabgrass has two to six fingerlike terminal spikes. Weed designation: CAN: federal, PQ.

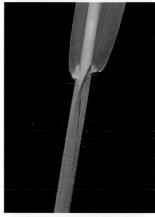

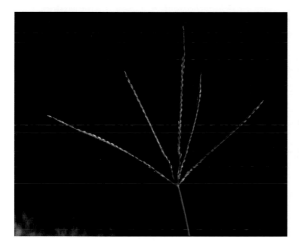

small crabgrass
(*Digitaria ischaemum*)

Echinochloa crus-galli (L.) Beauv.
BARNYARD GRASS

Also known as: cockspur grass, Japanese millet, cocksfoot panicum, barn grass, water grass, summer grass, billion-dollar grass

Scientific Synonym(s): *E. pungens* (Poir.) Rydb., *Panicum crus-galli* L.

QUICK ID: auricles and ligules absent; sheath open; stems reddish at base

Origin: native to Europe and India; first reported in California in 1825, Nova Scotia in 1829

Life Cycle: annual reproducing by seed; plants capable of producing up to 40,000 seeds

Weed Designation: USA: AR; CAN: MB, PQ; MEX: federal

DESCRIPTION
Seed: Elliptical (rounded on one side and flat on the other), whitish to grayish-brown, 3.5–4.2 mm long, 1.6–2.0 mm wide, surface with numerous short stiff hairs; awn 1–2 cm long.
Viable in storage for up to 9 years; commonly 6–30 months in soil.
Germination is shallow, as soil temperatures must be over 30°C for seedlings to emerge; optimal temperatures range from 20°C to 30°C; very little germination below 10°C; dormant up to 48 months.

Seedling: With one cotyledon, tip pointed, hairless, slightly red at base; stem somewhat flattened up until the third leaf stage; first tillers appear about 10 days after germination.

Leaves: Flat or V-shaped, 5–50 cm long, 0.5–2 cm wide, hairless; sheaths open, flattened, margins overlapping, slightly hairy; ligules and auricles absent.
Inflorescence a panicle 5–20 cm long; racemes greenish-red or purplish, 2–10 cm long; spikelets numerous, 3–4 mm long, borne in two to four rows; glumes 1–1.6 mm long, prominently three-nerved; lemmas 3–4 mm long, prominently five- to seven-nerved, awns 0–5 cm long.

Plant: Producing about 15 tillers; stems erect to prostrate, 20–150 cm long, forming large mats, base reddish-purple, lower nodes often rooting when in contact with soil; roots fibrous.

Fruit: Caryopsis (grain).

REASONS FOR CONCERN

Barnyard grass is a common weed of gardens and shelterbelts. Ranked third among the world's worst weeds, it has the potential to be a serious weed in row crops such as beets, potatoes, and corn. Barnyard grass may consume 60–80% of available soil nitrogen in one growing season. It also accumulates high levels of nitrates in plant tissues, which may be toxic to livestock. Barnyard grass is host for maize dwarf disease, barley stripe mosaic, wheat streak mosaic, and sugarcane mosaic viruses.

Similar species: awnless barnyard grass or jungle rice (*E. colonum* [L.] Link), native to India, is a serious weed of rice, cotton, corn, sorghum, and sugarcane. It is distinguished from barnyard grass by leaves 5–30 cm long and 2–6 mm wide. Racemes, 0.5–3 cm long, are borne in panicles less than 15 cm long.

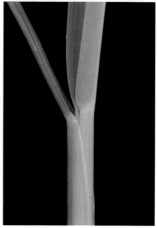

Eleusine indica (L.) Gaertn.
GOOSEGRASS

Also known as: silver
crabgrass, wire grass, yard
grass, crowfoot grass,
crows foot grass, bull
grass, Indian goosegrass,
dog's-tail
Scientific Synonym(s):
Cynosurus indicus L.

QUICK ID: leaf sheaths
keeled, hairy on margin;
flower cluster with a
windmill-like appearance
Origin: native to China, India,
Japan, Malaysia, or Tahiti
Life Cycle: annual
reproducing by seed;
large plants producing up
to 135,000 seeds; life cycle
completed in 120–180
days.
Weed Designation: none

DESCRIPTION
Seed: Elliptical, reddish-brown to black, 1–1.3 mm
long, 0.4–0.5 mm wide, surface with about 20
ridges
Viable in soil for up to 2 years.
Light increases germination. Germinates when
soil temperatures reach 18°C; 2–3 weeks later
than crabgrass.
Seedling: With one cotyledon; first blade about
4.5 cm long and 3–5 mm wide, hairless; sheaths
flattened, hairless, light green to white at base;
ligule less than 1 mm long, unevenly toothed;
auricles absent.
Leaves: Mostly basal, folded along midvein, 5–30
cm long, 3–8 mm wide; surfaces smooth to
sparingly hairy; margins rough; sheath open,
margins overlapping with a few hairs, whitish at
base; ligule membranous, 1–2 mm long, unevenly
toothed; auricles absent.
Inflorescence a panicle with two to 10 spikes at
top of stem; spikes 2.5–15 cm long, 3–3.5 mm
wide, fingerlike; spikelets 3–5 mm long, flattened,
borne in two rows, three- to nine-flowered;
glumes 1–3 mm long, one-veined; lemmas 2–3.6
mm long, three-nerved.
Plant: Prostrate and forming mats; stems 5–90 cm
tall, tufted, branched at base, flattened, hairless;
roots tough and fibrous, readily invading hard
compacted soils commonly found in high traffic
areas.
Fruit: Caryopsis (grain).

REASONS FOR CONCERN

A common weed of turf, nurseries, gardens, and roadsides, goosegrass tolerates drought conditions and close mowing. Mature leaves are extremely difficult to cut with a mower, leaving them frayed at the tip. It is a serious weed of corn, rice, sweet potatoes, sugarcane, and cotton. It is reported as a serious weed in 60 countries.

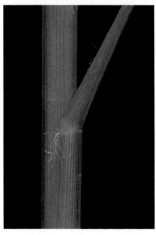

Elymus repens (L.) Gould
QUACK GRASS

Also known as: couch grass, twitch grass, quitch grass, scutch grass, quick grass, shelly grass, knotgrass, devil's grass, witchgrass, bluejoint, pondgrass, Colorado bluegrass, false wheat, dog grass, seargrass, quickens, stroil, wickens

Scientific Synonym(s): *Elytrigia repens* (L.) Nevski, *Elytrigia vaillantiana* (Wulfen & Schreb.) Beetle, *Agropyron repens* (L.) Beauv., *Triticum repens* L., *T. vaillantianum* Wulfen & Schreb.

QUICK ID: extensive creeping rhizome; auricles and ligules present; sheath open with overlapping margins

Origin: native to Europe; introduced as a contaminant in hay or straw; first report in North America in 1672

Life Cycle: perennial reproducing by rhizome and seeds; each stem producing up to 400 seeds, although 20–40 is common; rhizomes may spread up to 3 m per year, giving rise to 206 new shoots

Weed Designation: USA: AK, AZ, CA, CO, HI, IA, KS, OR, UT, WY; CAN: federal, BC, MB, PQ, SK

DESCRIPTION

Seed: Elliptical, pale yellow to brown, about 10 mm long, less than 1.7 mm wide; awns 1–10 mm long. Viability about 2% after 10 years in soil. Germination occurs at alternating temperatures between 15°C and 25°C; emergence from top 5 cm of soil, although 10 cm has been reported.

Seedling: With one cotyledon; first leaves slightly hairy, bright green, lower park of stem often pinkish-brown; first roots fibrous; rhizomes begin to form at the sixth to eighth leaf stage; tiller produced earlier at the to fourth to sixth leaf stage.

Leaves: Flat, 6–20 cm long, 4–10 mm wide; surface sparsely hairy above and hairless below; sheath open, margins overlapping with soft hairs; ligule less than 1 mm long, papery; auricles about 3 mm long, pointed.

Inflorescence a spike 5–25 cm long; spikelets numerous, 10–20 mm long, 3–6 mm wide, two- to nine-flowered, set edgewise to the stem, short-awned; glumes 7–12 mm long, prominently three- to seven-nerved; lemmas 8–13 mm long, five-veined, awns 0–10 mm long.

Plant: Forming clumps; stems 30–120 cm tall, three to five nodes; rhizomes extensive and creeping, yellowish-white, about 3 mm thick, up to 150 cm long, 5–20 cm deep, nodes with several brown scaly sheaths; root fibrous, appearing at each node.

Fruit: Caryopsis (grain); 15–400 per stem (commonly 25–40).

REASONS FOR CONCERN

In severely infested fields, quack grass rhizomes could weigh as much as 7–9 metric tons/acre. Rhizomes produce chemicals that suppress the growth of other plants and adversely affects the yield of all cultivated crops. It is often a contaminant of cultivated seed, which results in grade reduction. Over 40 countries report quack grass as a serious weed in 32 crops.

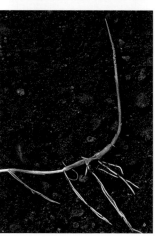

Hordeum jubatum L.
FOXTAIL BARLEY

Also known as: wild barley, foxtail, squirreltail barley, skunktail grass, flickertail grass, tickle grass, foxtail grass

Scientific Synonym(s): *Critesion jubatum* (L.) Nevski, *H. caespitosum* Scribn. ex Pammel

QUICK ID: bluish-green bunchgrass; ligule present; auricles absent

Origin: native to western North America

Life Cycle: perennial reproducing by seed; plants capable of producing over 180 seeds

Weed Designation: CAN: MB, PQ

DESCRIPTION

Seed: Elliptical, yellowish-brown, 3–4 mm long, surface with sharp backward-pointing barbs; four to eight awns, 1.5–6 cm long.
Viable in soil for up to 7 years.
Germination requires extended hours of darkness, which usually occurs in late August or early September; cannot emerge from depths of 7.5 cm or deeper.

Seedling: With one cotyledon; first leaf tall, narrow, bluish-green due to presence of short white hairs; sheaths open; seedlings capable of tolerating soils with 1% salt content.

Leaves: Flat or V-shaped, 5–15 cm long, 2–9 mm wide, grayish-green, surface with a sandpapery texture; sheaths split, margins overlapping with numerous soft hairs; ligule about 1 mm long, transparent; auricles very small or absent. Inflorescence a nodding raceme 5–12 cm long; three spikelets per node (one fertile, two sterile); fertile spikelets 4–7 mm long, glumes less than 7 mm long, awns 3.5–8 cm long, lemmas 4–7 mm long, five-veined, awns 3.5–8 cm long; sterile spikelets with glume awns 3.5–8 cm long, lemma absent or, if present, with awns 3.5–8 cm long; all awns purplish-green fading to straw-colored; spike breaking into several pieces at maturity.

Plant: Bunchgrass; stems 20–100 cm tall, bluish-green, brown nodes often swollen; roots fibrous.

Fruit: Caryopsis (grain).

REASONS FOR CONCERN

Foxtail barley is palatable to livestock in early summer, prior to flowering. In late summer, the sharp pointed awns may cause damage to the mouth, eyes, and skin of livestock that feed on this plant. Foxtail barley is host for wheat rust and barley stripe mosaic viruses.

Similar species: cultivated barley (*H. vulgare* L.) is distinguished from foxtail barley by shorter awns, 6–15 mm long, and the presence of auricles.

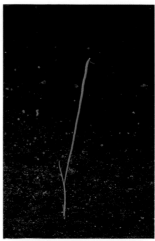

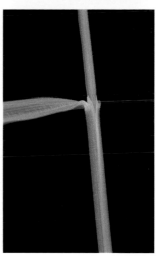

Lolium persicum Boiss. & Hohen. ex Boiss.

PERSIAN DARNEL

Also known as: darnel,
Persian ryegrass
Scientific Synonym(s):
L. dorei Boivin

QUICK ID: auricles and
ligules present; sheaths
closed; leaf surface shiny
Origin: native to
southwestern Asia; first
report from Canada in
1923
Life Cycle: annual
reproducing by seed
Weed Designation: CAN:
federal, MB, SK

DESCRIPTION

Seed: Light brown, 8–10 mm long, with a slightly
bent awn, 5–12 mm long.
Viability less than 2% after 3 years on the soil
surface.
Germination occurs with temperatures between
5°C and 26°C; less than 10% of seedlings emerge
from more than 4 cm depth; no emergence from
7.5 cm depth.

Seedling: With one cotyledon; first leaf long, narrow,
and folded, dark green, upper surface shiny; base
of first leaves often reddish-brown or purple;
auricles appearing by fifth leaf stage.

Leaves: Flat, twisted, 5–15 cm long, 2–6 mm wide;
upper surface rough; sheaths open, round,
margins overlapping and hairless, prominently
veined; ligules 1–2 mm long, papery; auricles
present.
Inflorescence a panicle 3–10 cm long; spikelets
10–20 mm long, 1.5–7 mm wide, five- to nine-
flowered, set edgewise to the stem; first glume
absent; second glume 7.5–23 mm long, five- to
nine-nerved; lemmas 9–10 mm long, five-nerved,
awns 5–15 mm long.

Plant: Similar in color to wheat; stems bright green,
three to four nodes, often branching from the
lower nodes, 15–75 cm tall; base of stem usually
reddish-purple; roots fibrous.

Fruit: Caryopsis (grain).

REASONS FOR CONCERN

Large populations of Persian darnel have been reported to cause significant decreases in crop yield. A problem pest in wheat, the seeds of Persian darnel lower the quality and grade of the crop. Seeds are similar in size to wheat grains, making it difficult to separate.

Similar species: perennial ryegrass (*L. perenne* L.) is an introduced grass cultivated in pastures, hayfields, and lawns. Stems are usually 20–80 cm tall. The panicle, 5–25 cm long, is composed of five to 10 flowered spikelets. Another species, poison darnel (*L. temulentum* L.), is an annual grass with stems to 1 m tall. It is distinguished from Persian darnel by its wider leaves (3–10 mm wide). Weed designation: USA: AR, SC.

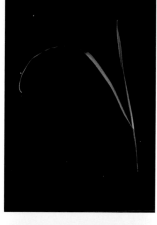

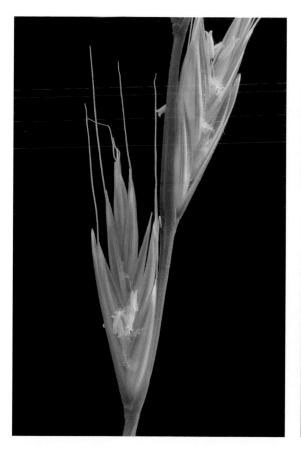

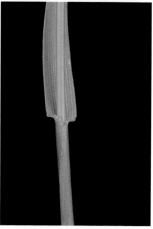

Nassella trichotoma Hackel ex Arech.
SERRATED TUSSOCK

Also known as: Yass River tussock, nassella tussock, nassella polgrass

Scientific Synonym(s): *Stipa trichotoma* Nees, *Urachne trichotoma* (Nees) Trin.

QUICK ID: panicle open, branches threadlike; ligule membranous; plants forming clumps

Origin: native to Argentina or Uruguay

Life Cycle: perennial (living up to 20 years) reproducing by seed; plants may produce up to 140,000 seeds; serious weed in Australia with over 1 million acres infested; accidental introduction there occurred in the early 1900s

Weed Designation: USA: federal, AL, AZ, AR, CA, FL, HI, MA, MN, NC, OR, SC, TX, VT; CAN: federal; MEX: federal

DESCRIPTION

Seed: Brown, about 1.5 mm long and 1 mm wide; awn about 25 mm long.

Viable in soil for up to 20 years; may pass through livestock digestive tracts unharmed.

Germination within 5 days at 20°C; optimal depth 5 mm.

Seedling: With one cotyledon; first leaves less than 3 cm tall and 1 mm wide.

Leaves: Threadlike, 15–45 cm long and less than 0.5 mm wide, leathery; ligule membranous. Inflorescence an open panicle, 8–20 cm long, branches threadlike, two to three per node; spikelets solitary, 4–8.5 mm long, one-flowered; glumes 4–8.5 mm long, three-veined; lemmas 1.5–2.5 mm long, awns 15–32 mm long and often twisted; panicles breaking off at maturity and forming tumbleweeds.

Plant: Forming clumps up to 25 cm wide at the base, drooping leaves extending the plant diameter to 75 cm; stems 20–50 cm tall; roots deep, fibrous.

Fruit: Caryopsis (grain); seeds light and easily wind-dispersed up to 20 km.

REASONS FOR CONCERN

Serrated tussock has a low protein and high fiber content, making it indigestible. Undigested balls of fiber form in the intestines of livestock and often lead to starvation and death. Serrated tussock infestations often go unnoticed as plants resemble native species of bluegrass (*Poa* L.), needlegrass (*Stipa* L.), and wild oatgrass (*Danthonia* Lam. & DC.).

Panicum capillare L.
ANNUAL WITCHGRASS

Also known as: old witchgrass, tickle grass, witches hair, tumbleweed grass, fool hay, tumble panic grass

Scientific Synonym(s): *P. barbipulvinatum* Nash

QUICK ID: ligule present; auricles absent; sheath open with hairy overlapping margins

Origin: native to North America

Life Cycle: annual reproducing by seed

Weed Designation: CAN: federal, PQ

DESCRIPTION

Seed: Elliptical, dull brown to dark gray, about 1.3 mm long and 0.8 mm wide.
Viable in soil for up to 47 years.
Germination occurs in the top 2.5 cm of soil; no emergence occurs from depths greater than 5 cm.

Seedling: With one cotyledon; first leaf 1–1.5 cm long, 3–4 mm wide; young stem often bent.

Leaves: Flat, 10–25 cm long, 5–15 mm wide, with a prominent white midrib; both surfaces densely haired; reduced in size upward; sheath open, strongly ribbed, margins overlapping with hairs 2–3 mm long; ligule a fringe of hairs, 1–2 mm long; auricles absent.
Inflorescence an open panicle, 15–50 cm long, profusely branched, breaking off at maturity and becoming a tumbleweed; spikelets 2–3 mm long; first glume 1–2 mm long, three-veined; second glume 2–3 mm long, five- to seven-veined; lemmas 2–3 mm long, five- to seven-nerved.

Plant: Plants 20–90 cm tall, becoming stiff and bristly with age; stems simple or branched; nodes bearded; roots fibrous.

Fruit: Caryopsis (grain.)

REASONS FOR CONCERN

A problem pest of cultivated fields, gardens, roadsides, and waste areas, witchgrass does not compete with other grasses in well-maintained pastures. Livestock that ingest large amounts of witchgrass are prone to photosensitization and nitrate poisoning. Witchgrass is host for the following viral diseases: barley stripe mosaic, barley yellow dwarf, wheat streak mosaic, and wheat striate mosaic.

Similar species: fall panicum (*Panicum dichotomiflorum* Michx.) is closely related to witchgrass. It is distinguished from witchgrass by the zigzag appearance of the stem, the absence of hairs on the sheath, and a coarser panicle that is 15–50 cm long. Weed designation: CAN: PQ.

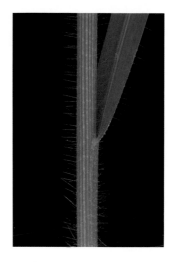

fall panicum
(*Panicum dichotomiflorum*)

Panicum miliaceum L.

WILD PROSO MILLET

Also known as: proso millet, broomcorn millet, blackseeded proso millet, hog millet, panic millet, wild millet, birdseed millet

Scientific Synonym(s): *Milium panicum* Mill.

QUICK ID: ligule a fringe of hairs; auricles absent; panicle arched and nodding

Origin: native to Europe; introduced into Canada as a cultivated crop in the 1700s; common constituent of birdseed; cultivated crop for at least 2000 years.

Life Cycle: annual reproducing by seed; heavy infestations producing up to 45,000 seeds per square meter

Weed Designation: USA: CO, OR; CAN: federal, ON

DESCRIPTION

Seed: Elliptical, pale orange to black, 2.9–3.2 mm long, 2–2.2 mm wide; surface shiny, faintly nerved.
Viable for up to 4 years in soil.
Germination occurs in the top 2.5 cm of soil; emergence from 7.5–13 cm depth depends on soil type.

Seedling: With one cotyledon, hairy, resembling young corn plants; seed remaining attached to the primary root.

Leaves: Flat, 10–30 cm long, 5–25 mm wide, surfaces smooth to somewhat hairy; sheath open, covered in stiff hairs, margins overlapped; ligule a fringe of hairs 2–5 mm long (fused at the base); auricles absent.
Inflorescence an open panicle, 8–30 cm long, erect, arched, or nodding; spikelets 4–5.5 mm long, two-flowered (one fertile); first glume 2–4 mm long, distinctly five-nerved, second seven- to nine-nerved; lemma 4–5.5 mm long, seven- to 11-nerved.

Plant: Plant 30–100 cm tall; stems coarse, erect to reclining; nodes hairless to bearded; roots fibrous, shallow.

Fruit: Caryopsis (grain).

REASONS FOR CONCERN

Wild proso millet is a serious weed of cultivated crops, especially corn. It is also common in forage crops, gardens, and farmyards. Wild proso millet is host for barley stripe mosaic, oat pseudo-rosette, panicum mosaic, rice dwarf mosaic, and wheat streak mosaic.

Similar species: wild proso millet is closely related to witchgrass (*P. capillare* L., p. 576). Witchgrass is distinguished by its erect panicle and smaller seeds. Fall panicum (*P. dichotomiflorum* [L.] Michx.) is distinguished from wild proso millet by its hairless leaf sheaths and larger spikelets.

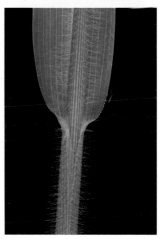

YELLOW FOXTAIL

Also known as: summer grass, golden foxtail, wild millet, pigeongrass, glaucous bristly foxtail, yellow bristle grass, pearl millet

Scientific Synonym(s): *P. americanum* (L.) Leeke, *P. typhoides* auct. non (Burm.) Stapf & C. E. Hubbard, *Setaria lutescens* (Weigel.) F. T. Hubbard, *S. glauca* (L.) Beauv., *Chaetochloa glauca* (L.) Scribn., *C. lutescens* (Weigel.) Stuntz, *Panicum americanum* L., *Panicum glaucum* L.

QUICK ID: ligule composed of a fringe of hairs; auricles absent; sheaths open with overlapping margins

Origin: native to Europe or Asia; first report in Canada from Montreal in 1821, spreading to British Columbia by 1889

Life Cycle: annual reproducing by seed; plants capable of producing over 8,400 seeds; each flower cluster producing up to 180 seeds

Weed Designation: USA: AL, CA, FL, MA, MN, NC, OR, SC, VT; CAN: MB, PQ

DESCRIPTION

Seed: Elliptical, dark grayish-brown, 2.3–2.5 mm long, 1.5–1.6 mm wide; surface dull, but somewhat shiny under low magnification. Viability in soil about 3% after 13 years. Germination does not require light; optimal depth is 1–2 cm; emergence from 10 cm depends on soil type; optimal temperature is 20°–35°C.

Seedling: With one cotyledon; early leaves arched, hairless; sheath-ridged, somewhat flattened; stem base often reddish.

Leaves: Flat or V-shaped, to 30 cm long, 0.5–1.2 cm wide, loosely twisted; surfaces hairless except for long silky white hairs (3–10 mm long) at the base; sheaths open, hairless, flattened by the third leaf stage, distinctly nerved, margins overlapping; ligule a fringe of hairs about 3 mm long; auricles absent.

Inflorescence a cylindrical panicle, 5–10 cm long, about 1 cm wide; spikelets 2–3 mm long, one-flowered, five to 20 bristles, yellowish-brown, 4–9 mm long; glumes and lemma 1–3 mm long.

Plant: Plants 40–130 cm tall; stems erect, solitary or tufted, often reddish at the base; nodes bearded.

Fruit: Caryopsis (grain).

REASONS FOR CONCERN

Yellow foxtail is a highly competitive weed that can drastically reduce crop yield and increase seed cleaning costs. Yellow foxtail is host for ergot, browning root rot, downy mildew, and smut.

Similar species: green foxtail (*S. viridis* [L.] Beauv., p. 590), another introduced annual grass, is distinguished from yellow foxtail by the presence of one to three greenish-purple bristles below each spikelet. Spikelets, 1.5–2 mm long, are smaller than those of yellow foxtail. Green foxtail does not have long white hairs at the base of the leaf blade.

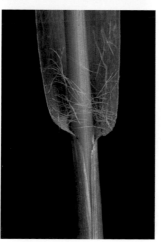

Phleum pratense L.
TIMOTHY

Also known as: herd's grass
Scientific Synonym(s):
   P. nodosum L.

QUICK ID: panicle spikelike; ligule 2–4 mm long; plants with a white bulbous base
Origin: native to Europe; introduced as a forage crop in New Hampshire prior to 1711
Life Cycle: perennial (living 4–7 years) reproducing by seed
Weed Designation: USA: NJ

DESCRIPTION
Seed: Dull to light brown, oval to oblong, 1.2–1.5 mm long.
   Viable for 4–5 years in soil.
   Germination occurs 3–4 weeks after seeds shed; optimal temperature range 5°–19°C.
Seedling: With one cotyledon; leaves rolled in bud; ligule present, 2–4 mm long, toothed; sheath fused.
Leaves: Flat, 10–30 cm long, 5–8 mm wide, upper blade shorter than others, erect with a clasping sheath; ligule membranous, 1–6 mm long. Inflorescence a dense cylindrical panicle, spikelike, 4–15 cm long, 5–20 mm across, spikelets numerous, 3–4 mm long, one-flowered, flattened, stiff-haired; glumes 3–3.8 mm long, three-veined, awn 1–2 mm long; lemmas 1.4–2.1 mm long, papery, five- to seven-veined.
Plant: With a white swollen bulblike base (formed in early summer), loose to densely tufted, intolerant of drought, tolerant of winter temperatures to –40°C, requires 90 days frost-free for seed production; stems 40–150 cm tall, smooth, three to six nodes; roots fibrous, shallow, extending to 120 cm deep.
Fruit: Caryopsis (grain).

REASONS FOR CONCERN

Once planted as a forage crop, timothy has now established itself along roadsides, pastures, and natural areas. The pollen of this species is suspected of being allelopathic to other grass species.

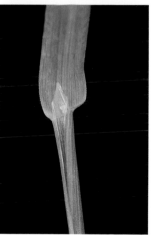

Phragmites australis (Cav.) Trin. ex Streud. ssp. *australis*
REED

Also known as: giant reed, common reed

Scientific Synonym(s):
P. communis Trin.,
P. phragmites (L.) Karst.,
Arundo phragmites L.

QUICK ID: plants reedlike, somewhat woody; panicle large; ligule membranous and fringed with hairs

Origin: introduced from Europe and Asia; native populations in North America referred to ssp. *americanus*

Life Cycle: perennial reproducing by seed, stolons, and creeping rhizomes; a single panicle may produce up to 2,000 seeds

Weed Designation: USA: AL, CT, MA, SC, VT, WA; Can: BC

DESCRIPTION

Seed: Dark brown, elliptical, 1.2–1.5 mm long; awn 6–7 mm long; rarely produced.
Viability less than 3 years.
Germination is inhibited by water depths more than 5 cm; occurs within 25 days at 16°–25°C.

Seedling: With one cotyledon; rarely produced in nature.

Leaves: Flat, 15–60 cm long, 1–6 cm wide; sheaths loose and overlapping, margins with minute soft hairs; ligule about 0.2 mm long, a yellowish-purple membrane topped with short white hairs (about 0.5 mm long) and firm white hairs (5–15 mm long), hairs falling off early; auricles absent, the yellowish-green collar resembling auricle-like shoulders.
Inflorescence a large terminal panicle, tawny to purple, 15–40 cm long, 6–15 cm wide, profusely branched, silky-haired; spikelets 10–17 mm long, three- to seven-seeded; first glume 2.5–7 mm long; second glume 5–12 mm long, three- to five-veined; lemma 8–15 mm long, three-nerved.

Plant: Reedlike, somewhat woody, 1.5–6 m tall, up to 10 mm in diameter; stems purplish-red near the base, growing 4–15 cm a day in early summer; rhizomes creeping, up to 2 m deep and 3 cm thick, growing about 40 cm per year; stolons spreading across the soil surface. Subspecies *americanus* has red internodes while ssp. *australis* has yellow to yellowish-brown internodes.

Fruit: Caryopsis (grain); rarely produced.

REASONS FOR CONCERN

The expansion of reed along the Gulf of Mexico and Atlantic coasts has been a concern for the past 30 years. Expansion of populations along the northern shorelines of Lake Ontario and Lake Erie in Ontario have been implicated in reduction of waterfowl habitats.

Poa annua L.
ANNUAL BLUEGRASS

Also known as: dwarf meadow grass, causeway grass, speargrass, sixweeks grass, walkgrass

Scientific Synonym(s): *P. triangularis* Gilib.

QUICK ID: sheath loose; leaf tips are boat-shaped; plants less than 30 cm tall

Origin: native to the Mediterranean area of southern Europe and Asia; first observed in California in 1797; New Brunswick in 1875, spreading to British Columbia by 1891

Life Cycle: annual or winter annual reproducing by seed; each plant capable of producing up to 2,250 seeds; 30–100 flower clusters per plant

Weed Designation: USA: none

DESCRIPTION

Seed: Straw-colored, oval, 2–3 mm long, 1–1.3 mm wide.
Viable in soil for 4 years.
Germination best at 2°–5°C; light required for germination.

Seedling: With one cotyledon; first leaf 5–9 mm long, 1–3 mm wide, young leaves with scattered soft hairs.

Leaves: Mostly basal, flat, 1–14 cm long, 1–4 mm wide, surface with soft hairs; sheaths flattened, loose, hairless, light green, tips boat-shaped; ligule membranous, 2–5 mm long.
Inflorescence an open panicle, pyramid-shaped, 2–12 cm long; spikelets 3–10 mm long, three- to 10-flowered; glumes 1.5–3 mm long, lance-shaped, one-veined; lemmas 2.5–4 mm long, elliptical, five-nerved.

Plant: Often rooting at lower nodes and forming large mats; stems 3–30 cm long, bright green; two to four nodes.

Fruit: Caryopsis (grain).

REASONS FOR CONCERN

Annual bluegrass is a serious weed in lawns, golf courses, and waste areas. In early summer, plants mature and die leaving brown patches of dead grass in turf. It is host for barley yellow dwarf and lucerne dwarf viruses.

Similar species: Canada bluegrass (*P. com-pressa* L.), another introduced species, is common in meadows and waste ground. It is distinguished from annual bluegrass by its creeping rhizomes and flattened stems. The panicle, 2–8 cm long, has several branches that occur in pairs.

Setaria faberi Herrm.
GIANT FOXTAIL

Also known as: Faber's foxtail, Chinese foxtail, Chinese millet, giant bristlegrass, Japanese bristlegrass, nodding foxtail, tall green bristlegrass

Scientific Synonym(s): none

QUICK ID: ligule a fringe of hairs; auricles absent; sheath open with hairy margins

Origin: native to China; introduced in 1931 as a contaminant in Chinese millet; first reported in Ontario in 1978

Life Cycle: annual reproducing by seed

Weed Designation: USA: AR, CA, KY; CAN: federal

DESCRIPTION
Seed: Elliptical, grayish-brown to pale green, 2.5–2.8 mm long, 1.4–1.5 mm wide; surface wrinkled. Viable up to 3 years in soil.
Germination occurs between 20°C and 35°C; optimal depth is top 5 cm of soil with little emergence from lower depths.

Seedling: With one cotyledon; first leaf 1.2–2.5 cm long, 2–4 mm wide.

Leaves: Flat, 30–50 cm long, 8–17 mm wide, upper surface rough-haired; sheaths open, 5–12 cm long; margins hairy; ligule a dense fringe of hairs 1–2 mm long; auricles absent.
Inflorescence a nodding spikelike panicle, 4–20 cm long, 1–3 cm thick; spikelets 2.5–3 mm long, 1.5–1.7 mm wide, with three to six yellowish-green bristles per spikelet; bristles 12–18 mm long; glumes 1–2.5 mm long, three-veined; lemmas 2.5–3 mm long, papery, seven-veined.

Plant: Often falling over if not supported by other plants; stems 80–200 cm tall, internodes 4–15 cm long; rooting at lower nodes.

Fruit: Caryopsis (grain).

REASONS FOR CONCERN

Giant foxtail is an aggressive weed in cultivated fields and waste areas of east central United States. It has been reported to reduce crop yields drastically. Germination occurs throughout the growing season, making control difficult. The roots of giant foxtail are believed to cause allelopathic affects to tomato and cabbage plants.

Similar species: green foxtail (*S. viridis* [L.] Beauv., p. 590) is distinguished from giant foxtail by its erect panicle, the absence of hairs on the upper surface of the leaves, and seeds that are not wrinkled. Green foxtail has shorter stems that rarely exceed 1 m in height.

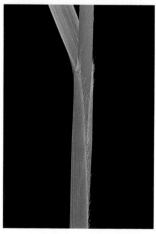

Setaria viridis (L.) Beauv.
GREEN FOXTAIL

Also known as: bottle grass, green brittlegrass, wild millet, pigeongrass, pussy grass

Scientific Synonym(s): *Chaetochloa viridis* Scribn., *Ixophorus viridis* (L.) Nash, *Panicum viride* L.

QUICK ID: ligule present; auricles absent; sheath open with overlapping margins

Origin: native to Europe; first reported in Montreal in 1821, spreading to Alberta by 1931

Life Cycle: annual reproducing by seed

Weed Designation: CAN: BC, MB, PQ, SK; MEX: federal

DESCRIPTION

Seed: Elliptical, pale green to grayish-brown, 1.5–1.9 mm long, about 1 mm wide, surface shiny under low magnification.
Viable for 33 months in dry storage at room temperature.
Germination depth optimal at 1.5–2.5 cm; no germination occurs from depths below 12 cm; germination occurs in late spring, usually after the crop has emerged. Optimal germination temperature 20°–30°C.

Seedling: With one cotyledon; first leaves arched, hairless; sheath margins hairy after emergence of the second leaf.

Leaves: Flat, 3–25 cm long, 5–15 mm wide, usually light green in color; sheath open, slightly flattened, margins overlapping, hairy; ligule a fringe of hairs 1.5–2 mm long; auricles absent. Inflorescence a cylindrical panicle, 3–11 cm long, 1–2.3 cm wide, purplish-green; spikelets numerous, one-flowered, 1.5–2 mm long, each with one to three bristles; bristles purplish-green, 6–12 mm long; glumes 1–2 mm long, one- to three-veined; lemmas 1–2 mm long, five- to seven-veined; seeds produced within 6 weeks of germination.

Plant: Tufted; stems 30–100 cm tall, surface with numerous upward-pointing hairs; three to five nodes, hairless; roots fibrous.

Fruit: Caryopsis (grain).

REASONS FOR CONCERN

Green foxtail, a serious weed of cultivated crops, gardens, and roadsides, causes dockage losses in wheat, barley, canola, and flax. The seeds of green foxtail germinate throughout the growing season when conditions are favorable. This factor allows many plants to survive or miss cultivation or herbicide applications. It is also host for the following viral diseases: barley stripe mosaic, corn new mosaic, soybean new disease, wheat spot mosaic, and wheat streak mosaic. It is a serious weed in 35 countries.

Similar species: yellow foxtail (*Pennisetum glaucum* [L.] R. Br., p. 580), another introduced annual grass, is more aggressive than green foxtail. Yellow foxtail has five to 12 yellowish-brown bristles below each spikelet, while green foxtail has one to three bristles. The ligule of yellow foxtail is composed of several long hairs.

Sorghum halepense (L.) Pers.

JOHNSONGRASS

Also known as: means grass, Egyptian grass, false Guinea grass, millet grass, Morocco millet, Cuba grass, St. Mary's grass, evergreen millet, maiden cane, aleppo milletgrass

Scientific Synonym(s): *Holcus halapensis* L., *S. miliaceum* (Roxb.) Snowden

QUICK ID: plants up to 3 m tall; panicle pyramid-shaped; ligule 2–5 mm long

Origin: native to the Mediterranean region of southern Europe; introduced from Turkey into South Carolina as a forage crop in 1830; first Canadian report from Ontario in 1959, with an additional 60 sites reported by 1980

Life Cycle: perennial reproducing by seed and creeping rhizomes; plants capable of producing up to 80,000 seeds; single plants may produce 60–90 m (about 5,000 nodes) of rhizome annually; new plants formed from fragments about 3 cm long

Weed Designation: USA: AR, CA, CO, DE, ID, IL, IN, KS, KY, MD, MO, NV, OH, OR, PA, SD, UT, WA, WV; CAN: federal, ON; MEX: federal

DESCRIPTION

Seed: Elliptical, pale yellow to purplish-brown, or mottled purple and yellow, 4.5–4.8 mm long, 1.5–2 mm wide, surface smooth and shiny. Viable in laboratory storage for up to 7 years. Germination occurs at 7–15 cm depth; optimal temperatures between 20°C and 35°C; light enhances germination when temperatures above 34°C and inhibits it when temperatures are below 22°C.

Seedling: With one cotyledon; first leaf 1.6–2.5 cm long, 4–6 mm wide, hairless; seedlings reach seventh leaf stage by day 50; young plants resemble corn or sorghum; rhizomes begin to appear at fifth leaf stage, well-developed by week 10.

Leaves: Flat, 20–90 cm long, 5–50 mm wide; midrib prominent and white; margins finely toothed; sheath open, margins overlapping and hairless; ligule papery 2–5 mm long; auricles absent. Inflorescence a pyramid-shaped panicle 15–50 cm long, 3–25 cm wide, hairy; branches up to 25 cm long, whorled at nodes; spikelets 35–350 per panicle; spikelets 4.5–5.5 mm long, two-flowered (one fertile, one sterile); sterile florets 4.5–6.5 mm long, glumes five-veined; fertile florets 4.5–5. 5 mm long, glumes leathery, yellow to dark brown or purplish-red to black, lemmas 2–2.5 mm long, one-veined, awns 10–16 mm long, often bent and twisted.

Plant: Plants 50–300 cm tall, up to 2 cm in diameter, shade-intolerant; stems hairless; nodes bearded; rhizomes up to 2 m long and 2 cm in diameter,

white with reddish-purple areas, rarely surviving winter soil temperatures of −17°C unless 20–25 cm deep; nodes with brown scalelike sheaths.
Fruit: Caryopsis (grain).

REASONS FOR CONCERN
Johnsongrass seeds are similar in size to maize and sorghum making separation difficult. Fresh or decaying leaves, rhizomes, and roots are reported to be allelopathic, allowing plants to form large colonies. Livestock grazing on large amounts of young plants are susceptible to hydrogen cyanide poisoning. It is reported to be a serious weed of 30 crops in 53 countries.

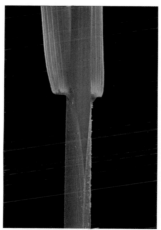

Taeniatherum caput-medusae (L.) Nevski
MEDUSAHEAD RYE

Also known as: medusahead

Scientific Synonym(s):
Elymus caput-medusae L.,
T. crinitum var. *caput-medusae* (L.) Wipff,
T. asperum auct. non (Simonkai)

QUICK ID: inflorescence a spike; awns twisted, stiff, and barbed; auricles present

Origin: native to Mediterranean region of southwestern Europe and northwestern Africa; first reported from Oregon in the early 1880s; not reported as weedy until 1950s

Life Cycle: annual or winter annual reproducing by seed; each flower cluster produces an average of five to 15 seeds

Weed Designation: USA: CA, CO, NV, OR, UT; CAN: AB

DESCRIPTION

Seed: Surface with stiff barbs; awns long, making them difficult to bury.
Viable in soil for 1–2 years.
Germination occurs in the top 2.5 cm of soil, emergence from 8 cm is possible; optimal temperatures between 10°C and 15°C; germination occurs in the fall when there is adequate moisture; dormancy about 3–4 months; awns contain a germination inhibitor, therefore must disintegrate before the seed germinates.

Seedling: With one cotyledon; first leaves about 5 mm wide; seedlings similar in appearance to downy brome (*Bromus tectorum*, p. 552); tolerant of desiccation of primary root, sprouting adventitious roots when moisture becomes available.

Leaves: Rolled, two to four per stem, 3–12 cm long, 2–5 mm wide, surface appearing glossy under magnification; sheath hairless to covered in minute hairs, open; upper leaves smaller; ligule less than 0.5 mm long, membranous; auricles up to 0.5 mm long, inconspicuous to sickle-shaped. Inflorescence a spike, 5–10 cm long; awns 2.5–10 cm long, twisted, stiff, barbed; spikelets crowded, bristly, two per node, each two-flowered; glumes 17–27 mm long, awn straight to slightly curved, 1–4 cm long; lemmas lance-shaped, 5–12 mm long, three-veined, surface with minute barbs, awns 2.5–10 cm long; inflorescence resembling foxtail barley (*Hordeum jubatum*, p. 570); commonly three to five spikes per plant, although 133 spikes have been reported.

Plant: Plants 15–60 cm tall, three to five tillers per plant, preferring climates with 25–50 cm rainfall; stems ascending; roots fibrous, up to 100 cm deep.

Fruit: Caryopsis (grain).

REASONS FOR CONCERN

Medusahead rye is an extremely aggressive plant that crowds out native vegetation. Heavily infested rangelands have suffered 40–75% reductions in grazing capacities. Livestock refuse to graze this species as it has a low forage value. The high silica content and barbed awns can cause injury to mouths, noses, and eyes of grazing livestock. Mature plants form a dense layer of litter and create fuel for wildfires.

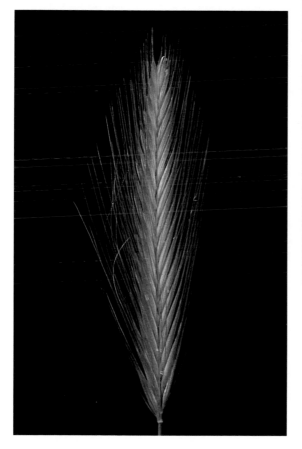

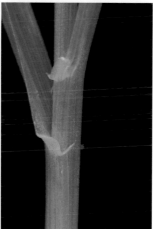

OTHER SPECIES OF CONCERN IN THE GRASS FAMILY

ARUNDO DONAX L. — GIANT REED

Native to India; perennial; rhizomes elongated; stems 2–6 m tall; blades drooping, 20–80 cm long, 10–70 mm wide; ligule a ciliate membrane; inflorescence an oblong panicle, 30–60 cm long, profusely branched; spikelets two- to seven-flowered, 8–15 mm long, breaking at maturity; glumes 7–14 mm long, three- to five-veined; lemma 6–12 mm long, five- to nine-veined, covered in hairs 5–8 mm long; palea 3–7 mm long; stamens 3, 2.5–3 mm long; caryopsis about 2 mm long. Weed designation: USA: TX.

Arundo donax

AVENA STERILIS L. — ANIMATED OAT

Native to Europe and Asia; annual; stems 30–180
cm long; two to five nodes; leaves 10–60 cm long,
4–18 mm wide, surface with a sandpapery texture;
ligule membranous, 3–8 mm long; inflorescence
an open panicle, nodding, 10–45 cm long, 5–25 cm
wide, branches threadlike, 5–35 mm long, drooping;
spikelets two- to five-flowered, wedge-shaped,
23–50 mm long; first glume 23–50 mm long, seven-
to 11-veined; second glume 23–50 mm long, seven-
to 11-veined; lemma 15–40 mm long, seven-veined,
awn 30–80 mm long, twisted. Weed designation:
USA: federal, AL, CA, FL, MA, MN, NC, OR, SC, VT; CAN:
federal.

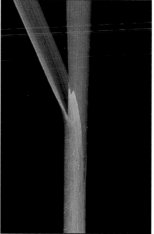

Avena sterilis

CHRYSOPOGON ACICULATUS (RETZ.) TRIN. — GOLDEN FALSE BEARDGRASS

Native to southeast Asia; mat-forming perennial, stolons present; stems 20–60 cm tall; leaves 1–9 cm long, 2–5 mm wide; surface hairless; ligule membranous; inflorescence an open panicle of racemes, 4–9 cm long; spikelets borne three per node (one fertile and stalkless, two sterile and stalked); sterile spikelets 4–5 mm long, glumes papery and enclosing the lemma; fertile spikelets 6–9 mm long; first glume 3–4 mm long; second glume lance-shaped, keeled; basal florets sterile, the lemma two-veined; upper florets oblong, lemma with awns 4–6 mm long. Weed designation: USA: federal, AL, CA, FL, MA, MN, NC, OR, SC, VT; MEX: federal.

DIGITARIA ABYSSINICA (HOCHST. EX A. RICH.) STAPF — AFRICAN COUCHGRASS

Native to Madagascar; mat-forming perennial, rhizomes creeping; stems weak and rambling, 5–60 cm long; leaves linear to lance-shaped, 2–12 cm long, 3–10 mm wide; ligule membranous; inflorescence of two to 25 racemes, 1–9 cm long; spikelets borne in pairs; spikelets with one sterile and one fertile floret, 1.5–2.5 mm long; first glume 0.1–0.8 mm long, zero- to one-veined; second glume 0.9–2.5 mm long, three- to seven-veined; lowest floret sterile; lemmas 1.5–2.5 mm long, seven-veined. Weed designation: USA: federal, AL, CA, FL, MA, MN, NC, OR, SC, VT.

IMPERATA BRASILIENSIS TRIN. — BRAZILIAN SATINTAIL

Native to South America; perennial spreading by scaly rhizomes; base with dead leaf bases; stems 25–75 cm tall, hairless; nodes bearded; leaves mostly basal, erect, 6–15 mm long, 8–10 mm wide, surface hairless except near the base, margins rough; sheaths hairless; ligule with brown hairs 0.3–1 mm long; inflorescence a spikelike panicle, 6–15 cm long, 1–2 cm wide, stalk 10–20 cm long; spikelets in pairs (lower spikelet sterile, upper fertile), 3.5–4 mm long, borne on threadlike stalks; callus hairs white, 8–12 mm long; glumes 3.5–4 mm long, three- to seven-

veined; lemma 0.5–1.1 mm long. Weed designation:
USA: federal, AL, CA, FL, MA, MN, MS, NC, OR, SC, VT.

IMPERATA CYLINDRICA (L.) BEAUV. — COGON GRASS, SATINTAIL

Native to southeast Asia, Philippines, China, and
Japan; introduced as a potential forage grass, which
turned out to be unpalatable to livestock; perennial
reproducing by seed and rhizomes, rhizomes
creeping, scaly, tips pointed, often dormant for long
periods of time; stems 60–120 cm tall, one to eight
nodes per culm; leaves mostly basal, flat, 3–150
cm long, 2–20 mm wide, erect; midvein whitish,
off-centered; tips sharp-pointed; margins finely
toothed; upper surface hairy, underside hairless;
sheaths short; ligule less than 1 mm long; auricles
absent; inflorescence a narrow panicle, silky white,
plumelike, 3–27 cm long, 1–3.5 cm wide; spikelets
3–6 cm long, crowded, borne on stalks of unequal
lengths (5–40 mm long); glumes 2–2.6 mm long,
three- to nine-nerved; lemma 1.5–4 mm long; fruit
a caryopsis (grain), brown, oblong, 1–1.3 mm long.
Weed designation: USA: federal, AL, CA, FL, HI, MN,
MS, NC, OR, SC, VT; MEX: federal.

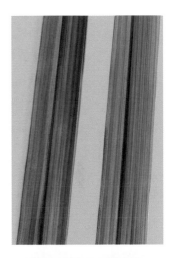

Imperata cylindrica

ISCHAEMUM RUGOSUM SALISB. — RIBBED MURAIN GRASS

Native to tropical Asia; annual; stems spreading,
10–100 cm long; leaves 5–30 cm long, 3–15 mm
wide; ligule membranous; inflorescence of terminal
and axillary racemes, paired, 3–12 cm long; spikelets
paired (lower sterile, upper fertile); sterile spikelet
2–6 mm long; fertile spikelet two-flowered (one
male, one perfect), 4–6 mm long; glumes 4–6 mm
long, keeled; lemma 4–6 mm long, awns 15–20 mm
long, twisted. Weed designation: USA: federal, AL,
CA, FL, MA, NC, OR, SC, VT; MEX: federal.

LEPTOCHLOA CHINENSIS (L.) NEES — CHINESE SPRANGLETOP

Native to tropical Asia; tufted annual; stems
ascending to spreading, 50–150 cm long,
occasionally rooting at lower nodes; leaves 25–50 cm
long, 3–7.5 mm wide, surface rough; sheath hairless;
ligule membranous; inflorescence of numerous

racemes, each 2–13 cm long; central axis 20–60 cm long; spikelets appearing in two rows; spikelets three- to six-flowered, elliptical to oblong, 2.1–3.2 mm long; first glume 0.6–1 mm long, one-veined; second glume 0.9–1.3 mm long, one-veined; lemmas elliptical to oblong, 0.8–1.4 mm long, three-veined; palea two-veined. Weed designation: USA: federal, AL, CA, FL, MA, MN, NC, OR, SC, VT; MEX: federal.

ORYZA LONGISTAMINATA A. CHEV. & ROEHR. — RED RICE, LONGSTAMEN RICE

Native to Madagascar; perennial with creeping rhizomes; stems ascending to spreading, 70–120 cm long, 5–10 mm diameter, spongy; four to 10 nodes, often rooted; leaves 10–75 cm long, 5–25 mm wide, surface and margins rough-textured; sheaths smooth and hairless; ligule membranous, 15–45 mm long; auricles 10–15 mm long; inflorescence an open panicle, 16–40 cm long, 1.5–8 cm wide; spikelets solitary, stalks 0.5–4 mm long; spikelets one-flowered, 7–12 mm long, 2–3 mm wide; glumes 2.5–3.8 mm long, one-veined; lemmas 7–12 mm long, five-veined, midvein spiny, awn 40–75 mm long; palea 6–11 mm long, three-veined; six anthers, 4.5–5.5 mm long. Weed designation: USA: federal, AL, CA, FL, MA, MN, NC, OR, SC, VT; MEX: federal.

ORYZA SATIVA L. — RICE

Native to southeast Asia; annual aquatic grass; stems 50–150 cm tall, often rooting at lower nodes; leaf sheaths inflated below; auricles sickle-shaped, hairy; blades 25–60 cm long and 5–20 mm wide with a sandpapery texture; ligule 10–40 mm long; panicle open, up to 30 cm long, nodding at maturity, one to three branches at lowest node, 2–12 cm long; spikelets 7–10 mm long; lemmas 1.5–4 mm long; fruit a caryopsis, 5–7 mm long, whitish-yellow to brown. Forms weedy hybrids with *O. rufipogon*. Weed designation: USA: AR, SC.

PANICUM REPENS L. — TORPEDO GRASS

Native to Africa; perennial with creeping rhizomes; stems 30–100 cm tall; leaves 7–25 cm long, 2–8 mm wide, stiff; ligule a ciliate membrane; inflorescence an open panicle, 5–20 cm long; spikelets 2.5–3 mm

Oryza sativa

long, one fertile and one sterile floret; lower glume about 1 mm long, one- to three-veined; upper glume 2.5–3 mm long, seven- to nine-veined; lemma 2.5–3 mm long, seven- to nine-veined. Weed designation: USA: AL, AZ, HI, TX;CAN: federal.

PASPALUM SCROBICULATUM L. — KODOMILLET

Native to tropical Asia; mat-forming or tufted perennial; stems erect to ascending, 10–150 cm long, 1–6 mm diameter; two to 17 nodes, often rooted; leaves 5–40 cm long, 3–15 mm wide; ligule membranous; inflorescence of one to 20 fingerlike racemes, each 2–15 cm long; spikelets borne in two rows; spikelets two-flowered (one sterile, one fertile), 1.4–3 mm long; first glume absent; second glume 1.4–3 mm long, dark brown, five- to seven-veined; lemma elliptical to round, dark brown, five-veined. Weed designation: USA: federal, AL, CA, FL, MA, MN, NC, OR, SC, VT; MEX: federal.

Panicum repens

PENNISETUM CLANDESTINUM HOCHST.
EX CHIOV — KIKUYU GRASS

Native to tropical east Africa; mat-forming perennial with creeping rhizomes; stolons present; stems trailing, 3–15 cm long, rooting at lower nodes, leaves 1–15 cm long, 1–5 mm wide; sheaths inflated; ligule a fringe of hairs; inflorescence with one to six spikelets, subtended by an inflated sheath; spikelets two-flowered (one sterile, one fertile), subtended by three to 15 bristles, 4–15 mm long; spikelets 10–20 mm long; glumes minute or absent; lemma 10–20 mm long, eight- to 12-veined; palea membranous, two- to four-veined; three anthers, 4–5 mm long; filaments 25–50 mm long. Weed designation: USA: federal, AL, CA, FL, MA, MN, NC, OR, SC, VT; MEX: federal.

Pennisetum clandestinum

PENNISETUM MACROURUM TRIN. —
AFRICAN FEATHERGRASS

Native to Africa and southwest Asia; perennial with creeping rhizomes; stems reedlike, 60–500 cm tall, nodes brown; leaves 10–45 cm long, 2–10 mm wide, leathery; ligule a fringe of hairs; inflorescence a spikelike panicle, 6–40 cm long; spikelets two-flowered (one sterile, one fertile), subtended by numerous bristles, 5–10 mm long; spikelets 2–6 mm long; first glume less than 1 mm or absent; second glume 0.2–1.2 mm long, zero- to three-veined; lemma 2–6 mm long, five to seven-veined; palea membranous; three anthers, 2–2.5 mm long. Weed designation: USA: federal, AL, CA, FL, MA, MN, NC, OR, SC, VT; MEX: federal.

ROTTBOELLIA COCHINCHINENSIS (LOUR.) CLAYTON—ITCHGRASS, GUINEA-FOWL GRASS

Native to Africa, Asia, and Australia; annual; stems erect, 30–300 cm tall; leaves 15–45 cm long, 5–20 mm wide; surface rough; sheath bristly haired; ligule membranous, about 1 mm long; inflorescence of axillary and terminal racemes, erect, 3–15 cm long; terminal spikelet sterile, four- to 14-flowered; spikelets borne in pairs; sterile spikelets 3–5 mm long; fertile spikelets two-flowered (one male, one perfect), 3.5–5 mm long; glumes 3.5–5 mm long, 11–13-veined; lemma 2.5–4.5 mm long, one- or five-veined; three anthers, about 2 mm long. Weed designation: USA: federal, AL, AR, CA, FL, MA, MN, MS, NC, OR, SC, TX, VT; MEX: federal.

SACCHARUM SPONTANEUM L.— WILD SUGARCANE

Native to tropical Africa and Asia; perennial with creeping rhizomes; stems erect, 2–4 m tall; leaves 50–200 cm long, 3–30 mm wide; ligule membranous; inflorescence an open panicle, 20–50 cm long, composed of racemes 3–15 cm long; spikelets two-flowered (one sterile, one fertile), 3.5–7 mm long, borne in pairs; glumes 3.5–7 mm long, one-veined; lemmas 1–3.5 mm long; palea absent. Weed designation: USA: federal, AL, CA, FL, MA, MN, NC, OR, SC, VT; MEX: federal.

SORGHUM ALMUM PARODI— COLUMBUS GRASS

Native to Argentina; perennial with short rhizomes; stems 2–3.2 m tall, 6–10 mm diameter; eight to 10 nodes, hairy; internodes 7–40 cm long; leaves 30–100 cm long, 25–40 mm wide, surface hairless, margins rough; sheaths hairless; ligule membranous, 2.5–3.5 mm long; inflorescence an open pyramidal panicle, 20–60 cm long, 8–25 cm wide; branches with four to nine divisions, 10–15 cm long, tipped by racemes; racemes 1.2–2.5 cm long, with one to five spikelets; spikelets borne in pairs on stalks 3–3.5 mm long; sterile spikelets (functionally male) about 5 mm long, nine-veined; fertile spikelets two-flowered (one sterile, one fertile), 5.5–6.5 mm long; glumes 3–3.5 mm long, dark brown, nine- to 11-veined; lemmas one- or two-veined, awns 8–10 mm long, twisted; three anthers, 2–2.5 mm long. Weed designation: USA: IL, IN, NV, UT.

SORGHUM BICOLOR (L.) MOENCH— BROOMCORN, GROS MIL, SORGHO, SORGO, DAZA

Native to North Africa; annual; stems erect, 1–6 m tall, 5–30 cm diameter; leaves 30–100 cm long, 5–10 mm wide; ligule membranous, 1–3 mm long; inflorescence an open panicle, 4–50 cm long, 2–20 cm wide; branches tipped by racemes; racemes with one to six fertile spikelets; spikelets borne in pairs; sterile spikelets functionally male; fertile spikelets two-flowered (one sterile, one fertile), 3–10 mm long; glumes 3–10 mm long, yellowish to red or black, five to seven-veined; lemmas 1–8 mm long, one- to five-veined, awns twisted. Weed designation: IN, IA, MD, NV, OH, PA.

UROCHLOA PANICOIDES BEAUV.— PANIC LIVERSEED GRASS

Native to southern Africa; tufted annual; stems ascending to spreading, 10–100 cm long, occasionally rooting at lower nodes; leaves 2–25 cm long, 5–18 mm wide; ligule a fringe of hairs; inflorescence of two to seven racemes, each 1–7 cm long; spikelets borne in two rows; spikelets two-flowered (one sterile, one fertile), 3.5–4.5 mm long; first glume 1.7–2.3 mm long, three- to five-veined; second glume 3.5–4.5 mm long, seven- to 11-veined; basal floret male or sterile; sterile floret lemmas 3.5–4.5 mm long, five- to seven-veined; fertile floret lemmas about 2 mm long, awns 0.3–1 mm long. Weed designation: USA: federal, AL, CA, FL, MA, MN, NC, OR, SC, VT; MEX: federal.

Polygonaceae

BUCKWHEAT FAMILY

Herbs, occasionally shrubs; stems with swollen nodes; leaves alternate, basal, opposite, or whorled, simple; sheath membranous (ocrea) and forming a collar around the node; inflorescence a raceme, head, spike, or panicle; flowers perfect, occasionally imperfect; three to six sepals, usually five, or two whorls of three, petal-like; petals absent; three to nine stamens, often in two whorls; one pistil; ovary superior; fruit a lens-shaped or three-sided achene. Worldwide distribution: 40 genera/800 species; found primarily in north temperate regions.

Distinguishing characteristics: herbs; stem nodes swollen; membranous sheath (ocrea) present at stem nodes.

Members of the buckwheat family are of little economic importance. Some well-known species are rhubarb and cultivated buckwheat. Many members of this family are weeds.

ocrea of water knotweed (*Persicaria amphibia*)

Fagopyrum tataricum (L.) Gaertn.
TARTARY BUCKWHEAT

Also known as: Tartarian buckwheat, sarrasin, Japanese buckwheat

Scientific Synonym(s):
F. sagittatum Gilib.,
F. vulgare Hill.,
Polygonum tataricum L.,
P. fagopyrum L.

QUICK ID: ocrea present; leaves triangular; flowers greenish-white with five petal-like sepals

Origin: native to China; introduced as a crop into Europe, then North America in the 1700s

Life Cycle: annual reproducing by seed; plants capable of producing up 1,100 seeds

Weed Designation: CAN: BC, MB, SK

DESCRIPTION

Seed: Oblong, three-sided, 5–7 mm long, 3–5 mm wide, dull gray to brown, wrinkled surface; similar in size to cereal grains, making it difficult to separate.
Viability unknown.
Germination occurs 2–5 months after seeds shed, due to dormancy; optimal temperatures 20°–30°C.

Seedling: With round cotyledons, stalked and notched at the base; first leaves triangular, the basal lobes rounded; papery sheath (ocrea) present.

Leaves: Alternate, triangular, 3–10 cm long, 2–8 cm wide, slightly longer than wide; ocrea papery, brownish, 5–11 mm long; lower leaf stalks 0.5–7 cm long, upper leaves stalkless.
Inflorescence an open axillary or terminal raceme; stalk 1–6 cm long.

Flower: Greenish-white, 4–6 mm wide; five sepals, petal-like, 1.5–3 mm long; eight stamens; one pistil; stalks 1–3 mm long; flowering starts within 4 weeks of germination; mature seeds produced 60–80 days after emergence.

Plant: Plant 20–80 cm tall; stems erect, light green, hairless; growth continues until killed by frost.

Fruit: Achene.

REASONS FOR CONCERN

Tartary buckwheat has been reported to reduce cereal crop yields by 50%, canola crops by 45%, and flax by 70%. Besides competing with the crop for moisture and nutrients, Tartary buckwheat can affect the grade and quality of the crop. Harvested crops containing large amounts of Tartary buckwheat seed cannot be used for flour, rolled oats, or malting processes.

Similar species: cultivated buckwheat (*F. esculentum* Moench) is quite similar to Tartary buckwheat. Cultivated buckwheat has stems up to 60 cm tall that turn reddish at maturity, brownish ocrea 2–8 mm long, and flower stalks 3–4 mm long.

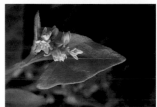

Fallopia convolvulus L. A. Love
WILD BUCKWHEAT

Also known as: black bindweed, dullseed cornbind, knot bindweed, bearbind, ivy bindweed, climbing knotweed, cornbind, devil's bindweed, blackbird bindweed, pink smartweed

Scientific Synonym(s): *Bilderdykia convolvulus* (L.) Dum., *Polygonum convolvulus* L., *Tiniaria convolvulus* (L.) Webb. & Moq.

QUICK ID: stems twining; leaves arrowhead-shaped; ocrea present

Origin: annual introduced from Europe as a contaminant in seed; first report in North America from California in 1860; first Canadian report from Manitoba in 1873

Life Cycle: annual reproducing by seed; large plants capable of producing over 30,000 seeds

Weed Designation: CAN: MB, PQ, SK; MEX: federal

DESCRIPTION

Seed: Oval, three-sided, dull black, 3–6 mm long, 1.8–2.3 mm wide.
Viable for at least 5 years in soil.
Germination occurs at 0.5–5 cm depth, although research has shown emergence from 19 cm depth; light not required for germination.

Seedling: With lance-shaped cotyledons, 7–33 mm long, less than 8 mm wide; attached at 120° from each other; stem below cotyledons often reddish-purple; first leaves arrowhead-shaped with downward-pointing lobes; ocrea present.

Leaves: Alternate, 2–6 cm long, 2–5 cm wide, arrowhead-shaped with backward-pointing basal lobes, tips pointed; stalks 0.5–5 cm long; ocrea tan to greenish-brown, 2–4 mm long. Inflorescence an axillary or terminal spikelike raceme or panicle, 2–10 cm long; two- to six-flowered; often drooping; stalk 1–100 mm long.

Flower: Greenish-white with a pink base, 4–6 mm across; four to six sepals, petal-like; three to nine stamens (commonly eight); one pistil; two or three styles; stalks 1–3 mm long.

Plant: Twining or trailing; stems 30–200 cm tall, branched, hairless, twining on vegetation or structures for support; taproot deep; roots fibrous.

Fruit: Achene; three-sided; enclosed by the dried green or brown sepals.

REASONS FOR CONCERN

Wild buckwheat reduces crop yields by competing for moisture, nutrients, and light. The twining nature of wild buckwheat entangles the crop and hampers harvesting. Wild buckwheat is reported to infest over 25 crops in 41 countries.

Similar species: field bindweed (*Convolvulus arvensis* L., p. 282) is often confused with wild buckwheat. Field bindweed has white or pink trumpet-shaped flowers, 1.5–2 cm long, and leaves with rounded tips.

Fallopia japonica var. japonica (Houtt.) Ronse Decraene
JAPANESE KNOTWEED

Also known as: Mexican bamboo, fleeceflower, Hancock's curse

Scientific Synonym(s): *Polygonum cuspidatum* Sieb. & Zucc., *P. sieboldii* de Vriese, *P. zuccarinii* (Small.) Small., *Reynoutria japonica* Houtt., *Pleuropterus zuccarinii* B. & B., Small., *Pleuropterus cuspidatus* (Sieb. & Zucc.) Moldenke

QUICK ID: plants with bamboo-like hollow stems, forming large colonies; flowers white, in axillary clusters; ocrea present

Origin: native to Japan, China, Korea, and Taiwan; introduced from Japan as a garden ornamental into the United Kingdom in 1825 and, from there, to North America in the late 1800s

Life Cycle: perennial reproducing by creeping rhizomes and seed; rhizome fragments can regenerate from 1 m depth, and emerge through 5 cm of asphalt; reproduction in Europe and North America primarily by rhizomes; abundant seed produced, but a large proportion nonviable; seed production in native range is up to 192,000 seeds per plant; northern limit in Europe coincides with sites with 120 frost-free days

Weed Designation: USA: AL, CA, CT, MA, NH, OR, WA, VT;CAN: AB

DESCRIPTION

Seed: Shiny black to brown, oblong, three-sided, 2–4 mm long, 1.4–1.9 mm wide; some report seeds rarely produced in northern climates: although viable seed production is abundant in southern Ontario.
Viable in dry storage for at least 4 years. Germination requires a cold period; greatest germination occurs on mineral soils; seedlings rarely establish under shady conditions.

Seedling: Rarely encountered in nature; seedling survival low; cotyledons narrowly elliptical to lance-shaped, 9–12 mm long, stalks 2–4 mm long; first leaves alternate, oval, 1–1.5 cm long; ocrea brownish, 4–10 mm long.

Leaves: Alternate, oval, 5–15 cm long, 5–12 cm wide, leathery, bases flat to wedge-shaped, tips pointed; stalks 1–3 cm long; ocrea 4–6 mm long, reddish to pale green and turning brown with age.
Inflorescence a panicle, 8–15 cm long; three- to eight-flowered per fascicle; borne in leaf axils; stalk 1–25 mm long.

Flower: Greenish-white, 2.5–3 mm long; male, female, or perfect; four or five sepals, petal-like, 2–8 mm long, borne in two whorls, outer three sepals enlarging in fruit; eight to 10 stamens; one pistil; three styles; seeds develop 2 weeks after pollination.

Plant: Plant 1–3 m tall, shrubby, arched near the top; stems bamboo-like and hollow, often mottled with red and purple; nodes swollen, usually reddish-brown at maturity; twigs often zigzagged; rhizomes thick and creeping, up to 20 m long, and forming stands up to 500 square meters; young rhizomes white with brown sheaths, older rhizomes brown and woody; plants tolerate high levels of sulfur dioxide.

Fruit: Achene; surrounded by enlarged calyx, 8–10 mm long; winged calyx assists in wind dispersal.

REASONS FOR CONCERN

The extensive creeping rhizome, ability to regenerate from small pieces of root, and seed viability make control of this plant difficult. Large colonies exclude native vegetation.

Similar species: a closely related species, Asian knotweed (F. *sachalinense* [F. Schmidt. Petrop.] Ronse Decr.) is similar in appearance to Japanese knotweed. A larger plant with angular stems 2–5 m tall, it has heart-shaped leaves 15–40 cm long and 7–25 cm wide, and ocrea 6–12 mm long. The green flowers, 4.5–6.5 mm long, are borne in panicles 2–8 cm long. Weed designation: USA: CA, CT, OR, WA;CAN: AB.

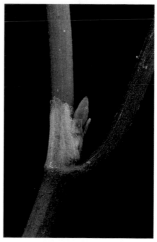

Persicaria lapathifolia (L.) S. F. Gray
PALE SMARTWEED

Also known as: pale persicaria, willow-weed, knotweed, curly top lady's thumb, dock leaf smartweed, nodding smartweed

Scientific Synonym(s): *Polygonum lapathifolium* L., *P. scabrum* Moench, *P. nodosum* Pers., *Persicaria incarnata* (Ell.) Small.

QUICK ID: leaves with a black "thumbprint"; ocrea present; flowers greenish-white to pink

Origin: annual native to western North America and South America

Life Cycle: annual reproducing by seed; large plants under optimal conditions may produce up to 19,000 seeds

Weed Designation: CAN: MB, PQ

DESCRIPTION

Seed: Heart-shaped, shiny black to dark brown, 2.5–3.5 mm long, 1.6–3 mm wide, concave on both sides.
Viable in dry storage for up to 6 years. Germination occurs in the top 4 cm of soil.

Seedling: With oval cotyledons, 4–15 mm long, less than 5 mm wide; upper surface rough, slightly hairy; first leaves alternate, underside densely haired; leaves after the fifth to seventh leaf stage hairless; ocrea present.

Leaves: Alternate, lance-shaped, 5–20 cm long, stalks 1–16 mm long; dark blotch or "thumbprint" near the middle of the leaf; upper leaves with sticky yellow dots or hairs on the underside; ocrea brownish, 4–24 mm long, the margins smooth and short-haired.
Inflorescence a dense terminal or axillary spike, 1–10 cm long, 5–12 mm wide; four- to 14-flowered; stalk 2–25 mm long.

Flower: Greenish-white to pink, 2.5–3 mm across; five sepals, petal-like; three to nine stamens (commonly five or six), anthers pink or red; one pistil; two or three styles; stalks 0.5–2.3 mm long.

Plant: Plants 20–250 cm tall; stems erect to spreading, branched, lower nodes often rooting when in contact with soil.

Fruit: Achene; remnants of the sepals often remaining attached after maturity.

REASONS FOR CONCERN

Green smartweed competes with cultivated crops for moisture and nutrients; in addition, it tends to dry slower than the crop and often delays harvest. Green smartweed is host for beet curly top virus.

Similar species: lady's thumb (*P. persicaria* L.), an introduced weed from Europe, is often confused with green smartweed. The flowers of lady's thumb are pink or purplish and borne in cylindrical-shaped terminal and axillary clusters. The leaves, 3–15 cm long, have a purplish-black spot on the leaflike green smartweed but does not have sticky yellow hairs on the underside. The ocrea has several stiff bristly hairs, each 1–2 mm long. Seeds are three-sided or lens-shaped. Weed designation: CAN: AB, MB, PQ.

Also known as: yard knotweed, knotgrass, doorweed, mat grass, pinkweed, birdgrass, stonegrass, waygrass, goosegrass, ninety-knots, bird's-tongue, cowgrass, road-spread, ovalleaf knotweed

Scientific Synonym(s):
P. aviculare L.
ssp. *arenastrum,*
P. montereyense Brenckle,
P. aequale Lindm.

QUICK ID: prostrate growth habit; ocrea present; flowers greenish appearing in leaf axils

Origin: native to Europe; first report from North America in 1809

Life Cycle: annual or short-lived perennial reproducing by seed; plants may produce up to 200 seeds

Weed Designation: CAN: PQ

DESCRIPTION

Seed: Egg-shaped, three-sided, dark reddish-brown, 2–2.5 mm long, surface smooth and shiny. Viability unknown.
Germination is at or near the soil surface; optimal soil temperatures 8°–12°C.

Seedling: With lance-shaped cotyledons, 6–15 mm long, 1–2 mm wide, united at base and forming a small cup; first leaves spatula-shaped, bluish-green; ocrea present.

Leaves: Alternate, oblong, 6–50 mm long, 1–8 mm wide, bluish-green, stalk 0.5–3 mm long; ocrea 3.5–12 mm long, silvery and translucent, margins torn.
Inflorescence a short cyme, one- to seven-flowered, borne in leaf axils.

Flower: Greenish-white, 1–3 mm across; five sepals, 2–3 mm long, green with pinkish-white margins; three to nine stamens; one pistil; two or three styles; stalks 1–2.5 mm long.

Plant: Prostrate; stems 10–100 cm long, profusely branched at most nodes and forming mats; taproot deep and wiry.

Fruit: Three-sided achene.

REASONS FOR CONCERN

Common knotweed, a weed of waste areas, has recently appeared in grain fields. It may become a problem because of its prolific seed production. It is also host for aster yellows, beet curly top, and beet ring spot viruses.

Rumex acetosella L.
COMMON SHEEP SORREL

Also known as: field sorrel, red sorrel, common sorrel, horse sorrel, sourgrass, redtop sorrel, cow sorrel, redweed, mountain sorrel

Scientific Synonym(s): *R. tenuifolius* (Wallr.) Löve, *Acetosella acetosella* (L.) Small., *Acetosella tenuifolia* (Wallr.) A. Löve, *Acetosella vulgaris* (Koch) Fourr., *R. angiocarpus* Murb., *R. pratensis* Mert. & Koch

QUICK ID: stems reddish-green; flowers borne in whorls of five to eight; leaves arrowhead-shaped

Origin: native to Europe or western Asia

Life Cycle: annual or perennial reproducing by rhizomes and seed

Weed Designation: USA: AR, CT, IA

DESCRIPTION

Seed: Three-angled, faces slightly concave, reddish-brown to golden brown, 0.9–1.5 mm long, 0.6–0.9 mm wide, surface shiny to slightly glossy. Viable for up to 25 years, with greatest decline occurring in the second year.
Germination occurs in the top 7 cm of soil; some germination occurs on the surface; optimal temperature of 20°–30°C.

Seedling: With lance-shaped cotyledons, 5–6.5 mm long, 1–1.5 mm wide, dull, united at the base and form a short tube; acidic tasting; stem below cotyledons shortens with age, pulling the young plant to the soil surface; first leaf oval, basal lobes appearing by the third to fifth leaf stage.

Leaves: Alternate, arrow-shaped with two basal lobes, 2.5–10 cm long; acidic tasting; ocrea ragged, colorless or pale greenish-yellow; lower leaves long-stalked, upper sessile and not lobed; all leaves often tinged with red; ocrea present. Inflorescence a terminal panicle, 10–25 cm long; flowers borne in whorls of five to eight; panicle red at maturity.

Flower: Nodding; male or female; male flowers yellowish-red, six sepals (three outer and three inner, the inner 1.5–2 mm long); six stamens; female flowers greenish, six sepals (three outer and three larger inner); one pistil; outer sepals lance-shaped; sepals do not enlarge as the fruit matures; flower stalks 1–3 mm long, not jointed.

Plant: Male or female; sour-tasting; highly variable species; stems reddish-green, 15–50 cm tall, several rising from the same crown; rhizomes creeping and up to 1.5 m deep, slender and branched, yellow, white buds producing new shoots; adventitious buds in top 8 cm of soil.

Fruit: Achene, three-sided.

REASONS FOR CONCERN

Once established, sheep sorrel is difficult to eradicate. Sheep sorrel is host for aster yellows, clover big vein, and tobacco mosaic viruses.

Similar species: garden sorrel (*Rumex acetosa* L.) may be mistaken for common sheep sorrel. The oblong to oval or lance-shaped leaves are 4–10 cm long and 1–4 cm wide and have a sagittate base. Flowers are borne in whorls of four to eight on stalks 2–5 mm long with a distinct joint at the middle. The dark brown to black seeds are 1.8–2.5 mm long and 1.2–1.5 mm wide.

Rumex crispus L.
CURLED DOCK

Also known as: yellow dock, curly dock, sour dock, narrowleaf dock

Scientific Synonym(s): *Lapathum crispum* (L.) Scopoli

QUICK ID: leaves with curled wavy edges; large terminal clusters of flowers that turn brown at maturity; taproot with a yellow core

Origin: native to Europe; first observed in North America in 1748

Life Cycle: perennial reproducing by seed; large plants capable of producing over 60,000 seeds

Weed Designation: USA: AR, IA; CAN: federal; MEX: federal

DESCRIPTION

Seed: Three-sided, reddish-brown, 2–3 mm long, 1.5–2 mm wide.
Viability about 52% in soil after 50 years; a few still viable after 80 years.
Germination requires light; emergence from a maximum depth of 3 cm.

Seedling: With oblong cotyledons, 7–16.3 mm long, 1.5–5.1 mm wide, hairless, dull green; first leaves lance-shaped, underside prominently veined; ocrea present.

Leaves: Alternate, lance-shaped, 10–30 cm long, margins wavy or crinkled, stalks 2.5–5 cm long; reduced in size upward; basal leaves 10–30 cm long; ocrea 1–5 cm long, turning brown and papery with age.
Inflorescence a panicle up to 60 cm long; flowers borne in whorls of 10–25; turning brown at maturity.

Flower: Greenish-red, 3–5 mm long; six sepals, in two whorls of three (outer green; inner red, 3.5–6 mm long); six stamens; one pistil; three styles; flower stalks 4–8 mm long, jointed.

Plant: Plant 40–160 cm tall, robust; stems reddish-green, enlarged at nodes; taproot thick and fleshy, yellow, up to 1.5 m deep.

Fruit: Achene; three-sided; enclosed by the enlarged inner sepals.

REASONS FOR CONCERN

Curled dock is a common weed in moist areas and low-lying depressions. It is a troublesome weed in cultivated cropland and pastures. The seeds and vegetation of curled dock are toxic to poultry. Curled dock is also host for the following viruses: beet curly top, cucumber mosaic, rhubarb ring spot, tobacco broad ring spot, tobacco mosaic, tobacco streak, and tobacco ring spot. It is reported to be a serious weed in 37 countries.

Similar species: Mexican dock (*R. salicifolius* var. *mexicanus* [Meisn.] C. L. Hitchc.) is often found growing alongside curled dock. Narrow-leaved dock has small side branches below the flower cluster. The leaves are 10 to 30 cm long with flat to wavy margins. CAN: federal.

OTHER SPECIES OF CONCERN IN THE BUCKWHEAT FAMILY

EMEX AUSTRALIS STEINH. — SOUTHERN THREECORNERJACK, DOUBLEGREE, SPINY EMEX

Native to South Africa; annual; stems trailing to spreading, 10–40 cm long, often reddish at base, branched; leaves alternate, elliptical to oval, 1–10 cm long, 0.5–6 cm wide; stalks 1–8 cm long; ocrea loose, hairless; inflorescence a raceme; flowers male or female; male flowers borne in groups of one to eight, sepals 1.5–2 mm long, four to six stamens with yellow or reddish anthers; female flowers borne in groups of one to four, outer sepals 4–6 mm long, inner sepals, 5–6 mm long; fruit an achene 4–6 mm long, subtended by enlarged sepals (7–9 mm long and 9–10 mm wide) with spines 5–10 mm long. Weed designation: USA: federal, AL, CA, FL, MA, MN, NC, OR, SC, VT; MEX: federal.

Emex spinosus

EMEX SPINOSUS (L.) CAMPD. — SPINY THREECORNERJACK, DEVIL'S THORN, LITTLE JACK, SPINY EMEX

Native to southern Europe and western Asia; annual; stems spreading to erect, 30–60 cm tall, base reddish, branched; leaves oval to oblong or triangular, 3–13 cm long, 1.1–12 cm wide, stalks 2–29 cm long, ocrea loose, hairless; inflorescence a raceme; flowers male or female; male flowers borne in groups of one to eight, sepals 1.5–2 mm long, four to six stamens; female flowers borne in groups of two to seven, outer sepals 4–6 mm long in fruit, inner sepals 5–6 mm long in fruit; fruit an achene 4–5 mm long, subtended by enlarged sepals with spines 2–4 mm long. Weed designation: USA: federal, AL, CA, FL, HI, MA, MN, NC, OR, SC, VT; MEX: federal.

PERSICARIA PERFOLIATA (L.) H. GROSS —
ASIATIC TEARTHUMB, MILE-A-MINUTE VINE

Native to southeast Asia; annual trailing vine with fibrous roots; stems reddish, up to 8 m long, surface with downward-pointing hooklike barbs; leaves alternate, triangular, 4–7 cm long, 5–9 cm wide, light green, stalks and leaf undersides with barbed hooks; ocrea saucerlike, 1–2 cm in diameter; inflorescence an axillary spike, 10–15-flowered; flowers white; fruit berry-like, blue and fleshy, about 5 mm in diameter; seeds black or reddish-black, glossy. Weed designation: USA: AL, CT, MA, NC, OH, PA, SC.

Pontederiaceae

PICKERELWEED OR WATER HYACINTH FAMILY

Aquatic herbs, rooted or floating; leaves opposite or whorled, the bases sheathing; leaf stalks often inflated; inflorescence a raceme or panicle, often with a spathe at the base; flowers perfect, usually regular; three sepals, petal-like; three petals; six stamens, occasionally one or three; one pistil; ovary superior; fruit a capsule or nut. Worldwide distribution: seven genera/30 species; primarily tropical and subtropical regions of the world.

Distinguishing characteristics: plants aquatic; flowers showy; three sepals, petal-like.

A few species are grown as ornamentals.

pickerelweed (*Pontederia cordata*)

Eichhornia crassipes (Mart.) Solms
WATER HYACINTH

Also known as: floating hyacinth, water-orchid

Scientific Synonym(s):
E. speciosa Kunth, *Piaropus crassipes* (Mart.) Raf., *Pontederia crassipes* Mart.

QUICK ID: aquatic floating plant; leaf bases bulb-shaped; flowers showy, lavender-blue

Origin: native to the Amazon basin of Brazil; introduced as an ornamental plant in 1884 at New Orleans; appearing in Florida by 1895, and California by 1904; by 1950 over 126,000 acres infested in Florida

Life Cycle: perennial reproducing by rhizomes, seed and offsets; plants may produce up to 10,000 seeds; primary mode of reproduction is by rhizome; experiments show that two plants produced over 1200 new plants within 4 months

Weed Designation: USA: AL, AZ, CA, CT, FL, SC, TX

DESCRIPTION

Seed: Ovoid, 1.1–2.1 mm long, 0.6–0.9 mm wide, surface with 12–15 longitudinal ridges. Viable 15–20 years in water or wet soil. Germination is initiated by high light intensity and alternating temperatures (5°–40°C); under ideal conditions seeds germinate within 3 days of being shed.

Seedling: With one cotyledon; first two or three leaves straplike, 5–15 mm long, produced 10 days after germination; by day 30, seven or eight leaves are produced; floats appearing by sixth leaf stage.

Leaves: Basal, oval to round, 2.5–10 cm long, 3.5–9.5 cm across, thick, succulent, waxy green, waterproof; leaf base heart-shaped; margins smooth; stalks inflated and bulblike 3.5–33 cm long, swelling or inflation one-sided; stipules 2.5–14 cm long.
Inflorescence a spike up to 30 cm long, four- to 25-flowered (commonly eight to 15); borne on flowering stalk 5–50 cm tall; spathe 4–11 cm long.

Flower: Showy, lavender to blue, 3.5–6 cm across; three sepals; three petals, lateral petals two-lipped, upper petal enlarged with a yellow spot ringed with blue; six stamens; anthers violet, 1.4–2.2 mm long; one pistil; three stigmas; two bracts, the lowest with a blade; flowers opening 2 hours after sunrise and wilting by nightfall; flowering occurs 100–120 days after germination; after flowering, the stem curves into the water to release the seeds; seeds produced about 18 days after pollination.

Plant: Aquatic to semiaquatic, mat-forming, anchored plants with fewer bulbous leaf bases; growing temperature range is 12°–35°C (optimal 25°–30°C); leaves and stems frost-intolerant, though rhizome tips have to freeze before the plant dies; flowering stems erect; rhizome branched, about 6 cm in diameter and 5–45 cm long; roots submerged, dark purple to black, feathery (about 70 lateral roots per cm); studies show over 50% of biomass is attributed to rhizome and roots; plants separated from mats are blown around by the wind as the leaves act as sails.

Fruit: Capsule, borne under water; zero to 20 fruit per plant; three to 450 seeds.

REASONS FOR CONCERN

Water hyacinth forms large floating mats that impede water flow and boat traffic. It has been reported to clog irrigation pumps and hydroelectric generators. Under ideal conditions populations may double in size every 6–18 days. These floating mats cause a reduction in oxygen levels and available sunlight for native aquatic plants and animals. Water hyacinth is a pest of rice paddy and taro fields.

OTHER SPECIES OF CONCERN IN THE PICKERELWEED OR WATER HYACINTH FAMILY

EICHHORNIA AZUREA KUNTH. — ROOTED WATER HYACINTH

Native to Central and South America; perennial rooting in mud; stems elongated and growing on water surface; leaves alternate; submersed leaves stalkless; emergent leaves round, 7–15 cm long, 2.3–16 cm wide, stalks 11–25 cm long, not inflated; stipules 7–13 cm long; inflorescence a raceme, 8–12 cm long, hairless; seven- to 60-flowered; spathes oblong to oval, 3–6 cm long, the stalk 1.9–15 cm long with orange hairs; flowers lasting 1 day, blue or white, dark blue at base with a yellow spot; sepals and petals, three each, 13–25 mm long, margins jagged; six stamens, 15–29 mm long; one pistil; fruit a capsule; 10–13 seeds, winged. Weed designation: USA: federal, AL, AZ, CA, FL, MA, NC, OR, SC, TX, VT.

Monochoria hastata

MONOCHORIA HASTATA (L.) SOLMS — ARROWLEAF FALSE PICKERELWEED

Native to southeast Asia; aquatic perennial; rhizomes well-developed; stems 30–100 cm tall; basal leaves 5–25 cm long, 4–20 cm wide, bases arrowhead-shaped, stalks sheathlike, 30–90 cm long; stem leaves with stalks 7–10 cm long; inflorescence a spikelike raceme, 15–60-flowered, opening at different times; lower flower stalks 15–30 mm long, upper stalks 7–20 mm long; flowers light blue; sepals and petals, three each, 15–18 mm long; six stamens (one stamen 5–6.5 mm long, other 5 3–4 mm long); one pistil; stalks 1–3 cm long; fruit a capsule about 10 mm long, enclosed by the persistent sepals and petals; seeds light brown. Weed designation: USA: federal, AL, CA, FL, MA, NC, OR, SC, VT.

MONOCHORIA VAGINALIS (BURM. F.)
K. PRESL EX KUNTH — HEARTSHAPE
FALSE PICKERELWEED

Native to Asia (India to Japan); annual rooting
in mud; rhizome short; flowering stems 8–17 cm
long; stem leaves heart-shaped, 4–8 cm long, 18–52
mm wide, stalkless; stipules 35–57 mm long; basal
leaves linear, 1–5 cm long, stalks 7–28 cm long;
inflorescence a panicle, three- to eight-flowered,
all opening on same day; spathe 19–40 cm long;
flowers blue or white; sepals and petals, three each,
10–14 mm long; six stamens (one stamen 4–6 mm
long and hooked, other 5 3–4 mm long); one pistil;
fruit a capsule; seeds winged, about 1 mm long.
Weed designation: USA: federal, AL, CA, FL, MA, NC,
OR, SC, VT.

Monochoria vaginalis

PONTEDERIA ROTUNDIFOLIA L. F. —
TROPICAL PICKERELWEED

Native to Central and South America; perennial
rooting in mud; rhizomes present; stems submersed
or growing to water surface; leaves basal, heart-
to kidney-shaped; stalked leaves floating or
submerged; inflorescence a spike, 50-or-more-
flowered; spathes folded; flowers opening 1 day,
funnel-shaped; three sepals; three petals; six
stamens (three long and three short), filaments
purple, anthers yellow; one pistil; fruit a utricle;
surface with spiny or toothed ridges. Weed
designation: USA: FL.

Pontederia rotundifolia

Portulacaceae
PURSLANE FAMILY

Herbs, often succulent; leaves alternate
or opposite, usually fleshy; inflorescence a
raceme or cyme; flowers perfect, regular; two
sepals; four to six petals; four to six or eight
to 12 stamens; one pistil; two to five styles;
ovary superior; fruit a capsule. Worldwide
distribution: 19 genera/580 species; found
throughout the world.

Distinguishing characteristics: plants
succulent; two sepals; four to six petals; fruit
a capsule.

Members of the purslane family are often
grown as ornamentals and include portulaca,
moss rose, pygmyroot, and miner's lettuce.

purslane (*Portulaca oleracea*)

Portulaca oleracea L.
PURSLANE

Also known as: wild
portulaca, pusley, low
pigweed, duckweed,
garden purslane, little
hogweed

Scientific Synonym(s):
P. neglecta Mackenzie &
Bush, *P. retusa* Engelm.

QUICK ID: leaves fleshy;
flowers yellow; plants
prostrate

Origin: native to southern
Europe or northern Africa;
introduced as a garden
plant; first reported in
Massachusetts in 1672;
Canada in 1863

Life Cycle: annual
reproducing by seed;
large plants producing up
to 242,500 seeds

Weed Designation: USA: AZ

DESCRIPTION

Seed: Oval to triangular, reddish-brown to black,
0.6–1 mm across, surface shiny.
Viable in soil for up to 40 years; seeds viable after
being exposed to temperatures of –15°C and
50°C.
Germination is enhanced by light and
temperatures over 24°C; emergence occurs
within 24 hours when soil temperatures are
30°–40°C, 2 days at 10°–20°C; light required for
germination.

Seedling: With succulent oblong cotyledons, 2–10
mm long, 1–2 mm wide, tinged with bright red;
stem below cotyledons bright red, succulent; first
pair of leaves alternate but appearing opposite,
oval, about 14 mm long and 7 mm wide.

Leaves: Alternate (appearing) opposite, oval to
wedge-shaped, 4–28 mm long, 2–20 mm wide,
succulent, green above and pale purple below,
hairless; crowded near the ends of branches.
Inflorescence a short raceme; one- to three-
flowered; borne in leaf axils.

Flower: Yellow, 3–10 mm across, inconspicuous, and
opening only on sunny mornings, stalkless; two
sepals, 3–4 mm long; four to six petals, falling off
soon after opening; six to 12 stamens; one pistil;
flowering begins 1 month after germination
(about the tenth to twelfth leaf stage); seeds
produced within 6 weeks of germination, and
7–12 days after pollination.

Plant: Prostrate, forming large mats up to 120 cm
across; stems succulent, red, branched, hairless,
often rooting at nodes; taproot 2–10 cm long;
uprooted plants able to ripen seeds from the
reserves in the stems and leaves, may survive
long enough to produce new roots.

Fruit: Globe-shaped capsule, 4–8 mm long, opening around the middle; few-seeded.

REASONS FOR CONCERN

Purslane is not shade-tolerant and rarely a weed of cultivated crops. It is more of a concern in row crops, home gardens, and flower beds. A prolific seed producer, it is not controlled by cultivation, as the stems and leaves store enough reserves to produce seeds even after being uprooted. Stem fragments may also root and produce more plants. It is reported weedy in 81 countries.

Similar species: prostrate pigweed (*Amaranthus blitoides* S. Wats., p. 4) is often confused with purslane. It is distinguished by nonfleshy spoon-shaped leaves and greenish-white flowers. The prostrate reddish stems, up to 120 cm long, do not root at leaf nodes.

OTHER SPECIES OF CONCERN IN THE PURSLANE FAMILY

TALINUM PANICULATUM (JACQ.) GAERTN. — PINK BABY'S BREATH, JEWELS OF OPAR

Native to western North America; perennial, hairless; roots tuberous, fleshy to somewhat woody; stems 20–150 cm tall, reddish; leaves alternate to subopposite, elliptical to oblong or oval, 1–12 cm long, 1–7 cm wide, somewhat succulent; inflorescence a panicle, few- to many-flowered; flowers red to pink (occasionally orangish, yellowish, or purplish); two sepals, falling off as the flower opens; five petals, 3–6 mm long, withering soon after opening; 15–20 stamens; one pistil; fruit a capsule, 3–5 mm long; seeds black about 1 mm across. Weed designation: none.

Potamogetonaceae

PONDWEED FAMILY

Aquatic herbs; rhizomes often producing tubers; stems jointed; leaves alternate or opposite, two-ranked, sheathing at the base and resembling stipules; stipules absent; inflorescence a spike, the flowers borne in whorls; flowers perfect or imperfect, regular; four tepals (sepals and petals similar in appearance); four stamens; four pistils; four styles; ovary superior; fruit a drupelet or achene. Worldwide distribution: two genera/100 species; predominantly found in North America.

Distinguishing characteristics: plants aquatic; stems jointed; flowers borne in spikes.

The pondweeds are of little economic importance.

plant, Richardson's pondweed
(*Potamogeton richardsonii*)

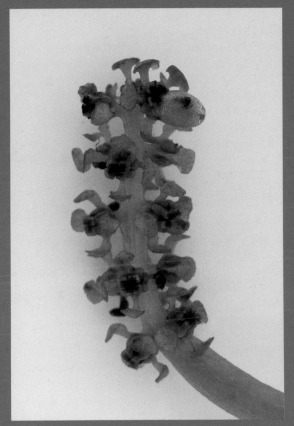

inflorescence of Richardson's pondweed
(*Potamogeton richardsonii*)

Potamogeton crispus L.
CURLY PONDWEED

Also known as: curly-leaf
pondweed
Scientific Synonym(s):
P. serratus Huds.

QUICK ID: aquatic plant;
leaves alternate; leaf
margins wavy and finely
toothed
Origin: native to Europe and
Asia; introduced into
North America about
1840
Life Cycle: perennial
reproducing by turions,
seeds, and stem
fragments; turions
overwintering or
producing new plants in
late summer and fall
Weed Designation: USA: AL,
CT, ME, MA, VT, WA CAN:
SK

DESCRIPTION

Seed: Oval and flattened, 3–5 mm long, red to
reddish-brown, surface pitted; beak 2–2.5 mm
long
Viability unknown.
Germination information unavailable.
Seedling: Information not available.
Leaves: Alternate, all submersed, two-ranked, one
to four per side, 1.2–9 cm long, 4–10 mm wide,
three- to five-nerved, stalkless, margins wavy
and finely toothed, midvein reddish; stipules
persistent, brown, 4–10 mm long, not fibrous or
shredding with age.
Inflorescence a cylindrical spike, 2.5–4 cm long;
stalk 2–10 cm long, curved.
Flower: Green; four tepals; four stamens; four pistils;
appearing in early spring.
Plant: Submersed, dormant during summer,
producing fruit and turions in spring; stems
20–100 cm long, 1–2 mm thick, flattened;
rhizomes absent; producing turions in leaf axils
in spring; turions 1.5–3 cm long, burlike and
spindle-shaped, consisting of three to seven
small, thickened leaves, germinating in fall and
overwintering as small plants.
Fruit: Achene.

REASONS FOR CONCERN

Curly pondweed may grow in dense stands and restrict recreational use of the waterway. It does not compete directly with native species due to its early spring growth and summer dormancy period.

Similar species: Richardson's pondweed (*P. richardsonii* [A. Bennett] Rydb., p. 638), a native pondweed, is distinguished by submersed oval to lance-shaped leaves, 1.6–13 cm long and 5–28 mm wide, with three to 35 veins, and smooth to crinkled margins.

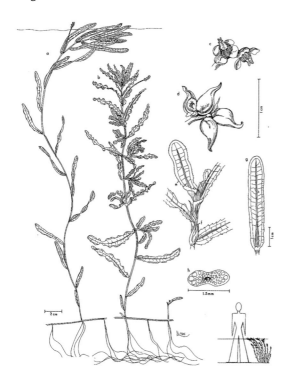

OTHER SPECIES OF CONCERN IN THE PONDWEED FAMILY

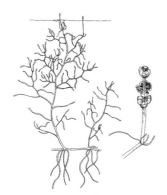

Potamogeton pusillus

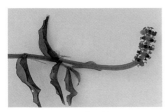

Potamogeton richardsonii

POTAMOGETON PUSILLUS L. — SMALL PONDWEED

Native to North America; rhizomes absent; stems 18–150 cm long and less than 0.5 mm diameter; turions 9–32 mm long, 0.3–1.8 mm wide; leaves alternate, two-ranked, all submersed, linear, 9–65 mm long, 0.2–2.5 mm wide, one- to three-veined; stalkless, margins smooth; stipules brown to green or white, 3–9 mm long, rarely fibrous or shredding; inflorescence a spike 0.5–6.2 cm long, 1.5–10 mm wide; fruit green to brown, 1.5–2.2 mm long, 1.2–1.6 mm wide; beak less than 0.6 mm long. Weed designation: none.

POTAMOGETON RICHARDSONII (A. BENNETT) RYDBERG — RICHARDSON'S PONDWEED

Native to North America; rhizomes present; stems 20–100 cm long; turions absent; leaves alternate, two-ranked, all submersed, oval to lance-shaped, 1.6–13 cm long, 5–28 mm wide, three- to 35-veined, stalkless, margins entire to crinkled; stipules white, 12–17 mm long, fibrous, and shredding; inflorescence a spike 1.5–14.8 cm long, 13–37 mm wide; fruit greenish-brown, 2.2–4.2 mm long, 1.7–2.9 mm wide; beak 0.4–0.7 mm long. Weed designation: none.

STUCKENIA PECTINATA (L.) BORNER — SAGO PONDWEED

Native to North America; rhizomes present; tubers present; turions absent; stems 20–75 cm long, branched; leaves alternate, linear, 5.6–9.2 cm long, 0.2–1 mm wide, one- to three-veined, stalkless, margins smooth; stipules sheathlike, not inflated, 8–11 mm long, fused to the leaf about two-thirds of the stipule length; ligule about 0.8 mm long; inflorescence a spike 14–22 mm long; flowers borne in three to five whorls; fruit yellowish-brown to brown, 3.8–4 mm long, 2.5–3.1 mm wide; beak 0.5–1.1 mm long. Weed designation: none.

Stuckenia pectinata

Primulaceae

PRIMROSE FAMILY

Herbs; leaves opposite, whorled, or basal, simple; stipules absent; inflorescence a raceme, spike, umbel, panicle, or solitary flower; flowers perfect, regular; five sepals; five petals, free or united; five stamens, opposite the petals; one pistil; one style; ovary superior or half inferior; fruit a capsule. Worldwide distribution: 28 genera/800 species; found throughout the world with the greatest diversity in the north temperate region.

Distinguishing characteristics: herbs; leaves opposite, whorled, or basal; five petals; five stamens, opposite the petals.

The family is of little economic importance. Several species are grown for ornamental value and include primroses, shooting stars, cyclamen, and pimpernel.

creeping Jenny (*Lysimachia nummularia*)

Anagallis arvensis L.
SCARLET PIMPERNEL

Also known as: poor
man's weather glass,
red chickweed, poison
chickweed, shepherd's
clock, eyebright, common
pimpernel, poisonweed,
wink a peep, bird's-eye

Scientific Synonym(s): from
Roman times to the late
1600s, it was believed the
red-flowered form was
male (*Anagallis mas*) and
the blue-flowered, female
(*Anagallis foemina*); now
accepted as subspecies

QUICK ID: leaves opposite;
leaf underside dotted;
flowers bell-shaped, red,
salmon, white, pink, or
blue

Origin: native to Europe and
the Mediterranean region

Life Cycle: annual, biennial,
or short-lived perennial
reproducing by seed;
plants in the wild
producing up to 900
seeds, while those in
greenhouses producing
up to 250,000 seeds

Weed Designation: none

DESCRIPTION

Seed: About 1.3 mm long, three-angled; elliptical,
dull brown to black; finely pitted.
Viable in soil for up to 10 years.
Germination controlled by a water-soluble
germination inhibitor; once broken down,
germination occurs within 6 or 7 days; optimal
temperature 7°–20°C; germination can occur at
2°C; shade-tolerant during early growth.

Seedling: With triangular to diamond-shaped to
elliptical or oval cotyledons, 1–6 mm long, 1–3
mm wide, dark green and shiny; stem below
the cotyledons pale green; first leaves opposite,
green with black or purple dots below, margins
smooth; roots threadlike.

Leaves: Opposite (occasionally in whorls of three),
egg-shaped to elliptical, 5–25 mm long, 4–18 mm
wide, both surfaces hairless, stalkless to clasping,
margins smooth; underside black to purple-
dotted.
Inflorescence a solitary flower borne on stalks 1–5
cm long, originating in leaf axils.

Flower: Salmon to brick red, purple, pink, blue, or
white, 8–14 mm across, opening around 8 A.M.
and closing by 2 P.M.; five sepals, green, awl-like,
3–5 mm long; five petals, 4–10 mm long; five
stamens, yellow, the filaments with pale bluish
hairs; one pistil.

Plant: Plants 10–40 cm tall (commonly less than 15
cm); can withstand temperatures of –10°C for
short periods; stems erect to spreading, four-
angled to square, hairless, branched from the
base, often rooting at the nodes; roots fibrous.

Fruit: Nodding, round capsule; 3–6 mm across;
many-seeded.

REASONS FOR CONCERN

Scarlet pimpernel is a common weed of turf and landscapes and is especially troublesome in sandy soils. It is also weedy in cereals, oilseeds, and vegetables. It is reported to be poisonous to dogs, calves, horses, sheep, and man. The toxic compound is saponin. Sheep are poisoned when 2% of their body weight in plant material has been consumed. Scarlet pimpernel is reported as a weed in 39 countries.

OTHER SPECIES OF CONCERN IN
THE PRIMROSE FAMILY

Androsace septentrionalis

Lysimachia nummularia

ANDROSACE SEPTENTRIONALIS L. —
FAIRY CANDELABRA, PYGMYFLOWER

Native to North America; annual or short-lived perennial; flowering stems 1–25 cm tall; leaves mostly basal, oblong to lance-shaped, 5–30 mm long, 1.5–5 mm wide, margins toothed near tip; inflorescence an umbel, three to 20 rays, bracts 2–3 mm long; flowers white or pinkish, borne on stalks 10–17 mm long (elongating 2–10 cm long in fruit); five sepals, five-veined; five petals, 1–1.2 mm long; five stamens; one pistil; fruit a capsule. Weed designation: none.

LYSIMACHIA NUMMULARIA L. —
CREEPING JENNY, MONEYWORT

Native to Europe and Asia; evergreen to semi-evergreen perennial; stems trailing, 10–40 cm long, branched from the base, winged; leaves opposite, round, 1.3–2.8 cm long and wide, dark green above and light green below, hairless, stalks about 5 mm long, margins smooth; inflorescence a single axillary flower, stalks about 2 cm long; flowers yellow-dotted with dark red, 2–3 cm across; five or six sepals, green, 7–9 mm long; five petals or 6, 5–14 mm long; five or six stamens, filaments and anthers yellow; one pistil; fruit a globe-shaped capsule, 7–9 mm long. Weed designation: USA: CT, MA.

LYSIMACHIA VULGARIS L. —
GARDEN YELLOW LOOSESTRIFE

Native to Europe and Asia; perennial spreading by
rhizomes and seed; stems erect, 60–150 cm tall,
densely glandular-haired; leaves opposite or in
whorls of three or four, elliptical to lance-shaped,
3–12 cm long, dotted with black or orange glands;
inflorescence a panicle, bracts leaflike; flowers
yellow, 8–15 mm wide; five sepals, about 5 mm long,
green with reddish-brown margins; five petals,
yellow; five stamens; one pistil; fruit a capsule. Weed
designation: USA: CT, WA.

Lysimachia vulgaris

Ranunculaceae

CROWFOOT OR BUTTERCUP FAMILY

Herbs, shrubs, and climbing vines; leaves alternate or opposite, simple or palmately compound; leaf bases sheathing; inflorescence a raceme, cyme, or solitary flower; flowers perfect, regular, or irregular; four to 15 sepals, often petal-like; five petals, occasionally absent; five to many stamens; three to many pistils; ovary superior; hypanthium absent; fruit a follicle, achene, berry, or capsule. Worldwide distribution: 35–70 genera/2,000 species; predominantly found in cooler regions of the northern hemisphere.

Distinguishing characteristics: herbs; leaves with sheathing bases; stamens and pistils numerous.

Several members of this family are grown as ornamentals. They include anemones, clematis, columbines, hellebores, and buttercups. Some species, while beautiful, are poisonous to livestock.

cursed crowfoot (*Ranunculus sceleratus*)

Delphinium bicolor Nutt.
LOW LARKSPUR

Also known as: plains
delphinium, little larkspur
Scientific Synonym(s): none

QUICK ID: flowers purple
to light blue; leaves
deeply lobed; stem and
leaves somewhat hairy
Origin: native to North
America
Life Cycle: perennial
reproducing by seed
Weed Designation: none

DESCRIPTION
Seed: Oval, brown, 1–3 mm long, margins with small
grayish wings.
Viability unknown.
Germination information not available.
Seedling: With oval cotyledons, hairless; first leaf
kidney-shaped, margins with rounded teeth; first
year's growth concentrated on root development;
basal rosette of three to five leaves produced by
year 3.
Leaves: Alternate, divided into a few narrow
segments, stalkless; basal leaves circular in
outline, 1–4 cm long, 1.5–7 cm wide, deeply
divided into three to 19 lobed segments, stalks
0.3–8 cm long; two to seven basal leaves and
three to six stem leaves at flowering; leaves
found on the lower third of the stem.
Inflorescence a raceme up to 20 cm long; three-
to 15-flowered.
Flower: Dark blue, irregular, 15–35 mm across; five
sepals, dark blue, hairy, spurred; four petals,
upper two light blue, the lower two dark blue;
several stamens; three pistils; spur 12–35 mm
long; stalks 10–40 mm long; flowering may occur
by year 3, but most often by year 5.
Plant: Of grassland habitats, 20–50 cm tall; stems
unbranched, solid, hairs when present yellow
and somewhat sticky, base often reddish; roots
fibrous and thick.
Fruit: Follicle (capsule), 15–25 mm long, 3–5 mm
wide, slightly hairy; several-seeded.

REASONS FOR CONCERN

Low larkspur is very poisonous. Under normal pasture conditions, horses and sheep are not affected by larkspurs. Cattle are often poisoned as they find it very palatable, especially the new growth in spring. It has been reported that a lethal dose is 0.7% of body weight. Death is usually caused by paralysis and asphyxiation. Constipation and bloating are common signs of larkspur poisoning. Cured hay containing larkspur is as toxic as fresh plant material, as the toxins are not destroyed during the drying process.

Similar species: tall larkspur (*D. glaucum* S. Wats.), a closely related species, has hollow leafy stems up to 2 m tall, and flower clusters more than 15 cm long. It is also very poisonous to livestock.

tall larkspur
(*Delphinium glaucum*)

Ranunculus acris L.
TALL BUTTERCUP

Also known as: meadow
buttercup, tall crowfoot,
blister plant, gold cup,
butter-rose, butter daisy,
horsegold, bachelor's
buttons
Scientific Synonym(s):
R. acer auct.

QUICK ID: flowers yellow;
stem leaves three-lobed;
plant with soft hairs
Origin: native to Europe
Life Cycle: perennial
reproducing by seed;
plants capable of
producing up to 250 seeds
Weed Designation: USA: MT;
CAN: AB, PQ

DESCRIPTION
Seed: Disc-shaped, reddish-brown, 2–3 mm long,
1.8–2.4 mm wide; beak short and hooked.
Viability unknown; some reports indicate viable
in soil for several years.
Germination occurs on the soil surface, or in the
top 1 cm of soil.
Seedling: With round to oval cotyledons, 4–11 mm
long, 1.5–3 mm wide, surface with three or five
veins; first leaves kidney-shaped, somewhat
hairy below, margin with rounded teeth; young
taproot often zigzagged with several lateral
white roots.
Leaves: Alternate, three-lobed, short-stalked; basal
leaves 1.8–5.2 cm long, 2.7–9.8 cm wide, deeply
divided into three to seven coarsely lobed
segments, long-stalked; all leaf surfaces with soft
hairs.
Inflorescence a solitary flower borne on long
stalks.
Flower: Golden yellow, 2–3.5 cm across; five sepals,
hairy, 4–8 mm long; five petals, shiny, 8–16 mm
long; stamens and pistils numerous.
Plant: Plants 30–100 cm tall, prefers heavy moist
soils; stems leafy below, branched above,
somewhat hairy; rootstalk thick and fibrous.
Fruit: Achene; borne in globe-shaped cluster 5–10
mm across.

REASONS FOR CONCERN

Tall buttercup is often found in overgrazed pastures as livestock find it unpalatable. Tall buttercup contains a bitter juice that causes inflammation of the mouth and intestinal tract when ingested. Hay containing tall buttercup does not cause these symptoms as the poisonous property is destroyed when the hay is cured. Tall buttercup is host for anemone mosaic and tomato spotted wilt viruses.

Ranunculus repens L.
CREEPING BUTTERCUP

Also known as: buttercup, crowfoot

Scientific Synonym(s): none

QUICK ID: plants with hairy creeping stolons; flowers yellow; compound leaves with three-toothed leaflets

Origin: native to Europe; common in Quebec by 1821

Life Cycle: perennial reproducing by stolons and seed; average plants producing 77–140 seeds

Weed Designation: USA: MA; CAN: PQ; MEX: federal

DESCRIPTION

Seed: Oblong to oval, light brown to blackish-brown, 2–3.5 mm long, 2–2.8 mm wide; beak triangular, 0.7–1.4 mm long.
Viable in soil for at least 80 years.
Germination enhanced by light; occurs at 0.5–3 cm depth; germination and early seedling growth can take place under waterlogged conditions.

Seedling: With oval cotyledons, 5–7 mm long; first leaves shallowly three-lobed; later leaves with several lobes, covered in soft hairs.

Leaves: Alternate, 3–8 cm across, dark green (often mottled with white), three-lobed, margins toothed; middle segment 0.5–5 cm long, stalked; terminal lobe 1.5–6 cm long; lower leaves long-stalked, upper leaves stalkless; basal leaves 1–8.5 cm long, 1.5–10 cm wide.
Inflorescence a corymb or solitary flower; borne on stalks up to 10 cm long.

Flower: Yellow, 2–3 cm across; five sepals, 4–6 mm long, hairy; five to nine petals, 6–17 mm long, 4–15 mm wide; stamens and pistils numerous; opening early morning and closing by midafternoon; not opening on rainy days, but already open flowers remain open during rainy weather.

Plant: Prostrate, prefers moist habitats; stems 10–30 long, often rooting at nodes, densely haired; roots thick and fibrous; stolons with three to six nodes, the internodes 4–8 cm long.

Fruit: Achene; borne in globe-shaped head 5–10 mm across with 20–25 achenes.

REASONS FOR CONCERN

The bitter sap of creeping buttercup may be poisonous to livestock. Creeping buttercup is host for anemone mosaic and tomato spotted wilt viruses. Serious weed of low-lying pastures and poorly drained cultivated fields.

OTHER SPECIES OF CONCERN IN THE CROWFOOT OR BUTTERCUP FAMILY

Actaea rubra

ACTAEA RUBRA (AIT.) WILLD. —
RED BANEBERRY

Native to North America; perennial; stems 30–100 cm tall; leaves alternate, stalked, pinnately compound with numerous leaflets; leaflets oval to elliptical, unlobed to three-lobed, margins irregularly toothed; inflorescence a pyramidal raceme, 2–17 cm long, 25 or more-flowered, bracts leaflike; flowers white, 0.7–1.2 mm across; three to five sepals, whitish-green, 2–4.5 mm long; four to 10 petals, cream-colored, 2–4.5 mm long; 15–50 stamens; one pistil; fruit a red or white berry, 5–11 mm long, stalks green; seeds 2.9–3.6 mm long, dark brown to reddish-brown. Poisonous. Weed designation: none.

CLEMATIS ORIENTALIS L. —
ORIENTAL VIRGIN'S BOWER

Native to Afghanistan; woody vine 2–8 m tall, climbing by tendril-like leaf stalks; leaves opposite, compound with five to seven leaflets; leaflets elliptical to oval, 1.5–5 cm long, 5–35 mm wide, often two- or three-lobed, margins smooth to coarsely toothed; inflorescence a cyme, three- to many-flowered; flowers greenish-yellow, stalks 1–11 cm long; four sepals, wide-spreading and recurved, 8–20 mm long; 20–40 stamens; 75–150 pistils; fruit an achene, beak 2–5 cm long. Weed designation: USA; CO.

Clematis tangutica

CLEMATIS TANGUTICA (MAXIM.) KORSH. —
YELLOW CLEMATIS

Native to eastern Asia; woody vine up to 5 m tall, climbing by tendril-like leaf stalks; leaves opposite, compound with five leaflets; leaflets diamond-shaped to oval, 1–6 cm long, 5–28 mm wide, often three-lobed at base; margins finely toothed; inflorescence a one- to three-flowered cyme; flowers 2–6 cm across, stalks 3.5–16.5 cm long; four sepals, yellow, 1.5–4 cm long, 0.6–1.4 cm wide; petals absent; stamens many, 5–11 mm long; pistils numerous; fruit an achene, styles elongating to 5 cm in fruit. Weed designation: CAN: AB.

CLEMATIS VITALBA L. — OLD-MAN'S BEARD, EVERGREEN CLEMATIS

Native to Europe and North Africa; woody vine 20–30 m tall, climbing by tendril-like leaf stalks; leaves opposite, compound with five leaflets; leaflets heart-shaped, 7–9 cm long, 3–5 cm wide, margins smooth to toothed; inflorescence a cyme, five- to 22-flowered; flowers greenish-white, 15–25 mm across, stalks 1–1.5 cm long; four sepals, wide-spreading; about 50 stamens; 20 or more pistils; fruit an achene, beak about 3.5 cm long. Weed designation: USA: OR, WA.

RANUNCULUS FICARIA L. — FIG BUTTERCUP, LESSER CELANDINE

Ranunculus ficaria

Native to Europe; perennial flowering in March–April and disappearing by late May; stems erect to spreading; roots tuberous; leaves basal, round to kidney- or heart-shaped, 1.8–3.7 cm long, 2–4 cm wide, dark green and shiny, base heart-shaped, margins smooth or scalloped; inflorescence a solitary flower; flowers yellow, 1–2.5 cm wide; five sepals, 4–9 mm long; eight petals, 10–15 mm long; stamens and pistils numerous; fruit a cluster of achenes 4–5 mm long, 6–13 mm wide; achenes 2.6–2.8 mm long, beak absent. Weed designation: USA: CT, MA.

RANUNCULUS SCELERATUS L. — CURSED CROWFOOT

Ranunculus sceleratus

Native to North America; annual; stems hairless, often rooting at lower nodes; leaves alternate, kidney-shaped to semicircular in outline, three-lobed, 1–5 cm long, 1.6–7 cm wide, margins with rounded teeth; flowers yellow; three to five sepals, bent backward, 2–5 mm long; three to five petals, 2–5 mm long; stamens and pistils numerous; fruit a cylindrical or elliptical head of achene, 5–13 mm long, 3–7 mm wide; achenes 1–1.2 mm long, beak triangular. Poisonous. Weed designation: none.

Rhamnaceae

BUCKTHORN FAMILY

Trees, shrubs, and vines; leaves alternate
or opposite, simple; stipules present;
inflorescence a cyme, umbel, spike or head;
flowers perfect or imperfect; five sepals;
five petals, occasionally four or absent; five
stamens, opposite the petals; one pistil;
ovary superior; fruit a capsule or drupelike
berry. Worldwide distribution: 58 genera/900
species; found throughout the world.

Distinguishing characteristics: trees and
shrubs; flowers with stamens opposite the
petals; fruit a berry.

A few members of this family are grown for
their ornamental value.

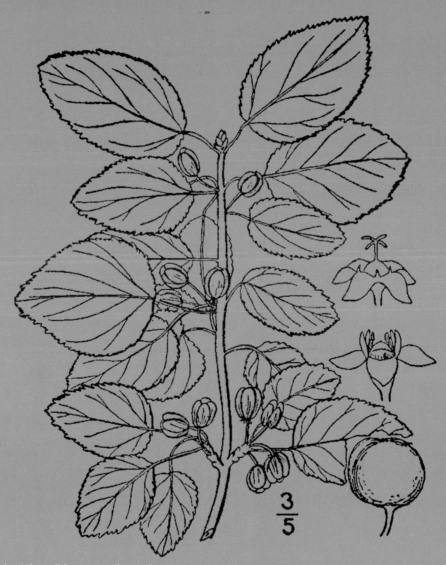

$\frac{3}{5}$

buckthorn (*Rhamnus cathartica*)

Rhamnus cathartica L.
EUROPEAN BUCKTHORN

Also known as: common
buckthorn, Christ's-
thorn, French berries,
Hart's thorn, purging
buckthorn, rainberry
thorn, rhineberry thorn,
waythorn, Carolina
buckthorn

Scientific Synonym(s):
Cervispina cathartica (L.)
Moench

QUICK ID: tree or shrub
with alternate leaves;
flowers green; fruit black,
several-seeded

Origin: native to Europe;
introduced in the early
1800s as an ornamental
shrub for hedges and
shelter belts; by late 1890s
reported to have escaped
cultivation and to have
begun spreading into the
natural environment

Life Cycle: perennial
reproducing by seed;
fruit production begins
at 5 years of age

Weed Designation: USA: CT,
IA, MA, MN, NH, VT; CAN:
AB, MB, ON

DESCRIPTION

Seed: Oval, greenish-gray to black, 4–5 mm long, tips
pointed.
Viable in soil for 2–3 years.
Germination requires the seed coat to
disintegrate, as well as a cold period to break
dormancy.

Seedling: With heart- to kidney- or wedge-shaped
cotyledons; first leaves opposite, elliptical,
stipules present; not shade-tolerant.

Leaves: Opposite, elliptical to oblong, 3–6 cm long,
upper surface dark green, underside yellowish-
green; two to four pairs of veins that curve
toward the tip; margins with 6–8 teeth per cm;
stipules 3–5 mm long, toothed, often absent;
remaining green late into fall.
Inflorescence a cyme, borne in leaf axils; male
flowers borne in groups of two to seven; female
flowers borne in groups of two to 15.

Flower: Greenish-yellow, fragrant, 3–5 mm wide;
male or female; male flowers with four sepals,
four petals, and four stamens; female flowers
with four sepals, four petals, and one pistil; petals
lance-shaped, 1–1.5 mm long.

Plant: Small tree or shrub up to 6 m tall, trunk up
to 25 cm across, tolerant of a variety of soils and
moisture regimes; male or female; bark light gray,
slightly scaly; inner bark yellow; heartwood pink
or orange; branches tipped with a short thorn.

Fruit: Juicy berry, purplish-black, 5–9 mm across;
four-seeded.

REASONS FOR CONCERN

Human consumption of the berries should be avoided, as ingestion of 20 berries will lead to abdominal pain, vomiting, kidney damage, and muscular convulsions. European buckthorn is host for leaf and crown rust of oats, which is responsible for large crop losses in oat fields. This shrub's ability to grow on a variety of soils, its prolific seed production, and its quick germination allow the plant to invade natural ecosystems displacing native species. Some reports indicate that the fruit and leaves of common buckthorn may be allelopathic, thereby reducing competition from adjacent plants.

female flowers

male flowers

OTHER SPECIES OF CONCERN IN THE BUCKTHORN FAMILY

Frangula alnus

FRANGULA ALNUS P. MILL. — GLOSSY BUCKTHORN

Native to Europe, Asia, and North Africa; tall shrub or small tree up to 7 m tall; bark grayish with lenticels; branches brownish-green; leaves alternate, glossy, 4–8 cm long, 2–5 cm wide, oblong to oval or elliptical, stalks about 2 cm long, margins wavy-toothed, five to 10 prominent veins per side; inflorescence an umbel, one- to six-flowered; flowers yellowish-green, bell-shaped, inconspicuous, 2–6 mm wide, stalks 3–10 mm long; five sepals; five petals, 1–1.4 mm long; five stamens; one pistil; fruit a red to black berry-like drupe, about 8 mm diameter; two or three seeds. Weed designation: USA: CT, MA, MN, NH, VT; CAN: MB.

COLUBRINA ASIATICA (L.) BRONGN. — LATHERLEAF, ASIAN NAKEDWOOD

Native to tropical Asia; vinelike evergreen shrub with scrambling branches up to 15 m long; leaves alternate, oval, 4–9 cm long, 2.5–5 cm wide, shiny dark green above, stalks 1–2 cm long, margins toothed; inflorescence a cyme; flowers greenish-white; five sepals; five petals; five stamens; one pistil; fruit a globe-shaped capsule 7–9 mm long, green and fleshy turning brown and dry at maturity; three seeds, gray. Weed designation: none.

Colubrina asiatica

Rosaceae

ROSE FAMILY

Trees, shrubs, and herbs; leaves alternate, simple or compound; stipules often present; inflorescence racemes, spikes, panicles, and solitary flowers; flowers perfect, regular; five sepals; five petals, occasionally absent; five to many stamens; one, five, or many pistils; ovary superior or inferior; hypanthium present; fruit an achene, aggregate, drupe, pome, or follicle. Worldwide distribution: 100 genera/3,000 species; found throughout the world.

Distinguishing characteristics: trees, shrubs, and herbs; stipules present; hypanthium present.

The rose family is very important economically. Members of the family include roses, raspberries, strawberries, apples, peaches, plums, cherries, cinquefoil, hawthorn, and mountain ash. Several species are weedy.

prairie rose (*Rosa arkansana*)

Also known as: tall five-finger, Norwegian cinquefoil, upright cinquefoil

Scientific Synonym(s):
P. grossa Dougl,
P. monspeliensis L.

QUICK ID: flowers yellow; leaves compound with three leaflets; stems and leaves covered in stiff hairs

Origin: native to North America; some introduced subspecies may be present

Life Cycle: annual, biennial, or short-lived perennial; large plants producing up to 13,000 seeds

Weed Designation: none

DESCRIPTION

Seed: Oval, yellowish-brown, 0.6–1.3 mm long, less than 0.8 mm wide.
Viability about 5% after 10 years in soil; about 2% at 20 years.
Germination requires light.

Seedling: With round to oval cotyledons, 1.3–4 mm long, 1–2 mm wide; first leaf oval, three-lobed, surface with soft hairs; compound leaves with three leaflets appear by the fifth leaf stage.

Leaves: Alternate, 2–10 cm long, compound with three (rarely five) leaflets, surfaces with long hairs; leaflets oval, 1.5–7 cm long, 0.8–4 cm wide, margins coarsely toothed, the tips often reddish; basal leaves with stalks up to 10 cm long; leaves reduced in size upward; stipules four- to five-lobed, 1–2.5 cm long.
Inflorescence a leafy terminal cyme; stalks up to 9 mm long, hairy.

Flower: Yellow, 7–12 mm across; five sepals, 4–6 mm long; five petals, 2.5–4 mm long, notched at the tip; 15–20 stamens; pistils numerous; five bractlets, alternating with the sepals, 4–5 mm long, tips reddish.

Plant: Robust, 20–90 cm tall, range climatically controlled where rainfall exceeds 500 mm; stem erect, green to red, covered in long stiff hairs.

Fruit: Achene; several produced per flower; enclosed by the enlarged sepals.

REASONS FOR CONCERN

Rough cinquefoil is a problem weed in pastures, roadsides, gardens, and row crops. It is occasionally found in cultivated fields.

Similar species: the basal leaves of rough cinquefoil are often mistaken for strawberry (*Fragaria* ssp.). Strawberry leaves have teeth on the lower half of the leaflets, unlike those of rough cinquefoil, which are toothed to the tip of the leaflet. The leaves of strawberry have very few hairs. Silverweed (*Argentina anserina* L.), a native species to North America, has seven to 21 leaflets, silver-colored below and green above. Plants produce several runners that root and form leaf clusters at the nodes.

Potentilla recta L.
SULPHUR CINQUEFOIL

Also known as: five-finger cinquefoil, rough-fruited cinquefoil, upright cinquefoil, yellow cinquefoil, tormentil

Scientific Synonym(s): *Hypargyrium recta* Fourr., *P. obscura* Nestler, *P. pilosa* Willd., *P. sulphurea* Lam. & DC.

QUICK ID: leaves palmately compound with five to nine leaflets; flowers yellow; stems densely haired

Origin: native to Europe; introduced into Ontario prior to 1897; first infestations believed to originated from the Azores

Life Cycle: perennial reproducing by seed; plants capable of producing up to 6,000 seeds

Weed Designation: USA: CO, MT, NV, OR, WA; CAN: AB, BC

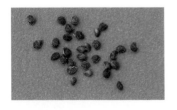

DESCRIPTION
Seed: Blackish-brown, about 2.5 mm long, beak hooked, margins slightly winged; under magnification, surface wrinkled.
Viable after 28 months in soil; possibly up to 4 years.
Germination occurs when soil temperatures reach 20°C; no germination above 35°C; light required for germination as less than 5% of seeds germinate in darkness.

Seedling: With oval cotyledons, 2.8–4 mm long, 1–1.5 mm wide, faintly three-nerved; margins and stem with gland-tipped hairs; first two leaves with toothed margins, surface covered in soft hairs; later leaves with longer bristly hairs; compound leaves with three leaflets appearing at the fifth leaf stage.

Leaves: Alternate, palmately compound with five to nine leaflets; leaflets oblong to lance-shaped, 3–14 cm long, hairy, margins with seven to 17 teeth, pale beneath; lower leaves long-stalked; upper leaves stalkless, three leaflets; stipules present.
Inflorescence a compact terminal corymb; 15–35-flowered.

Flower: Sulfur yellow, 2–2.5 cm across; five sepals, hairy; five petals, heart-shaped, about 1 cm long; 30 stamens; pistils numerous; five bractlets, small and alternating with the sepals; average stems with 25 flowers.

Plant: Long-lived perennial (up to 20 years), range coincides where rainfall is between 750 and 1,250 mm; stems 20–80 cm tall, leafy, hairy; multiple stems (up to 21) originating from the crown; crown shoots produce new plants as the central crown dies.

Fruit: Achene; an average of 62 fruits per flower.

REASONS FOR CONCERN

Sulphur cinquefoil is host for rose mosaic virus. Plants contain a high tanin content, making them unpalatable to livestock and wildlife. An aggressive competitor, it has been reported to crowd out native grasses and herbs.

Rosa multiflora Thunb. ex Murr.
MULTIFLORA ROSE

Also known as: baby rose, Japanese rose, seven sisters rose, bramble-flowered rose

Scientific Synonym(s): *R. cathayensis* (Rehd. & Wilson) Bailey

QUICK ID: shrub with alternate compound leaves; stipules present and featherlike; flowers white in clusters of 25–100

Origin: native to Japan, Korea, and China; introduced into the eastern United States in 1886 as rootstock for ornamental roses; northern range limited by inability to tolerate winter temperatures below −30°C

Life Cycle: perennial reproducing by layering, suckering, and seed; plants may produce up to 1 million seeds each year

Weed Designation: USA: AL, CT, IN, IA, KY, MA, MO, NH, PA, SD, WV, WI

DESCRIPTION

Seed: Oval to oblong, somewhat flattened, yellowish to tan, 4–4.5 mm long, 2–2.6 mm wide. Viable in soil up to 20 years. Germination is enhanced when fruit pass through the digestive tract of birds, the primary vector for dispersal.

Seedling: With oval cotyledons, 4–6 mm long, 3–5 mm wide, margin with prickles; stem below cotyledons prickly; first leaf compound with three oval leaflets (largest being terminal), margins coarsely toothed; leaves with five leaflets appearing by the fifth leaf stage; stipules appearing by the third leaf stage; seedling grow unnoticed for 1–2 years as a low trailing stem.

Leaves: Alternate, 8–11 cm long, stalks 1–1.3 cm long, compound with five to 11 (usually seven or nine) leaflets; leaflets 2.5–4 cm long, margins with five to eight teeth per centimeter, upper surface hairless, lower surface slightly hairy; stipules feathery or comblike, often glandular. Inflorescence a pyramidal panicle of 25–100 flowers; stalk 1–1.5 cm long, hairy, often glandular-haired; one or two bracts.

Flower: White or pinkish-white, 13–25 mm across; five sepals; five petals; stamens and pistils numerous.

Plant: Shrub, forming dense, impenetrable clumps up to 10 m in diameter; bark grayish-brown to brown, hairless; stems 1.5–4.6 m tall, branched, erect to arching or trailing on the ground, stem tips rooting when in contact with soil (layering), surface covered in stout, recurved prickles; twigs reddish-green, hairless; taproot deep, resprouting from the crown.

Fruit: Achene; borne in globe-shaped hips; hips red, 5–8 mm across, surface leathery, remaining on the stem throughout the winter, readily eaten by birds.

REASONS FOR CONCERN

Reproduction by seed, suckering, and layering has allowed the plant to invade pastures, rangeland, abandoned farms, and forest edges, thereby displacing native vegetation. Livestock avoid grazing near the plants due to their prickly nature.

Similar species: several other species of rose have escaped cultivation and become part of the natural environment. Multiflora rose is distinguished from other native and introduced roses by the presence of feathery stipules.

stipule

OTHER SPECIES OF CONCERN IN THE ROSE FAMILY

Crataegus monogyna

***CRATAEGUS MONOGYNA* JACQ. — ENGLISH HAWTHORN, SINGLE-SEED HAWTHORN**
Native to Europe, Asia, and North Africa; deciduous shrub to small tree up to 6 m tall; bark pale gray, smooth; stems with stout spines up to 12 mm long; leaves alternate, oval to oblong, 1.5–3.5 cm long, three- to seven-lobed; stalked, margins mostly smooth; inflorescence a corymb; flowers white fading to pink, 10–15 mm across; five sepals; five petals; 20 stamens; one pistil; fruit a red to purple pome, about 12 mm across; one seed. Weed designation: none.

RUBUS FRUTICOSUS L. (SPECIES COMPLEX INCLUDING R. ARMENIACUS FOCKE, R. DISCOLOR WEIHE & NEES, AND R. LACINIATUS WILLD.) — BLACKBERRY, HIMALAYAN BLACKBERRY

Native to western Europe and North Africa; scrambling shrub up to 2 m tall, canes spreading to trailing, 3–7 m long, often rooting at the tips when in contact with soil; rhizomes creeping; stems arched, green to reddish or purplish, ribbed, prickles straight or backward-pointing and curved, 6–10 mm long; leaves alternate, stalks prickly, palmately compound with three to five leaflets; leaflets oval, dark green above, midribs prickly; inflorescence a panicle; flowers white to pink, 2–3 cm across; five sepals; five petals; stamens and pistils numerous; fruit a globe-shaped aggregate of drupelets, 1–1.5 cm long, 1–3 cm across, green turning to red then black; 20–52 drupelets. Weed designation: USA: federal, AL, CA, FL, MA, MN, NC, OR, SC, VT; MEX: federal.

Rubus fruticosus

Rubiaceae

**BEDSTRAW OR
MADDER FAMILY**

Herbs and shrubs; leaves opposite or whorled, simple; stipules present, often leaflike; inflorescence a cyme; flowers perfect, regular; four or five sepals; four or five petals, united; four or five stamens, borne on the petals; one pistil; ovary inferior; fruit a capsule, berry, or schizocarp. Worldwide distribution: 500 genera/6,000–7,000 species; predominantly found throughout the tropics.

Distinguishing characteristics: herbs; leaves whorled or opposite; stipules present; flowers funnel-shaped.

Economically important members of this family include coffee, gardenias, and the quinine-producing genus of *Cinchona*. A few species are weedy.

yellow spring bedstraw (*Galium verum*)

Galium aparine L.
CLEAVERS

Also known as: bedstraw, white hedge, valiant's cleavers, catchweed, scratch grass, grip grass, goosegrass, catchweed bedstraw, spring cleavers, stickywilly

Scientific Synonym(s): *G. vaillantii* DC., *G. spurium* var. *vaillantii* (DC.) G. Beck, *G. agreste* var. *echinospermum* Wallr.

QUICK ID: leaves in whorls of six to eight; stem square; plants clinging to vegetation by downward-pointing hairs

Origin: native to central Europe and western Asia; first Canadian report in 1870

Life Cycle: annual or winter annual reproducing by seed; plants may produce up to 400 seeds

Weed Designation: USA: NY; CAN: federal, AB, BC, MB, SK

DESCRIPTION

Seed: Round, grayish-brown, 1.9–3 mm in diameter, surface rough-haired, the hairs bristly and hooked, less than 0.6 mm long, often attach themselves to clothing and fur.
Viable in soil for up to 6 years.
Germination is shallow with optimum depth at 8–15 mm; seedlings do not emerge from more than 4 cm depth; germination decreases when soil temperatures exceed 20°C.

Seedling: With oval cotyledons, 10–30 mm long, 5–14 mm wide, notched at the tip; stem below cotyledons often splotched with brown or purple; first leaves in whorl of four, spine-tipped; roots 5–6 cm long at this leaf stage.

Leaves: In whorls of six to eight, 1–8 cm long, 2–3 mm wide, prominently one-veined, stalkless, margins and midrib with bristly hairs assisting the plant in clinging to surrounding vegetation. Inflorescence a cyme, borne in groups of three to nine in leaf axils.

Flower: Greenish-white, about 2 mm across, often overlooked because of their size; four sepals, minute, green; four petals, greenish-white; four stamens; two pistils; flowers appear within 8 weeks of germination.

Plant: Weak and trailing; stems, square, 30–150 cm long, nodes and corners of stem with numerous backward-pointing bristly hairs, enabling the plant to cling to surrounding vegetation for support; roots shallow, branches numerous, short.

Fruit: Nutlet, 3–4 mm long, covered in hooked bristles; borne in pairs.

REASONS FOR CONCERN

Cleavers is an aggressive weed found in cultivated fields, gardens, and roadsides. It is a weed of particular concern to canola growers. The seeds are similar in size to canola and difficult to separate. This results in the downgrading of crop quality. The bristly hairs of cleavers allow the trailing stems to cling to the crop, making harvesting difficult. It has been reported as a principal weed in 31 countries.

Similar species: false cleavers (*G. spurium* L.), another introduced weed, is often confused with cleavers. Its greenish-yellow flowers, 1–1.5 mm wide, produce fruit less than 2.8 mm wide. The leaves appear in whorls of six to eight and are less than 4 cm long. Unlike cleavers, there are very few hairs present at the leaf node. Weed designation: CAN: federal, SK.

Paederia foetida

PAEDERIA FOETIDA L. — SKUNK VINE

Native to eastern Asia; twining vine, ill-scented; stems 2–7 m long; leaves opposite or in whorls of three, oval to oblong or lance-shaped, 2–14 cm long, 2–5 cm wide, lower surface with tufts of hairs in axils of veins, stalks 2–5 cm long, margins smooth; stipules 1.5–1.7 mm long, fringed to hairy; inflorescence a cyme, 2–12 cm long; flowers white to pale yellow with a purple center, 12–16 mm long; five sepals; five petals; five stamens; one pistil; fruit a yellowish-brown to red capsule, globe-shaped, 4–6 mm across. Weed designation: USA: AL, FL.

Salviniaceae

**FLOATING WATER
FERN FAMILY**

Free-floating aquatic herbs; roots absent; three
leaves per node, two floating on surface, one
dissected and resembling roots; sporangia
produced in sporocarps borne at the base of
submersed leaf. Worldwide distribution: one
genus/10 species; tropical and temperate
regions throughout the world.

Distinguishing characteristics: plants
aquatic, floating; two leaves and roots
appearing at the same node; flowers absent.

A few species of this family are serious
weeds of lakes, ponds, and irrigation canals.

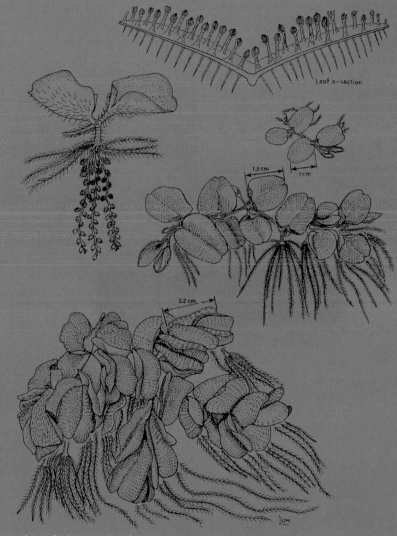

giant salvinia (*Salvinia molesta*)

Salvinia molesta D. S. Mitchell
GIANT SALVINIA

Also known as: kariba weed
Scientific Synonym(s):
  Salvinia auriculata Auctt.
  non Aubl.

QUICK ID: free-floating
  aquatic; leaves in whorls
  of three (two green, one
  brown and resembling
  roots)
Origin: native to Brazil and
  Argentina; introduced
  ornamental for ponds and
  aquaria in 1990s
Life Cycle: perennial
  reproducing by stem
  fragments, lateral root
  buds, and spores; primary
  mode of reproduction is
  by fragmentation
Weed Designation: USA:
  federal, AL, AZ, CA, CO, CT,
  FL, MA, MS, NC, NV, OR,
  SC, TX, VT

DESCRIPTION
Seed: Not produced by this species; reproduction is
  by spores.
  Viability of spores unknown.
  Germination of spores unknown.
Seedling: Not produced by this species.
Leaves: In whorls of three; two floating leaves,
  opposite each other, round to oblong, 5–60
  mm across, upper surface with rows of hairs
  with four branches fused at the tip (resembling
  eggbeaters), acting as a water repellent;
  submersed leaf brown, resembling roots,
  feathery and branched, up to 25 cm long.
  Spores produced by this species; sporocarps egg-
  shaped, 2–3 mm across, borne on stalks less than
  1 mm long, spirally arranged along the main axis
  of the submersed filaments.
Plant: Aquatic, up to 30 cm long, preferring tropical
  to subtropical areas, and still to slowly moving
  water, optimal water temperatures between
  20°C and 30°C; rhizome horizontal and below
  water surface; fronds light green, often with
  brownish edges, distinctly folded in the center;
  buds intolerant of temperatures below −3°C and
  above 43°C; under ideal conditions, plants double
  in size every 2–4 days, and biomass doubled every
  10 days; dense mats don't form when water
  temperatures are below 10°C.
Fruit: Not produced by this species; sporocarps
  borne in chainlike clusters along the submersed
  frond (roots); sporangia numerous.

REASONS FOR CONCERN

Recognized as one of the world's worst aquatic weeds. Its rapid growth causes it to form dense mats that reduce water flow, light and oxygen levels, and species diversity. It has been observed to clog dams and machinery associated with hydroelectric generation. It has been reported as a serious weed in 22 countries.

OTHER SPECIES OF CONCERN IN THE FLOATING WATER FERN FAMILY

Salvinia auriculata

***SALVINIA AURICULATA* AUBL. (SPECIES COMPLEX INCLUDING *S. BILOBA* RADDI AND *S. HERZOGII* DE LA SOTA) — EARED WATERMOSS**
Native to South America; plants aquatic; stems horizontal, just below water surface; leaves in whorls of three; two floating leaves, 1.5–2 cm long, 1.8–2.5 cm wide, upper surface with eggbeater-shaped hairs, 2–4 mm long; submersed leaf resembling roots, stalks less than 1.4 cm long, branches 8–60 mm long; sporocarps globe-shaped, borne in chainlike clusters, stalks 1–12 mm long. Weed designation: USA: federal, AL, CA, FL, MA, NC, OR, SC, TX, VT.

SALVINIA MINIMA BAKER—ROUND-LEAVED SALVINIA, WATER SPANGLES

Native to Mexico, Central America, and South America; free-floating aquatic; stems horizontal 1–6 cm long; leaves whorled (two floating, one submersed); submersed leaf dissected, appearing rootlike; floating leaves round to oval, 4–20 mm long, emerald green (turning brown at maturity), base heart-shaped; upper surface with rows of white bristly hairs, each with four branches at the tip and repelling water; underside with long brown hairs; sporocarps, 2–3 mm long, nutlike, borne in groups of four to eight, spirally arranged along the main axis of the submersed filaments. Weed designation: USA: FL, NC, TX.

Salvinia minima

Sapindaceae

SOAPBERRY FAMILY

Trees, shrubs, and vines; leaves alternate, simple or pinnately compound; stipules absent; inflorescence a raceme or cyme; flowers imperfect, regular, or irregular; five sepals; five petals, occasionally absent; 10 stamens (in two whorls of five), or numerous; one pistil; ovary superior; fruit a capsule, nut, berry, drupe, schizocarp, or samara; seeds often with arils. Worldwide distribution: 150 genera/2,000 species; tropical and subtropical regions of the world.

Distinguishing characteristics: trees, shrubs, and vines; stipules absent; 10 stamens, in two whorls.

A few species are grown for their ornamental value.

heartseed (*Cardiospermum halicacabum*)

Cardiospermum halicacabum L.
HEARTSEED

Also known as: balloon vine, love in a puff, heart pea, winter cherry

Scientific Synonym(s): *C. microcarpum* Kunth.

QUICK ID: climbing vine; leaves compound with three leaflets; seeds with a heart-shaped attachment

Origin: native to tropical North America

Life Cycle: annual, biennial, or perennial reproducing by seed

Weed Designation: USA: AL, AR, SC, TX

DESCRIPTION

Seed: Round, grayish-black, 4.7–5.1 mm across, surface smooth; attachment face heart-shaped and buff colored, covering a third or more of seed face.
Viable up to 18 months in soil.
Germination occurs from 20°C to 35°C; optimal depth 1–5 cm.

Seedling: With rectangular cotyledons; first leaf three-lobed, the terminal lobe also three-lobed.

Leaves: Alternate, stalks 10–57 mm long, compound with three leaflets; leaflets elliptical to oblong, oval, or triangular, 2.8–6.2 mm long, 1–4.3 cm wide, margins coarsely toothed to lobed, surfaces with simple hairs; tendrils present.
Inflorescence a raceme.

Flower: Greenish-white, 2–4 mm across; four sepals, 0.8–4 mm long; four petals, 1.5–5 mm long; stamens 8 (4 long, 4 shorter); one pistil; stalks 1–4 mm long.

Plant: Climbing vine; stems branched, 2–8 m long.

Fruit: Inflated capsule, three-sided, 1.2–4.1 cm long, 1.5–4.4 cm wide.

REASONS FOR CONCERN

Heartseed is a fast-growing vine that quickly covers shrubs and natural vegetation. It is also reported as a problem weed in soybean fields in Texas.

OTHER SPECIES OF CONCERN IN
THE SOAPBERRY FAMILY

CUPANIOPSIS ANACARDIOIDES (A. RICH.) RADLK. — CARROTWOOD

Native to Australia; evergreen tree up to 10 m tall; bark dark gray, inner back orange; leaves alternate, stalks swollen at base, compound with four to 12 leaflets; leaflets oblong, up to 20 cm long and 7.5 cm wide, leathery, shiny yellowish-green, stalked, margins smooth; inflorescence a panicle up to 35 cm long; flowers white to greenish-yellow, about 8 mm wide; five sepals; five petals; six to eight stamens; one pistil; fruit a woody capsule, about 2.2 cm across, yellowish-orange when ripe and turning brown at maturity; three seeds, black, covered with a yellowish-red aril. Weed designation: USA: FL.

Scrophulariaceae

FIGWORT FAMILY

Herbs or subshrubs; leaves alternate, opposite, or whorled, simple to pinnately dissected; stipules absent; inflorescence a spike or raceme; flowers perfect, regular, or irregular; four or five sepals, united; four or five petals; four stamens, occasionally two or five; sterile stamen (staminode) often present; one pistil; ovary superior; fruit a capsule. Worldwide distribution: 65 genera/1,800 species; found throughout the world.

Distinguishing characteristics: herbs; leaves alternate or opposite; corolla often two-lipped.

Scrophulariaceae is of limited economic importance. A few species are noxious weeds.

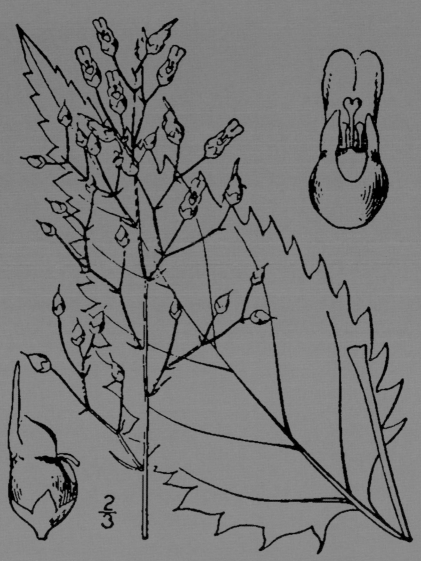

lanceleaved figwort (*Scophularia lanceolata*)

Verbascum thapsus L.
COMMON MULLEIN

Also known as: velvet dock, big taper, candle-wick, flannel leaf, torches, Jacob's-staff, blanket leaf, ice leaf, velvetleaf, hedge taper, Aaron's rod, devil's-tobacco, flannel plant

Scientific Synonym(s): none

QUICK ID: flowers yellow; stem and leaves covered in soft feltlike hairs; flowering stem with reduced leaves

Origin: native to Greece; introduced into North America as a medicinal plant and as a fish-control agent; first report from Virginia in the mid-1700s, spreading to the Pacific coast by 1876

Life Cycle: biennial reproducing by seed; plants capable of producing up to 180,000 seeds

Weed Designation: USA: CO, HI; CAN: AB

DESCRIPTION
Seed: Oblong, dark gray to brown, 0.7–0.9 mm long, 0.4–0.5 mm wide, surface wrinkled with several intersecting ridges.
Viable in soil for at least 100 years.
Germination requires light and temperatures above 10°C for germination; establishment occurs more readily on bare soil than vegetated areas.

Seedling: With egg-shaped cotyledons, 1–6 mm long, 1–3.5 mm wide, surface slightly hairy; first leaves opposite, slightly hairy; leaves after the third leaf stage covered in dense woolly hairs; rosettes less than 15 cm across not surviving winter temperatures.

Leaves: Alternate, 10–40 cm long, reduced in size upward, leaf bases continuing down the stem; basal leaves elliptical to oblong, 15–45 cm long, 2–10 cm wide; all leaf surfaces with dense woolly branched hairs.
Inflorescence a dense spike, 20–50 cm long, about 3 cm across.

Flower: Yellow, about 2.5 cm across, stalkless; five sepals, united, five petals, united; five stamens (three short, two long), covered in white or yellow hairs; one pistil.

Plant: Range climatically controlled as growing season must be at least 140 days; flowering stems 50–250 cm tall, covered in woolly hairs, occasionally short-branched; taproot deep.

Fruit: Oval capsule, 3–10 mm long, surface with woolly hairs; capsule averaging of 600 seeds; average stems produce 180–250 capsules.

692 ■ FIGWORT FAMILY

REASONS FOR CONCERN

Common mullein increases on overgrazed pastures as livestock find the woolly leaves unpalatable. The large number of seeds and their longevity make control of this weed difficult. It is often found growing in gravelly soils. Common mullein is host for Prunus B & G, tobacco mosaic, and tobacco streak viruses.

Similar species: black mullein or black torch (*V. nigrum* L.) is distinguished from common mullein by stamens that are covered in purple hairs. Another species, orange mullein (*V. phlomoides* L., p. 695), has leaf bases that do not continue down the flowering stem as they do in common mullein.

OTHER SPECIES OF CONCERN IN
THE FIGWORT OR SNAPDRAGON FAMILY

Verbascum blattaria

VERBASCUM BLATTARIA L. — MOTH MULLEIN
Native to eastern Europe and Russia; biennial, flowering stem unbranched, basal leaves 3–10 cm long, 1–4 cm wide, margins with rounded teeth to lobed; stem leaves reduced in size upward, margins irregularly toothed; inflorescence 10–50 cm long; flowers yellow, on stalks 5–10 mm long; five sepals, 5–6 mm long; five petals, the upper three lobes woolly at base; five stamens, the filaments purple and woolly; one pistil; fruit a capsule 7–8 mm long. Weed designation: CO.

VERBASCUM PHLOMOIDES L. —
ORANGE MULLEIN, AARON'S ROD

Native to Europe and Asia; biennial; stems 60–120 cm tall, covered in simple nonglandular hairs; leaves alternate, lance-shaped, clasping, the bases running slightly down the stem; flowers yellow, 2–5.5 cm across, borne in a dense spikelike stem; five sepals; five petals, the lower three slightly larger than the upper two; five stamens, bearded; one pistil; fruit a capsule, 5–8 mm long; seeds brown. Weed designation: none.

Verbascum phlomoides

Simaroubaceae

QUASSIA FAMILY

Trees and shrubs; male or female; leaves alternate, pinnately compound; inflorescence a raceme or panicle; flowers imperfect, regular; three to eight sepals; three to eight petals; six to 16 stamens; one pistil; ovary superior; fruit a schizocarp (breaking into one to five mericarps), capsule or aggregate of drupes or samaras. Worldwide distribution: 20–32 genera/100–200 species; predominantly tropical and subtropical Asia and Australia.

Distinguishing characteristics: tree; leaves compound; fruit with a papery wing.

A few species are grown for their ornamental value.

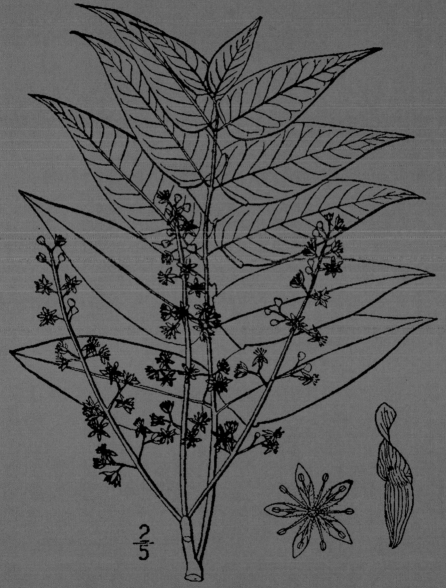

$\dfrac{2}{5}$

tree of heaven (*Ailanthus altissima*)

Ailanthus altissima (P. Mill.) Swingle
TREE OF HEAVEN

Also known as: ailanthus, copal tree, paradise tree, Chinese sumac, stinking sumac, varnish tree

Scientific Synonym(s):
A. glandulosa Desf.

QUICK ID: tree; leaves alternate, compound with 11–41 leaflets; leaves ill-scented when crushed

Origin: native to China; introduced into England in 1751 mistakenly as a lacquer tree and, later, into Pennsylvania in 1784; once widely planted along city streets

Life Cycle: perennial (living to 50 years) reproducing by resprouting and seed; plants may produce up to 350,000 seeds per year; most abundant seed production on trees aged 12–20 years old

Weed Designation: USA: CT, MA, NH, VT

DESCRIPTION

Seed: Elliptical, flattened, brown, 4.5–7 mm long. Viable less than 1 year in soil.
Germination can occur on compacted soils with a high salt content; requires a cold period to break dormancy.

Seedling: With round to oval cotyledons; first leaves opposite, compound with three leaflets (terminal being largest); first season of growth 1–2 m; well-formed taproot at 3 months of age; resprouts distinguished by leaves with more than three leaflets and thick ropelike roots; resprouts may emerge up to 15 m from the nearest stem, and grow 3–4 m per year.

Leaves: Alternate, 30–100 cm long, dark green, compound with 11–41 leaflets; leaflets lance-shaped, 7.6–12 cm long, 2.5–5 cm wide; base with one or two teeth, gland-bearing; both surfaces with minute hairs and glands; leaf scars conspicuous, triangular; ill-scented when crushed; turning yellow in autumn.
Inflorescence a panicle 10–20 cm long, borne on new growth; each plant producing hundreds of panicles.

Flower: Greenish-yellow, 6–8 mm long; male or female; male flowers ill-scented; five or six sepals; five or six petals, 1.5–2.5 mm long; six to 16 stamens; one pistil; flowering generally commences around 10 years of age, although first-year seedlings have been reported producing flowers; 2-year-old resprouts reported to produce flowers.

Plant: Deciduous tree up to 30 m tall, growing 1–1.5 m per year; bark gray, smooth; twigs light brown, covered in fine hairs when young; shade-intolerant; surviving winter temperatures of −35°C; male or female, male trees producing three to four times more flowers than female trees.

Fruit: Yellowish-green to reddish-brown schizocarp; breaking into two to five winged mericarps (samara-like); mericarps 3–5 cm long, 7–13 mm wide, twisted, wind-dispersed, single-seeded; fruiting clusters up to 30 cm across.

REASONS FOR CONCERN

The leaves of trees of heaven may cause skin irritation in sensitive individuals. If the main stem is damaged, suckers are produced and may grow 3–4 m per year. The seeds are wind-dispersed and germinate readily. In addition to its prolific vegetative and sexual reproduction, ailanthus has allelopathic effects on other tree species. Studies have shown effects on 34 native hardwood species, except white ash (*Fraxinus americana* L.).

Similar species: staghorn sumac (*Rhus typhina* L., p. 700) may be mistaken for young specimens of tree of heaven. This shrub, 2–10 m tall, has compound leaves with toothed leaflets. During winter it can be distinguished by red cone-shaped clusters of fruit at branch tips. *Ailanthus* may be confused with other trees having alternate compound leaves with several leaflets. Tree of heaven is distinguished from other tree species by ill-scented foliage.

The following pages illustrate trees and large shrubs that have alternate compound leaves with seven or more leaflets and that may be confused with tree of heaven.

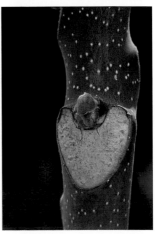

staghorn sumac (*Rhus typhina*)

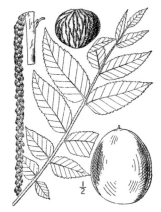

black walnut (*Juglans nigra*)

poison sumac
(*Toxicodendron vernix*)

butternut (*Juglans cinerea*)

American mountain-ash
(*Sorbus americana*)

black locust
(*Robinia pseudoacacia*)

bitternut hickory
(*Carya cordiformis*)

Solanaceae
NIGHTSHADE OR POTATO FAMILY

Herbs, shrubs, vines, and trees; leaves alternate, simple to compound; stipules absent; inflorescence a cyme; flowers perfect, regular; five sepals, united; five petals, united; five stamens, borne on the petals; one pistil; ovary superior; fruit a berry or capsule. Worldwide distribution: 85 genera/2,300 species; tropical and subtropical regions of the world with the greatest diversity in Central and South America.

Distinguishing characteristics: herbs; leaves alternate; flowers regular; fruit berry-like or capsule.

The nightshade family is a very important group of plants. Many species are grown for food and include potatoes, tomatoes, eggplants, and peppers. Several other species are grown as ornamental plants. These include petunias, chinese lanterns, and angel's trumpets. Tobacco is another important crop species of this family. The poisonous members of this family include henbane and belladona.

wild tomato (*Solanum triflorum*)

Datura stramonium L.
JIMSONWEED

Also known as: Jamestown weed, thorn apple, mad apple, stinkwort, angel's trumpet, devil's trumpet, dewtry, whiteman's-weed, moonflower

Scientific Synonym(s): *D. tatula* L.

QUICK ID: flowers white, trumpet-shaped; fruit a spiny capsule; leaves alternate

Origin: native to India; first North American report in 1676; first Canadian report from Montreal in 1821

Life Cycle: annual reproducing by seed; plants capable of producing up to 35,000 seeds

Weed Designation: USA: CT, PA; CAN: federal, MB, NS

DESCRIPTION

Seed: Black, disc-shaped, 3.4–3.8 mm long, 2.8–3.2 mm wide, less than 1.5 mm thick; surface dull, slightly wrinkled; extremely poisonous. Viable for at least 40 years in soil. Germination occurs in the top 8 cm of soil.

Seedling: With lance-shaped cotyledons, 20–40 mm long, and 4 mm wide, midrib prominent; stem below cotyledons often purplish; first leaves opposite, egg-shaped, upper surface with scattered hairs, underside prominently veined; cotyledons wither and remain attached to the developing stem.

Leaves: Alternate, oval to triangular, 7–20 cm long, long-stalked, margins irregularly toothed; ill-scented when crushed. Inflorescence a solitary flower borne in leaf axils.

Flower: White or purplish, funnel- to trumpet-shaped, 5–12.5 cm long and 5 cm across; five sepals, fused into a tube; petals, united; five stamens; one pistil.

Plant: Ill-scented; stems 70–200 cm tall, 3–5 cm diameter, branched near the top, often purplish-green; roots thick and shallow.

Fruit: Spiny capsule, egg-shaped, 3–6 cm wide; four-celled; 600–700-seeded; plants may produce up to 50 fruits.

REASONS FOR CONCERN

Jimsonweed is a serious weed of cultivated fields, row crops, and farmyards. It has been reported to reduce yields by 45% in beans and 77% in tomatoes. All parts of the plant are poisonous, especially the seeds. It is reported that the plant has a narcotic affect that may result in death to livestock or humans who ingest it. Less than 0.1% of body weight is considered a lethal dose. The nectar of jimsonweed flowers contains several alkaloid compounds that are reported to contaminate honey. Over 55 viral diseases have been identified with jimsonweed, including beet curly top, cucumber mosaic, and several potato, tobacco, and tomato viruses.

Similar species: datura or thorn apple (*D. inoxia* Small.), an ornamental plant, is closely related to jimsonweed. It is distinguished by flowers 12–20 cm long, stems up to 60 cm tall, and reddish-brown disc-shaped seeds.

datura (*Datura inoxia*)

Hyoscyamus niger L.
BLACK HENBANE

Also known as: fetid
nightshade, henbane
Scientific Synonym(s): none

QUICK ID: flowers
yellowish-green with
purple veins; leaves
alternate, sticky haired;
fruit with bristle-tipped
calyx
Origin: native to Europe;
introduced as a medicinal
herb
Life Cycle: annual or biennial
reproducing by seed; each
plant producing up to
100,000 seeds
Weed Designation: USA: CA,
CO, ID, NV, NM, WA; CAN:
AB

DESCRIPTION
Seed: Egg- to pear-shaped, light grayish-brown,
1.5 mm long, 1.2–1.4 mm wide, less than 0.8
mm thick; under magnification surface with a
network of raised veins.
Viable in soil for up to 5 years.
Germination of fresh seed does not require light;
dormant seeds require light for germination.
Seedling: With oval cotyledons, 3.5–5 mm long,
1–1.5 mm wide, with a few hairs near the base;
first leaves elliptical, stalks with numerous hairs,
clammy; ill-scented if crushed.
Leaves: Alternate, 7–20 cm long, stalkless and
clasping the stem, margins irregularly lobed,
reduced in size upward; basal leaves 10–45 cm
long, 3–15 cm wide, margins wavy to shallowly
lobed; all leaves covered in sticky hairs; ill-
scented if crushed.
Inflorescence a solitary flower borne in leaf axils;
flowers often appearing on one side of stem.
Flower: Yellowish-green and purple-veined, funnel-
shaped, 25–45 mm long, 25–35 mm wide, short-
stalked; five sepals, united; five petals, united;
five stamens, purple; one pistil.
Plant: Ill-scented; stems branched, 40–100 cm tall,
surface with long sticky hairs; taproot robust,
spindle-shaped.
Fruit: Globe-shaped capsule, 1–1.5 cm long,
50–200-seeded; persistent calyx, 2–2.5 cm long,
urn-shaped and bristle-tipped, surrounding the
capsule.

REASONS FOR CONCERN

Black henbane is a common weed of waste areas, roadsides, and pastures. It is rarely found in cultivated fields. Black henbane is toxic to livestock and man. Poisonings are rare as the sticky hairs make it unpalatable. The poisonous properties are retained in cured hay. An early symptom is depression of the central nervous system by hyoscyamine. Levels of this toxin in the plant range from 0.08% to 0.3%. Ingestion of sufficient plant material will cause convulsions, increase heart rate, decrease respiration, and eventually cause death. It is host for cucumber mosaic, lucerne mosaic, and several potato, tobacco, and tomato viruses.

Nicandra physalodes (L.) Gaertn.
APPLE OF PERU

Also known as: shoofly plant
Scientific Synonym(s): *Atropa physalodes* L., *Physalodes physalodes* (L.) Britt.

QUICK ID: stems angular; flowers pale blue or white; fruit a yellowish berry surrounded by the papery sepals
Origin: native to Peru; introduced as an ornamental plant
Life Cycle: annual reproducing by seed; large plants producing up to 44,000 seeds
Weed Designation: none

DESCRIPTION

Seed: Round, light brown, 1.25–1.5 mm across, surface smooth with fine lines.
Viable in soil (7.5 cm depth) for at least 5 years. Germination occurs within 5 days; studies show seeds may remain dormant for up to 28 years.
Seedling: With lance-shaped cotyledons, 1–4.5 mm wide, 3.5–14 mm long, unequal, surface smooth; stem below cotyledons purple-tinged, often with hairlike projections; first leaves oval to oblong, upper surface with cone-shaped hairs; later leaves sparsely haired.
Leaves: Alternate, oval to oblong, 6–25 cm long, 2.5–18 cm wide, stalks 1–20 cm long, margins irregularly or shallowly lobed; both surfaces bright green with a few scattered hairs.
Inflorescence a solitary flower borne in leaf axils.
Flower: Pale blue, white, or blue with a white center, bell-shaped, 1–4 cm across, erect during flowering and nodding later, stalks 1–4 cm long; five sepals, united, 1–3 cm long, membranous; five petals, united; five stamens, one pistil; flowering occurs 43–54 days after germination.
Plant: Plant ill-scented; stems 40–200 cm tall, erect, hollow, angular, grooved or furrowed, branched.
Fruit: Globe-shaped berry, yellowish, 1–1.5 cm across; many-seeded; calyx persistent, enlarging and enclosing the mature fruit, green turning brown and papery with age; fruit appearing 53–64 days after germination.

REASONS FOR CONCERN

The seeds of apple of Peru germinate early in spring and then sporadically through the growing season making control difficult. The seeds are also similar in shape to grasses and legumes, making cleaning difficult. It is a serious weed of pastures, pineapple, sugarcane, fodder beets, lettuce, papaya, peppers, and sweet potatoes.

Solanum carolinense L.
HORSE NETTLE

Also known as: bull nettle, Carolina horse nettle, apple of Sodom, wild tomato, sand briar, tread-softly, devil's tomato; devil's potato

Scientific Synonym(s): none

QUICK ID: plants prickly; flowers white; fruit a yellow berry

Origin: native to southeastern United States and northern Mexico

Life Cycle: perennial reproducing by rhizome and seed; large plants producing up to 5,000 seeds

Weed Designation: USA: AK, AZ, AR, CA, HI, IA, NV; CAN: federal, MB; MEX: federal

DESCRIPTION

Seed: Disc-shaped, yellow to orange brown, 1.5–3 mm long, 1.3–2.2 mm wide, surface smooth and glossy.
Viable up to 10 years in soil; seeds lose viability when exposed to temperatures exceeding 50°C. Germination enhanced by alternating temperatures; emergence from soil depths to 10 cm in light-textured soils (optimal 1–5 cm).

Seedling: With elliptical to lance-shaped cotyledons, 10–12 mm long and 1.5 mm wide, shiny green above and pale beneath; surfaces with stiff prickly hairs, margins hairy; stem below cotyledons purplish, surface with a few short hairs; first leaves alternate, elliptical, surface with a few short stiff hairs; third leaf stage onward, leaves with star-shaped hairs, margins wavy or lobed.

Leaves: Alternate, oval to egg-shaped, 7–15 cm long, 3–6 cm wide, dark green, tips pointed, margins sharply lobed (two to five lobes per side) or toothed; stalks, veins, and midribs covered in rough star-shaped hairs and prickly spines; spines yellow, flattened, 2–5 mm long.
Inflorescence a one-sided raceme-like cyme; five- to 20-flowered; stalk prickly

Flower: White to violet, 1.5–2.5 cm across; lower flowers perfect, upper flowers male; five sepals, united; five petals, united; five stamens, about 6 mm long; anthers 7–9 mm long; one pistil; flowering commences 1 month after new shoots emerge; berries begin to mature 4–8 weeks later.

Plant: Plants 20–120 cm tall; stems erect, branched, covered in star-shaped hairs (four- to eight-rayed) and yellow spines (6–12 mm long); rhizome deep, fleshy, extending up to 1 m before producing new shoots; vertical roots up to 2 m deep.

Fruit: Globe-shaped berry, 9–15 mm across, green
turning yellow at maturity; 40–170-seeded
(average 85); calyx persistent and spiny, partially
surrounding the fruit.

REASONS FOR CONCERN

Fresh plants and hay containing horse nettle
is poisonous to livestock. Toxins are often
most concentrated in the fall. Symptoms
of acute poisoning include gastrointestinal
irritation and nervous effects such as apathy,
drowsiness, salivation, trembling, breathing
difficulties, progressive weakness, paralysis, and
unconsciousness. Chronic poisoning includes
appetite loss, emaciation, rough coat, and
constipation. Sheep and goats are more resistant to
poisoning than are cattle or horses. A common weed
of orchards, pastures, and corn and vegetable crops,
it is especially difficult to control in tomato, potato,
and tobacco crops. It also harbors leaf spot fungus
of tomatoes, mosaic virus of tomatoes and potatoes,
verticillium wilt of eggplant, potato stalk borer,
potato flea beetle, potato psyllid, and onion thrips.

Solanum dulcamara L.
BITTERSWEET NIGHTSHADE

Also known as: climbing
nightshade, deadly
nightshade, bittersweet,
bitter nightshade,
European bittersweet,
blue nightshade, woody
nightshade, poison berry,
scarlet berry, fellenwort,
dogwood

Scientific Synonym(s):
Dulcamara flexuosa
Moench

QUICK ID: climbing vine
with an unpleasant odor;
flowers purple; fruit a red
berry

Origin: native to Europe and
Asia

Life Cycle: perennial
reproducing by seed and
creeping rhizomes

Weed Designation: USA: CT;
CAN: MB

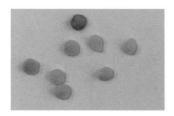

DESCRIPTION

Seed: Disc-shaped, pale yellow, 2.1–2.5 mm across,
surface pitted.
Viable for at least 4 years in laboratory storage.
Germination occurs when soil temperatures
reach 20°–35°C; emergence from top 7.5 cm of
soil.

Seedling: With elliptical to oval cotyledons; first
leaves oblong to elliptical or oval, margins
smooth, distinctly veined.

Leaves: Alternate, oval to heart-shaped, 4–10 cm
long, dark green often with purplish tinge; leaf
shape variable, often on the same plant; upper
leaves with one or two small lobes or leaflets.
Inflorescence a raceme; borne on stalks on
internodes or opposite the leaves.

Flower: Purple, wheel-shaped, 8–12 mm across; five
sepals; five petals; five stamens, yellow; one
pistil.

Plant: Semiwoody vine up to 4 m tall, ill-scented;
bark light gray, shredding with age; stems
branched; rhizomes creeping.

Fruit: Elliptical berry, 8–12 mm long, bright red at
maturity, 40–60-seeded; often remaining on the
stem throughout the winter.

REASONS FOR CONCERN

The leaves and berries of bittersweet nightshade contain toxic alkaloids reported to be poisonous to livestock and humans. The bitter and sweet flavor of the berries is responsible for poisoning of small children. This species is host of viral diseases that affect potatoes, tomatoes, and tobacco.

Solanum nigrum L.
BLACK NIGHTSHADE

Also known as: deadly nightshade, poisonberry, garden nightshade, hound's berry, garden huckleberry, West Indian nightshade

Scientific Synonym(s):
S. rubrum auct. non L.

QUICK ID: leaves alternate; flowers white with yellow centers; fruit a black berry

Origin: native to South America; introduced into North America; first Canadian report from New Brunswick in 1877

Life Cycle: annual reproducing by seed; large plants producing up to 178,000 seeds

Weed Designation: CAN: MB; MEX: federal

DESCRIPTION

Seed: Disc-shaped, light brown, 1.7–2.1 mm long, 1.4–1.6 mm wide, less than 0.7 mm thick. Viable in soil for 8 years. Germination occurs between 20°C and 30°C.

Seedling: With lance-shaped cotyledons, 3–9.5 mm long, 1–4 mm wide, upper surface and midvein below with minute hairs; stem below cotyledons with short sticky hairs; first leaves egg-shaped, upper surface hairy at first and becoming hairless with age, underside remaining hairy.

Leaves: Alternate, oval to triangular, 3–8 cm long, 1–5 cm wide, dark green, long-stalked, margins wavy or irregularly toothed.
Inflorescence a raceme; five- to 10-flowered; borne on short stalk connected directly to the stem.

Flower: White with a yellow center, 3–8 mm across; five sepals, united; five petals, united; five stamens; one pistil.

Plant: Somewhat hairy, ill-scented; stems erect, 10–100 cm tall, widely branched; roots fibrous.

Fruit: Globe-shaped berry, 9–15 mm across; green and turning bluish-black at maturity; many-seeded; plants producing up to 100 berries.

REASONS FOR CONCERN

A weed of cultivated fields and waste areas, black nightshade is not easily controlled by herbicides. The fruits are often harvested with beans and peas, which lowers the quality of the crop. The unripe fruit of black nightshade are poisonous to humans and livestock. It is also host for the following viral diseases: cucumber mosaic, tobacco mosaic, tomato spotted wilt, beet curly top, chili mosaic, petunia mosaic, and potato A. It is a serious weed in 61 countries.

Similar species: hairy nightshade (*S. physalifolium* Rusby), another introduced species, is often confused with black nightshade. It is distinguished from black nightshade by bristly and sticky hairs on the stem and fruits that are yellowish-green at maturity.

Solanum rostratum Dunal
BUFFALO BUR

Also known as: beaked
  nightshade, sandbur,
  Colorado bur, Kansas
  thistle, Mexican thistle,
  Texas thistle, horned
  nightshade, prickly potato
Scientific Synonym(s):
  S. cornutum auct. non
  Lam., *Androcera rostrata*
  (Dunal) Rydb.

QUICK ID: plants prickly;
  flowers yellow; fruit a
  brown or black berry
  surrounded by prickly
  calyx
Origin: native to central and
  southwestern United
  States
Life Cycle: annual
  reproducing by seed
Weed Designation: USA: ID,
  OR, WA; CAN: MB

DESCRIPTION
Seed: Disc-shaped, brownish-black to dark red-
  brown, 2.4–2.7 mm long, 2–2.1 mm wide, surface
  coarsely pitted and wrinkled.
  Viability unknown.
  Germination occurs throughout the growing
  season.
Seedling: With lance-shaped cotyledons, 1.5–3 mm
  wide, 9–18 mm long; upper surface, margins,
  and underside midveins with gland-tipped hairs;
  stem below cotyledons succulent, purple-tinged;
  first leaf with a few needle-like hairs; second and
  later leaves with star-shaped bristly hairs.
Leaves: Alternate, oval in outline, 2–12 cm long,
  stalked, margins irregularly five- to seven-lobed,
  the lobes often two- to five-lobed, all surfaces
  with yellow starlike hairs; leaves resembling
  those of watermelon.
  Inflorescence a raceme; stalk prickly.
Flower: Yellow, wheel-shaped, 2–3.5 cm across; five
  sepals, united, prickly and becoming burlike in
  fruit; five petals, united; five stamens (one long,
  four short); flowering continues until killed by
  frost.
Plant: Plants 30–100 cm tall, drought-tolerant;
  stems branched, covered in yellow spines and
  star-shaped hairs; spines 1–15 mm long, shorter
  spines with star-shaped tips with five to seven
  rays.
Fruit: Dry berry, dark brown or black, 8–12 mm
  across, enclosed by the burlike calyx, 5–20 mm
  long; 50–120-seeded.

REASONS FOR CONCERN

The prickly nature of buffalo bur hampers handpicking of fruit and vegetables. Livestock often avoid areas where the plant is growing, thereby reducing the usable acreage of the pasture. The leaves and fruit also contain the poisonous alkaloid solanine. Buffalo bur also serves as host for the Colorado potato beetle.

Similar species: a closely related perennial species, horse nettle (*S. carolinense* L., p. 710) resembles buffalo bur. It is densely prickly with purplish to white flowers, a smooth yellow spineless fruit, and leaves with shallow lobes.

Also known as: Sodom apple
Scientific Synonym(s):
  S. khasianum var.
  chatterjeeanum Sen
  Gupta

QUICK ID: plants bushy
  and prickly; flowers white;
  fruit a yellow berry, 2–3
  cm wide

Origin: native to Brazil,
  Paraguay, and Argentina;
  first observed in Glades
  County, Florida in 1987;
  now infesting over
  1 million acres in 29
  counties; dispersion has
  been through transport
  of livestock, hay, and
  sod and through wildlife
  migration

Life Cycle: perennial
  reproducing by seed;
  plants producing up to
  90,000 seeds

Weed Designation: USA:
  federal, AL, AZ, CA, FL, MA,
  MN, MS, NC, OR, SC, TN,
  TX, VT; MEX: federal

DESCRIPTION
Seed: Light reddish-brown, 2–3 mm across, flattened.
  Viable for 2 years in soil.
  Germination normally exceeding 75%; emergence
  from 8 cm soil depth.
Seedling: With oblong to oval or lance-shaped
  cotyledons; first leaf round to oval, surface and
  margins finely haired.
Leaves: Alternate, oval to triangular, 5–20 cm long,
  2–15 cm wide, stalks prickly, margins with angular
  lobe, surface with soft hairs; prickles 5–20 mm
  long, borne on veins.
  Inflorescence a raceme, stalk prickly; borne on
  stems below the leaves.
Flower: White, 10–15 mm across; five sepals, prickly,
  five petals, recurved; five stamens, yellow; one
  pistil.
Plant: Bushy and prickly, 30–200 cm tall; roots
  2.5–10 mm diameter and extending 2 m
  radius, producing new shoots; stems erect,
  branched, surface with straight downward-
  pointing prickles; maturity about 105 days after
  germination.
Fruit: Globe-shaped berry; 2–3 cm wide; green with
  dark veins when young, dull yellow when ripe;
  averaging 400 seeds per berry; plants producing
  125–400 fruits.

REASONS FOR CONCERN

Tropical soda apple is a serious weed of rangeland, where it displaces forage and native vegetation. It creates dense stands that livestock avoid due to the prickly nature of the plant. Once the fruits become ripe, leaves are shed, and cattle readily consume the fruit, thereby spreading the seeds in their manure. Seeds are also dispersed by birds and other wildlife.

OTHER SPECIES OF CONCERN IN THE NIGHTSHADE OR POTATO FAMILY

LYCIUM FEROCISSIMUM MIERS — AFRICAN BOXTHORN

Native to South Africa; shrub up to 5 m tall; lateral branches ending in spines up to 15 cm long; branches numerous; leaves alternate (often appearing clustered), elliptical to oblong or linear, 5–43 mm long, 3–12 mm wide, somewhat fleshy, short-stalked, margins smooth; inflorescence an axillary cyme (two-flowered) or solitary flower; flowers white with purple marks, tube-shaped, 10–13 mm long, 10–12 mm across, fragrant; five sepals, 4–8 mm long; five petals; five stamens, exceeding the petals; one pistil; stalks 5 mm long and elongating to 20 mm at fruiting; fruit a red or orange berry, 5–12 mm across, egg- to globe-shaped; seeds dull yellow, about 2.5 mm across. Weed designation: USA: federal, AL, CA, FL, MA, MN, NC, OR, SC, VT; MEX: federal.

Solanum elaeagnifolium

SOLANUM ELAEAGNIFOLIUM CAV. — SILVERLEAF NIGHTSHADE, TROMPILLO

Origin uncertain, probably southern North America to South America; perennial; roots to 2 m deep; stems 30–120 cm tall; stem and foliage with slender red to yellow or orange spines; spines 2–5 mm long; leaves lance-shaped, 2–10 cm long, 5–20 mm wide, margins wavy, surface with silvery white star-shaped hairs; inflorescence a cyme, three- to five-flowered; flowers blue to violet or white, 1–4 cm long; five sepals; five petals; five stamens, bright yellow; one pistil; stalks about 1 cm, elongating 2–3 cm long at fruiting; fruit a green berry, turning yellow to brown at maturity, 8–15 mm across; seeds brown, 2.5–4 mm long. Weed designation: USA: AR, CA, HI, ID, NV, OR, WA.

SOLANUM TAMPICENSE DUNAL — SCRAMBLING OR WETLAND NIGHTSHADE

Native to Mexico and Central America; shrub; branches sprawling, 1–5 m long, armed with recurved prickles; prickles white to tan, about 5 mm long; leaves alternate, oblong to oval, 5.5–25 cm long, 2–7 cm wide, stalks 5–15 mm long, often prickly, margin irregularly lobed, surface with star-shaped hairs and prickles; inflorescence an axillary cyme, three- to 11-flowered, borne opposite the leaves; flowers white to yellowish-white, 6–11 mm long; five sepals, 1.5–2 mm long, prickly; five petals, 12–14 mm across; five stamens; one pistil; stalks 5–8 mm long, recurved with maturity; fruit a globe-shaped green berry, 6–10 mm across, turning orange to red at maturity; seeds 2–2.5 mm long. Weed designation: USA: federal, AL, CA, FL, MA, NC, OR, SC, VT.

SOLANUM TORVUM SW. — TURKEY BERRY

Native to tropical America; shrub 1–5 m tall, profusely branched; stems woolly haired, armed with scattered prickles; leaves alternate, oval to elliptical, 5–25 cm long, 5–18 cm wide, margins shallowly to irregularly lobed, upper surface with sandpaper textures, lower surface and stalks woolly haired; inflorescence a corymb, many-flowered; flowers white, 12–25 mm long; five sepals, woolly haired; five petals; five stamens, yellow; one pistil; fruit a green or yellow berry, globe-shaped, 10–15 mm across; seeds 2–3 mm long. Weed designation: USA: federal, AL, CA, FL, HI, MA, MN, NC, OR, SC, VT.

Tamaricaceae

TAMARISK FAMILY

Trees and shrubs, often growing in dry alkaline areas; leaves alternate, small and scalelike; stipules absent; inflorescence a spike, panicle, or solitary flower; flowers perfect, regular; four or five sepals; four or five petals; eight, 10, or more stamens, borne in bundles of five; one pistil; ovary superior; fruit a capsule; seeds with a tuft of hair at one end. Worldwide distribution: four genera/120 species; temperate to subtropical regions of the world.

Distinguishing characteristics: trees and shrubs with coniferlike foliage; flowers borne in spikes or panicles.

Species from the genus *Tamarix* are often grown as ornamentals. Some species have escaped cultivation and become weedy.

salt cedar (*Tamarix ramosissima*)

Tamarix ramosissima Ledeb.
SALT CEDAR

Also known as: tamarix, tamarisk

Scientific Synonym(s): *T. pallasii* var. *brachystachys* Bunge

QUICK ID: shrub or small tree; leaves scalelike; flowers pink

Origin: native to Ukraine, Iraq, China, and Korea; ornamental shrub introduced into the western United States as a windbreak in 1823

Life Cycle: perennial (living 75–100 years) reproducing by resprouts, rhizomes, and seed; average-sized trees producing over 2.5 billion seeds per year

Weed Designation: USA: CO, MT, NM, NE, NV, ND, OR, SD, TX, WA, WY; CAN: AB, SK

DESCRIPTION

Seed: Less than 0.5 mm diameter, tufts of hairs at apex aid in wind dispersal.
Viable up to 45 days; most seeds losing viability within 14 days.
Germination occurs within 24 hours of contact with water or wet soil.

Seedling: Tiny; drought-intolerant; capable of floating and relocating to new sites; may survive submergence for 24 hours; at 8 weeks, plants are 11–12 cm tall with roots about 15 cm long; taproot appears 20 hours after germination, cotyledons within 24 hours.

Leaves: Alternate, scalelike, 1.5–3.5 mm long; salt-secreting glands present, often covered in a salty bloom.
Inflorescence a panicle, the branches 2–8 cm long, 3–5 mm wide; about 20 flowers per 2.5 cm of raceme; one bract.

Flower: Pink; five sepals; five petals, 1.1–1.8 mm long; four to 10 stamens (commonly five); one pistil; flowers may appear during the first season's growth but commonly by the third year.

Plant: Shrub or small tree, 4–8 m tall, capable of tolerating soils with 50,000 ppm salts; taprooted until the water table is reached, then profusely branched; bark reddish-brown.

Fruit: Oval to lance-shaped capsule, 3–5 mm long; eight- to 20-seeded.

REASONS FOR CONCERN

Salt cedar is the dominant riverbank tree along the Colorado River. Tamarisk is a heavy user of groundwater in riparian areas. It also has salt-secreting glands that turn the soil alkaline, thereby reducing species diversity.

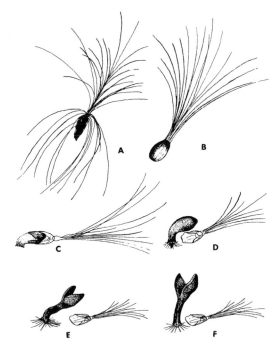

Germination of salt cedar seed

OTHER SPECIES OF CONCERN IN THE TAMARISK FAMILY

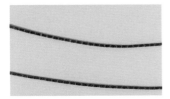

Tamarix aphylla

Tamarix chinensis

TAMARIX APHYLLA (L.) KARST. (SYNONYM: *T. ARTICULATA* VAHL) — ATHEL PINE, TAMARIS, TARAY

Native to northeastern Africa and western Asia; evergreen tree 2–15 m tall; young bark light gray, older bark dark gray, rough; branches jointed; leaves scalelike, dusty gray, 1.5–4 mm long, sheathing the stem, sharp-pointed; inflorescence a raceme, 4–6 cm long, 3–4 mm wide; bracts sheathing, less than 1 mm long; flowers pinkish-white, about 3 mm long; five sepals, pinkish-green, about 1 mm long; five petals, about 2 mm long; five stamens, white; one pistil; fruit a capsule, 4–5 mm long; seeds brown, apical hairs about 3 mm long. Weed designation: USA: MT, NM, SD, TX, WY.

TAMARIX CHINENSIS LOUREIRO (SYNONYM: *T. PENTANDRA* PALLAS) — FIVE-STAMEN SALT CEDAR

Native to eastern Asia; tree up to 6 m tall, diameter up to 10 cm; bark reddish-brown to blackish; branches green to purplish or reddish; leaves scalelike 1.5–3 mm long, oblong to lance-shaped, dull bluish-green, sharp-pointed; inflorescence a raceme, 1.5–7 cm long; bract linear to lance-shaped; flowers pink to whitish, less than 3 mm long; five sepals, 0.5–1.5 mm long; five petals, 1.5–2 mm long; five stamens; one pistil; fruit a reddish-brown capsule about 3 mm long; seeds hairy. Weed designation: USA: CO, MT, ND, NM, SD, TX, WY; CAN: AB, SK.

TAMARIX PARVIFLORA DC. —
SMALL-FLOWERED SALT CEDAR

Native to southeastern Europe; evergreen shrub
or small tree 1.5–5 m tall; bark reddish-brown to
purplish-gray; branches greenish-purple; leaves
scalelike, 1–3 mm long, stalkless; inflorescence a
raceme 1.5–5 cm long; flowers pink, about 3 mm
across; four sepals, 1–1.5 mm long; four petals,
about 2 mm long; four stamens; one pistil; fruit a
brown capsule about 3 mm long; seeds hairy. Weed
designation: USA: CO, MT, NE, NM, NV, ND, SD, TX,
WY; CAN: AB, SK.

Urticaceae
STINGING NETTLE FAMILY

Herbs, often with stinging hairs; leaves alternate or opposite, simple; stipules present; inflorescence a spike, raceme, or panicle; flowers imperfect, regular; four or five sepals; petals absent; four or five stamens, opposite the sepals; one pistil; ovary superior or inferior; fruit an achene or drupe. Worldwide distribution: 45 genera/550 species; temperate and tropical regions throughout the world.

Distinguishing characteristics: herbs with stinging hairs; flowers green, nonshowy.

The stinging nettle family is of little economic importance. A few species without stinging hairs are grown as ornamental plants.

stinging hairs of stinging nettle (*Urtica dioica*)

Urtica dioica L.
STINGING NETTLE

Also known as: big-sting nettle, tall nettle, slender nettle, California nettle

Scientific Synonym(s):
U. lyallii S. Wats., *U. gracilis* Ait.

QUICK ID: leaves opposite; stems with stinging hairs; stems four-sided

Origin: native to North America; introduced European subspecies with male and female plants

Life Cycle: perennial reproducing by rhizomes and seed; plants may produce up to 20,000 seeds; rhizome fragments may produce new plants with a rhizome radius of 2.5 m in a single season

Weed Designation: CAN: MB, PQ

DESCRIPTION

Seed: Oval, dull orange to pale brown, 1–1.5 mm long, 0.7–1 mm wide; narrow wing encircling the seed.
Viable in soil for about 10 years.
Germination occurs 5–10 days after being shed.

Seedling: With oblong cotyledons, 2–4 mm long and 1 mm wide, midrib distinct; first pair of leaves with shallowly toothed margins and a few stinging hairs, each tooth tipped with a long hair; stinging hairs present at second leaf stage; seedlings surviving only in shade.

Leaves: Opposite, heart- to lance-shaped, 5–20 cm long, 2–13 cm wide, prominently three- to seven-nerved, stalks 10–15 mm long, margins toothed, surfaces covered in stinging hairs; two stipules, brown and papery, 5–12 mm long.
Inflorescence a panicle; racemes with male or female flowers.

Flower: Less than 2 mm across; male or female; male flowers greenish-yellow, four sepals (1–2 mm long), four stamens; female flowers green, four sepals, the inner two, 1.4–1.8 mm long and surrounding the ovary, the outer two, 0.8–1.2 mm long; one pistil; wind-pollinated.

Plant: Plants 50–300 cm tall, long-lived (up to 50 years); stems four-sided, covered in stinging hairs; rhizome creeping and spreading up to 2.5 m in diameter per year; small pink buds giving rise to new shoots.

Fruit: Achene; surrounded by the papery inner sepals.

REASONS FOR CONCERN

Stinging nettle is a weed of disturbed areas, farmyards, roadsides, and pastures. It is rarely a problem in cultivated fields or row crops. Stinging hairs located on the stem and leaves cause a type of dermatitis. The sharp, pointed hairs have a bulbous base filled with an irritating fluid. These hairs break when penetrating the skin and release fluid beneath the surface of the skin causing the irritation.

OTHER SPECIES OF CONCERN IN
THE STINGING NETTLE FAMILY

Laportea canadensis

Parietaria pensylvanica

LAPORTEA CANADENSIS (L.) WEDDELL —
WOOD NETTLE
Native to North America; perennial with rhizomes and tuberous roots; stems 30–150 cm tall, covered with stinging hairs and nonstinging hairs; leaves alternate, oval to round, 6–30 cm long, 3–18 cm wide, margins toothed; stipules present; inflorescence a panicle of cymes, male or female; male flowers 1–1.5 mm across, five tepals; five stamens; female flowers about 0.5 mm across, two to four tepals, one pistil; fruit an achene. Weed designation: none.

PARIETARIA PENSYLVANICA MUHL.
EX WILLD. — PELLITORY
Native to North America; annual without stinging hairs; stems 4–60 cm tall, often covered in hooked and straight hairs; leaves alternate, elliptical to oblong, oval, or lance-shaped, 2–9 cm long, 4–30 mm wide, margins smooth; stipules absent; inflorescence an axillary cyme of perfect, male and/ or female flowers; bracts 1.8–5 mm long; flowers green; perfect flowers with four tepals, 1.5–2 mm long, four stamens, one pistil; fruit a light brown achene 0.9–1.2 mm long. Weed designation: none.

URTICA URENS L. — DOG NETTLE, SMALL NETTLE

Native to Europe and Asia; annual without stinging hairs; stems 10–80 cm tall, often branched; leaves opposite, elliptical to oval, 1.8–9 cm long, 1.2–4.5 cm wide, margins coarsely toothed; stipules present; inflorescence a spikelike panicle; flowers green, male or female; male flowers with four tepals and four stamens; female flowers with four tepals (two inner and larger, two outer and smaller) and one pistil; fruit an achene, 1.5–1.8 mm long. Weed designation: none.

Urtica urens

Verbenaceae

VERVAIN FAMILY

Herbs, shrubs, and trees; stems often four-sided; leaves opposite or whorled; stipules absent; inflorescence a cyme; flowers perfect, irregular and somewhat tubular or two-lipped; five sepals, fused; five petals, fused; four stamens; one pistil; one style, terminal; ovary superior; fruit a drupe, nutlet, or capsule. Worldwide distribution: 75 genera/3,000 species; primarily tropical and subtropical regions of the world.

Distinguishing characteristics: leaves opposite; flowers irregular; style borne at tip of ovary.

The vervain family contains the highly prized teak tree (*Tectona grandis*). Several species are grown as ornamental plants, while a few are recognized weeds.

lantana (*Lantana camara*)

Lantana camara L.
LANTANA

Also known as: largeleaf
lantana, shrub verbena,
red-flowered sage, tick
berry, wild sage, white
sage, confetti bush
Scientific Synonym(s):
L. aculeata L., *L. tilifolia*
auct. non Cham.

QUICK ID: shrub with ill-
scented leaves; flowers
multicolored; fruit a black
drupe
Origin: native to the West
Indies; introduced as an
ornamental in the early
1800s
Life Cycle: perennial
reproducing by seed,
layering, and resprouting;
large plants producing up
to 12,000 seeds
Weed Designation: none

DESCRIPTION
Seed: Yellow to pale brown, 1.5–4 mm long.
Viable up to 5 years in soil; 6 months in dry
storage.
Germination enhanced by exposure to light;
adequate moisture required during germination.
Seedling: With round cotyledons, tip shallowly
notched; first leaves opposite, margins coarsely
toothed, surface prominently veined and rough-
haired.
Leaves: Opposite, oval, 5–15 cm long, 2–7.5 cm wide,
surface rough-haired, stalks up to 2 cm long,
margins blunt-toothed; strongly aromatic when
crushed.
Inflorescence a dense flat-topped cluster 2–4 cm
across; 20–40-flowered.
Flower: Multicolored (white, pink, lavender, orange,
yellow, or red), tube-shaped and four-lobed, 5–9
mm long; five sepals; four petals, united; four
stamens; one pistil.
Plant: Deciduous shrub 2–5 m tall, ill-scented,
somewhat shade-intolerant, does not survive
temperatures below 5°C; stems four-angled,
branched, surface with bristly hairs and small
prickles.
Fruit: Fleshy drupe, 4–8 mm across, one-seeded;
green turning purple then bluish-black at
maturity.

REASONS FOR CONCERN

Reports indicate that the roots of lantana release chemicals that kill or suppress the growth of other species. This feature, in addition to large seed production, allows lantana to displace native vegetation along roadsides, pastures, woodlands, and orchards. Over 4 million hectares are infested in Australia; 160,000 hectares in Hawaii. The leaves and fruit of lantana are reported to be highly toxic to grazing livestock. There have been reports of the berries being poisonous to children. It is a serious weed in 14 crops in 47 countries.

OTHER SPECIES OF CONCERN IN
THE VERVAIN FAMILY

Phyla nodiflora

PHYLA NODIFLORA (L.) GREENE — MAT GRASS, TURKEY TANGLE FOGFRUIT

Native to tropical America; perennial; stems creeping and rooting at nodes, four-angled, internodes less than 4 cm long, branches commonly less than 15 cm long; leaves opposite, oblong to oval or spatula-shaped, 0.5–4 cm long, 5–20 mm wide, stalks 5–6 mm long, margins toothed near the tip, surface with numerous hairs and giving the plant a grayish-green appearance; inflorescence a spikelike head 1–2.5 cm long, stalks 1–9 cm long; bracts oblong to oval; flowers white to pale violet, about 2 mm long; four sepals, fused and two-lipped; five petals, fused and two-lipped; four stamens; one pistil; fruit a capsule about 1.8 mm long and 1.5 mm wide, two-seeded. Weed designation: none.

VERBENA BRASILIENSIS VELL. — BRAZILIAN VERVAIN

Native to South America; perennial; stems 40–150 cm tall, four-angled, often bristly on the angles; leaves opposite, oval to lance-shaped, 3–10 cm long, short-stalked to stalkless, margins toothed; inflorescence an axillary spike, 1–7 cm long, borne on long stalks; bracts 3–3.5 mm long; flowers violet to pink; five sepals, fused; five petals, fused; four stamens; one pistil; fruit a nutlet, 1–1.5 mm long. Weed designation: none.

Violaceae

VIOLET FAMILY

Herbs; leaves alternate, simple; stipules present; inflorescence a raceme, cyme, or solitary flower; flowers perfect, irregular; some flowers cleistogamous, small and petal-less, never opening but setting seed; five sepals; five petals, the lowermost often enlarged and spurred; five stamens; one pistil; ovary superior; fruit a capsule, berry, nut, or samara. Worldwide distribution: 16 genera/850 species; found throughout the world.

Distinguishing characteristics: herbs; flowers pansy-like; fruit a capsule.

A few species of violets are grown as ornamentals.

Johnny-jump-up (*Viola tricolor*)

Viola arvensis Murr.
FIELD VIOLET

Also known as: wild pansy, field pansy, heartsease, European field pansy

Scientific Synonym(s): *V. agrestis* Jord., *V. arvatica* Jord., *V. deseglisei* Jord., *V. lloydii* Jord., *V. pentela* Jord., *V. ruralis* Jord., *V. timbali* Jord., *V. tricolor* ssp. *arvensis* L. (Murray) Gaudin

QUICK ID: flowers pansy-like; stipules leaflike; leaves alternate

Origin: native to Europe; introduced for use as a garden ornamental, medicinal, and culinary uses.

Life Cycle: annual or winter annual reproducing by seed; large plants producing up to 2,500 seeds

Weed Designation: none

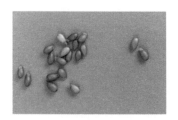

DESCRIPTION

Seed: Oval to kidney-shaped, yellowish-brown to dark brown, 1–2 mm long, surface glossy. Viable in soil for 6–11 years; European excavations have shown viability in 400-year-old seed. Germination occurs at temperatures between 5°C and 15°C; optimal depth 5–10 mm.

Seedling: With oval to squarish cotyledons, 3–5 mm long, 3–4 mm wide, blunt-tipped; first leaves oval to spatula-shaped, blunt-pointed, long-stalked, margins with rounded teeth; later leaves alternate but developing into a basal rosette.

Leaves: Alternate, oblong to lance-shaped, 2–8 cm long, 1–1.5 cm wide, margins with rounded teeth; stipules leaflike, divided into five to nine narrow segments; basal leaves rounded to oblong, margins toothed, 1–1.5 cm long, 7–12 mm wide, stalks 1–2 cm long, veins on underside sparsely haired; stipules small.

Inflorescence a solitary flower borne on long stalks originating in leaf axils.

Flower: Pale yellow to white with purple or mauve markings, irregular, 1–1.5 cm long, about 1 cm wide; five sepals, united, about the same length or longer than the petals; five petals, the lower spurred, laterals bearded; five stamens; one pistil; spur 2–4 mm long; stalks 2–4 cm long.

Plant: Erect to ascending, partially shade-tolerant; stems prostrate to ascending, 10–35 cm tall, branched, hairy on the angles; roots fibrous, wintergreen-scented when crushed.

Fruit: Spherical capsule, 5–10 mm long, tan-colored; about 75-seeded; ejected with force at maturity up to 2.1 m from parent.

REASONS FOR CONCERN

Field violet is a common weed of nursery crops, gardens, and strawberry fields. It is a problem in winter wheat crops, as the seeds are shed prior to the crop being harvested. It has been determined that 250 seedlings per square meter will result in a 5% crop loss.

OTHER SPECIES OF CONCERN IN
THE VIOLET FAMILY

Viola odorata

VIOLA ODORATA L. — SWEET VIOLET, ENGLISH VIOLET

Native to Europe and Asia; perennial with rhizomes 5–20 mm diameter and leafy stolons rooting at nodes; flowering stems 6–24 cm tall; leaves all basal; oval to heart-shaped, 1.5–6 cm long, thick and somewhat leathery, stalks 2–17 cm long, margins toothed; inflorescence a solitary flower, borne on stalks 2–15 cm tall; flowers purple with a white throat to pale violet or white, fragrant; five sepals; five petals, laterals bearded, lowest spurred; five stamens; one pistil; style hooked; fruit a capsule, 5–8 mm long, purplish-green, finely haired; seeds cream-colored. Weed designation: none.

VIOLA TRICOLOR L. — JOHNNY-JUMP-UP

Native to Europe, escape from cultivation; annual, biennial, or perennial; stems 10–40 cm tall, erect, angular; leaves alternate and basal; basal leaves oval to lance-shaped, long-stalked; stipules leaflike, 1–4 cm long, deeply dissected; stem leaves smaller, oblong to round to lance-shaped, margins with rounded teeth; flowers solitary in leaf axils, usually tricolored with purple, white, and yellow, 3.5–6 cm across; five sepals, 1.2–2.2 cm long; five petals, lateral and lower purple-striped, lower petals with spur 5–8 mm long; five stamens; one pistil; stalks with two small bracts; fruit a capsule, 8–12 mm long. Weed designation: none.

Viola tricolor

Zygophyllaceae

CALTROP FAMILY

Herbs and shrubs, often growing in dry alkaline areas; leaves opposite, pinnately compound; stipules present, modified; inflorescence a cyme or solitary flower; flowers perfect, regular; four or five sepals; four or five petals; five, 10, or 15 stamens; one pistil; ovary superior; fruit a capsule. Worldwide distribution: 30 genera/250 species; primarily tropical and subtropical areas of the world.

Distinguishing characteristics: herbs; leaves opposite and compound, stipules present.

Members of this family include lignum vitae (the densest and hardest wood), creosote bush, and Syrian rue.

false puncture vine (*Tribulus cistoides*)

Tribulus terrestris L.
PUNCTURE VINE

Also known as: caltrop, ground burnut, tackweed, bullhead, goathead, Mexican sandbur, Texas sandbur

Scientific Synonym(s): *T. terrester* Landolt

QUICK ID: plants prostrate; leaves opposite and compound; fruit with two to four sharp spines

Origin: native to the Mediterranean region of southern Europe and the northern fringes of the Sahara Desert in Africa; spread throughout the world in sheep wool

Life Cycle: annual reproducing by seed; plants producing up to 10,000 seeds

Weed Designation: USA: AZ, CA, CO, ID, IA, NV, NC, OR, WA; CAN: AB, BC.

DESCRIPTION

Seed: Wedge-shaped, tan, 4.7–5.7 mm long, 3.5–6 mm wide; one or two short spines about 3–5 mm long radiate from the midpoint of seed; surface covered in ridges and prominent white bristles. Viable in soil for 5 years.
Germination requires a dormant period of 6 months to 1 year; optimal depth in top 5 cm of soil.

Seedling: With oblong cotyledons, 8–15 mm long, 3–4 mm wide, peach to green in color, thick, tip notched, upper surface distinctly creased and rough-textured, thick, notched at the tip; stem below cotyledons salmon-colored, surface with short hairs; first leaf compound with six to 10 leaflets, surface bristly with sharp pointed hairs.

Leaves: Opposite, about 6 cm long, compound with eight to 16 leaflets; leaflets, 5–15 mm long, about 5 mm wide, surface slightly to densely haired; stipules up to 1 cm long.
Inflorescence a solitary flower borne in leaf axils.

Flower: Pale yellow, about 2 cm across; five sepals, 3–5 mm long; five petals, 3–12 mm long; 10 stamens; one pistil; flowering occurs within 2 weeks of germination.

Plant: Prostrate and forming mats up to 5 m across, preferring semiarid regions; stems green to reddish-brown, up to 2.4 m long, surface covered in dense bristly hairs; taproot woody, extending to 2.6 m deep and 6.6 m in diameter.

Fruit: Schizocarp (podlike), about 1.2 cm long, breaking into five triangular woody pieces (mericarps) at maturity, each one- to four-seeded; each segment with two to four sharp spines; fruit produced within 5 weeks of germination.

REASONS FOR CONCERN

Puncture vine is reported to be poisonous to livestock. Fair-colored livestock eating this plant become highly sensitive to bright light and should be provided shade to escape the rays of the sun. It is also reported to cause bighead in sheep, a condition that causes the head to swell. Puncture vine may also cause irreversible high quarter-stagger in sheep. In late summer, livestock often refuse to graze in areas where this plant is abundant. The spiny fruit can cause injury to livestock and humans and damage to vehicle tires. It is reported to be weedy in 37 countries.

OTHER SPECIES OF CONCERN IN
THE CALTROP FAMILY

***PEGANUM HARMALA* L.—HARMAL PEGANUM**
Native to Mediterranean region of Europe, Asia, and Africa; perennial reproducing by seed and lateral roots; stems stiff, branched, 15–80 cm tall; leaves alternate, succulent, bright green, 2–8 cm long, divided three or more times into narrow segments, the segments 3–5 cm long, 2–5 mm wide; stipules bristle-like, 1.5–2.5 mm long; inflorescence a solitary flower borne on stalks 2–5 cm long originating in leaf axils; flowers white, 2–2.5 cm across; five sepals, 10–18 mm long; five petals, 10–16 mm long; 15 stamens; one pistil; stalks about 1.2 cm long, threadlike; fruit a leathery capsule, 6–10 mm diameter, orangish-brown at maturity; seeds black, 2–4 mm long. Weed designation: USA: AZ, CA, CO, NV, NM, OR.

Tribulus cistoides

***TRIBULUS CISTOIDES* L.—FALSE PUNCTURE VINE, JAMAICAN FEVER PLANT**
Native to tropical America; perennial reproducing by seeds; stems trailing; leaves alternate, compound with 12–16 leaflets; leaflets 6–22 mm long, 3–9 mm wide, silky-haired, margins smooth; inflorescence a solitary axillary flower; flowers yellow; five sepals, silky-haired; five petals, 10–20 mm long; 10 stamens, five short and five long; one pistil; five stigmas; fruit a schizocarp, horned and woody; mericarps one-seeded. Weed designation: none.

ZYGOPHYLLUM FABAGO L. — SYRIAN BEANCAPER

Native to southwestern Asia; perennial with a thick woody crown; stems 50–100 cm tall, hairless, branched; leaves opposite, stalk 12–15 mm long, compound with two leaflets; leaflets oblong to oval, thick and waxy, 1–4 cm long, 10–18 mm wide, short-awned between the leaflets; stipules oval to elliptical, 5–10 mm long; inflorescence a solitary flower borne in leaf axils; flowers yellowish or copper-colored, about 10 mm across, stalks 6–8 mm long; five sepals, 6–7 mm long, margins membranous; five petals, 6–8 mm long, orangish-red at base; eight to 10 stamens, filaments 10 mm long; one pistil; fruit a capsule, 20–35 mm long, 4–6 mm wide, hairless, borne on bent stalks; seeds oblong, about 3 mm long. Weed designation: USA: CA, ID, OR, WA.

Glossary

achene
dry, one-seeded fruit that does not open when ripe

allelopathic
a plant that produces a chemical that suppresses the growth of other plants

annual
plant with a life cycle that is completed in one growing season

anther
the pollen-producing sacs of the stamens

apex
tip of a leaf or petal

apical
at the tip

auricle
earlike appendages; in grasses, auricles are found at the junction of the blade and the sheath

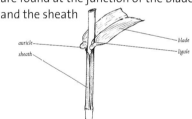

awn
a stiff bristle

axil
the angle formed between the leaf and the stem

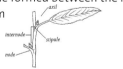

axillary
found in the axil of leaves

basal
from the base; often used in reference to leaves

berry
a fleshy fruit with one to many seeds

biennial
a plant that takes two years to complete one life cycle

bract
leaflike structure found below a flower or flower cluster

bracteole
a small bract

bulb
a short vertical underground stem

bulbil
a fleshy reproductive structure borne on the aerial part of the plant

calyx
a collective term for the sepals of a flower

capitulum
a head

capsule
a dry fruit that opens when mature

caryopsis
the fruit or grain of a grass

cleft
deeply lobed

collar
in grasses, part of the leaf where the blade and sheath are connected

compound leaf
a leaf composed of two or more leaflets

coma
a tuft of hairs

corolla
a collective term for the petals of a flower

corona
an appendage derived from petals and stamen filaments

cotyledons
the seed leaves

cyathium
a specialized inflorescence type found only in the spurge family (Euphorbiaceae)

cyme
a flower cluster having the oldest flowers appearing at the tip of the stem

deciduous
having leaves that fall in autumn; also refers to seeds that are shed at maturity

dicotyledon
term given to plants whose seeds produce two cotyledons

disc floret
small flowers, usually tube-shaped, in the aster family

dormancy
in reference to seeds, the period between seed maturity and germination

drupe
a fleshy fruit; usually one-seeded (i.e., a cherry)

drupelet
one part of an aggregate fruit, like those of raspberries

elliptical
shaped like an ellipse

emergent
a plant whose stem rises out of the water

family
a group of plants having similar characteristics

fibrous
threadlike roots of plants

filament
the stalk of the stamen

floret
a small flower

fruit
the part of the plant that contains seeds

genus (pl. genera)
a group of species having similar characteristics

glabrous
without hair

gland
a structure that usually produces nectar or another sticky substance

glume
in grasses, a small bract at the base of a spikelet

herb
a plant having no woody stems

host
an organism on which another organism lives as a parasite

hypanthium
a structure formed by the fusion of the lower parts of the sepals, petals, and stamens

inflorescence
flower cluster

internode
the region of stem between two leaves

involucre

a bract or group of bracts below a flower cluster

legume

fruit from the pea family that opens along two suture lines to release seeds

lemma

in grasses, one of two small bracts that enclose the flower (see glume)

ligule

in grasses, a small projection from the top of the sheath (see auricle)

loment

a podlike fruit that breaks into one-seeded segments at maturity

margin

the edge of a leaf

mericarp

a one-carpel unit of a schizocarp

monocot

term given to plants whose seeds produce one cotyledon

nerve

a vein of a leaf, petal, sepal, or other structure

node

the point at which the leaf is attached to the stem

noxious

a plant capable of causing damage to agricultural crops

nutlet

a small nut

ocrea

a sheath found at leaf nodes derived from the fusion of two stipules

ovary

part of the pistil containing the ovules

ovate

egg-shaped and broader at the base, usually in reference to leaves

palea

the innermost bract surrounding a grass flower (see glume)

panicle

a branching flower cluster that is often pyramid-shaped

pappus

hairs or bristles that are attached to the seed in the aster family

parasite

a plant that obtains food and nutrients from another living plant

perennial

a plant, or part of a plant, living more than two growing seasons

perfect

term given to flowers that have both stamens and pistils present

petal

part of a flower that is usually colored to attract insects

petiole

the stalk of a leaf

pinna (pl. pinnae)

the primary division of a compound leaf

pinnate

compound leaf with leaflets arranged on both sides of the stalk (see p. xiii)

pistil

the female part of the flower

pod

a dry fruit that releases its seeds when mature

prickle

a spiny structure on the surface of a plant

prothallus
in ferns and fern allies, a small structure producing male and female sex organs

raceme
flower cluster with flowers attached along a central stalk with flower stalks of equal length

ray floret
straplike, often marginal flower type in the aster family (see disc floret)

recurved
bent or curled backward

rhizome
an elongated, underground stem

rosette
a cluster of basal leaves

schizocarp
a fruit that breaks into sections that contain one or more seeds

scorpioid
resembling a scorpion's tail, often used to describe a cyme

semiaquatic
a plant capable of living in water and on wet shorelines

sepals
the outermost part of the flower; usually green and leaflike

serrated
leaf margin having jagged edges

sessile
stalkless

sheath
the base of the leaf that surrounds the stem

shrub
woody plant having many stems arising from the root

silicle
a podlike fruit of the mustard family (Brassicaceae) that is less than twice as long as wide

silique
a podlike fruit of the mustard family (Brassicaceae) that is more than twice as long as wide

simple leaf
a leaf with a single blade

sorus (pl. sori)
a cluster of sporangia

spadix
a thick fleshy spike, often associated with a spathe

spathe
a large bract that encloses a flower cluster

spatulate
spoon-shaped leaves

species
a group of similar plants capable of interbreeding to produce offspring like themselves

spike
an unbranched flower cluster where flowers are attached directly to the main stem

spikelet
a small spike; often used to refer to members of the grass and sedge family

spikelet

spore
a reproductive structure in ferns and fern allies

spur
a slender projection from the corolla or calyx

stamen
the male part of the flower

staminate
flowers containing male sex organs only

staminode
a sterile stamen

sterile
a flower without functional reproductive structures

stigma
the receptive surface to which pollen grains attach

stipulate
stems with stipules at the base of the leaf

stipules
a pair of membranous structures found at the base of a leaf

stolon
a horizontal stem that roots at nodes

strobilus (pl. strobili)
a cone

style
the elongated part of the pistil between the stigma and the ovary

subspecies
in taxonomy, the rank below species; often used to distinguish between characteristics within the same species

sucker
a stem that grows from the underground root of a plant

taproot
the main root, like those of carrots

tendril
a clasping or twining part of a leaf

tepal
a collective term used to describe sepals and petals that are similar in size, appearance, and color

terminal
the end of the stem or leaf

thallus (pl. thalli)
plants without differentiation between stems and leaves

tillers
in grasses, stems or branches originating at the base of the first stem

tree
a woody plant having one stem arising from the root

true flower
flower possessing stamens and/or pistils

tuber
a short, swollen underground stem

umbel
flower arrangement where the flower stalks originate from one point

utricle
an achene with a loose fragile covering, often bladderlike

venation
pattern of veins on a leaf, petal, etc.

verticel
a whorl

winter annual
plants that germinate and produce a leafy rosette in the fall, produce seeds, and die the following growing season

Illustration Credits

All color photos in this guide are by France Royer and copyrighted in her name.

The illustrations on pages xiii and 753–57 are the copyright of Sundew Productions Ltd., Edmonton, Alberta, Canada.

We would like to acknowledge the copyright-free illustrations of plant species from the United States Department of Agriculture Plants Database. For information about this excellent resource, please visit: http://plants.usda.gov; USDA, NRCS. 2012. The Plants Database. National Plant Data Team, Greensboro, NC 27401–4901 USA.

We also wish to acknowledge the copyright-free illustrations from *Aquatic, Wetland and Invasive Plants in Pen-and-Ink DVD* provided by the Center for Aquatic and Invasive Plants at the University of Florida.

Bibliography

Agri-Growth Research, Inc. 1993. *Southern Weed Seedling Identification*. Agri-Growth Research, Inc., Hollandale, MN.

Alberta Agriculture. 1983. *Weeds of Alberta*. Alberta Agriculture and Alberta Environmental Centre, Edmonton.

———. 1986. *Weed Seedling Identification*. Agdex 640-3. Alberta Agriculture, Edmonton.

Alex, J. F. 1992. *Ontario Weeds: Descriptions, Illustrations and Keys to Their Identification*. Ontario Ministry of Agriculture and Food, Publication 505. Agdex 640. Consumer Information Centre, Ontario Ministry of Agriculture and Food, Toronto.

Alex, Jack F., Richard Cayouette, and Gerald A. Mulligan. 1982. *Common and Botanical Names of Weeds in Canada*. Canadian Government Publishing Centre, Ottawa.

Askew, Shawn. 2001-3. "Turf Weeds." Turfgrass Weed Science, Virginia Polytechnic Institute and State University. http://turfweeds.contentsrvr.net/category_browse.vesh.

Auld, B. A., and R. W. Medd. 1987. *Weeds: An Illustrated Botanical Guide to Weeds of Australia*. Inkata Press, Melbourne.

Australian Government. 2013. "Invasive Species." Department Sustainability, Environment, Water, Population and Communities. Last updated May 8, 2013. http://www.environment.gov.au/biodiversity/invasive.

Best, K. F., and J. Baden Campbell. 1971. *Prairie Grasses, Identified and Described by Vegetative Characters*. Canadian Department of Agriculture Publication 1413. Department of Agriculture, Ottawa.

British Columbia. 2011. "B.C. Weed Control Act: Noxious Weeds in B.C." Ministry of Agriculture. Updated August 8, 2011. http://www.agf.gov.bc.ca/cropprot/noxious.htm.

———. 2002. "Browse." Weeds BC. http://www.weedsbc.ca/browse.html.

California Department of Food and Agriculture. 2012. "Welcome to Integrated Pest Control." Modified May 22, 2012. http://www.cdfa.ca.gov/phpps/ipc.

California Invasive Plant Council. 2007. "California Invasive Plant Inventory Database." Accessed February 15, 2012, http://www.cal-ipc.org/paf/.

Canadian Food Inspection Agency. 2008. *Invasive Species in Canada: Summary Report*. Canadian Food Inspection Agency, Canada.

Cavers, P. B., ed. 1995. *The Biology of Canadian Weeds: Contributions 62-83*. Agriculture Institute of Canada, Ottawa.

Center for Aquatic and Invasive Plants. 2007. "Plants by Scientific Name." Accessed February 15, 2012, http://plants.ifas.ufl.edu/node/22.

Clayton, W. D., M. S. Vorontsova, K. T. Harman, and H. Williamson. "GrassBase—the Online World Grass Flora: Descriptions.". Royal Botanic Gardens, Kew. 2002 onward. http://www.kew.org/data/grasses-db.html.

Colorado Weed Management Association. 2007. "Noxious Weed Information." Accessed February 15, 2012, http://www.cwma.org/noxweeds.html.

Conabio Programa de Especies Invasoras. 2003. "Especies Invasoras—Plantas." Last modified July 20, 2011. http://www.conabio.gob.mx/invasoras/index.php/Especies_invasoras_-_Plantas.

Davis, Linda W. 1993. *Weed Seeds of the Great Plains: A Handbook for Identification*. University Press of Kansas, Lawrence.

DiTommaso, A., Lawlor, F. M., and Darbyshire, S. J. 2005. "The Biology of Invasive Alien Plants in Canada. Pt. 2. Vincetoxicum rossicum (Kleopow) Borhidi (= *Vincetoxicum rossicum* [Kleopow] Barbar.) and *Cynanchum louiseae* (L.) Kartesz & Gandhi (= Vincetoxicum nigrum [L.] Moench)." *Canadian Journal of Plant Sciences* 85:243-63.

Dickinson, Richard B., and France Royer. 1996. *Weeds of Canada and the Northern United States*. University of Alberta Press, Edmonton.

Flora of China Editorial Committee, eds. 2002. "Flora of China." http://www.efloras.org/flora_page.aspx?flora_id=2.

Flora of North America Editorial Committee, eds.1993–. "Flora of North America." http://www.efloras.org/flora_page.aspx?flora_id=1.

Flora of Pakistan. 2013. "Pakistan Plant Database." http://www.tropicos.org/Project/Pakistan.

Florida Exotic Pest Plant Council. 2006. "Florida Exotic Pest Plant Council Invasive Plant Lists." Last updated November 8, 2011. http://www.fleppc.org/list/list.htm.

Frankton, Clarence, and Gerald A. Mulligan. 1970. *Weeds of Canada*. Canada Department of Agriculture, Publication 948. Canada Department of Agriculture, Ottawa.

Fuller, T. C., and E. McClintock. 1986. *Poisonous Plants of California*. University of California Press, Berkeley.

Gilkey, Helen M. 1957. *Weeds of the Pacific Northwest*. Oregon State College, [Corvallis].

Gleason, Henry A. 1963. *The New Britton and Brown Illustrated Flora of the Northeastern United States and Adjacent Canada*. Hafner Publishing, New York.

Global Invasive Species Database. 2005. "Welcome to the Global Invasive Species Database." Accessed February 15, 2012, http://www.issg.org/database/welcome.

Government of Australia, Department of Sustainability, Environment, Water, Population and Communities. "Weeds Australia: Weed Identification Tool." 1998. http://www.environment.gov.au/cgi-bin/biodiversity/invasive/weeds/weedidtool.pl.

Government of Canada. 2013. "Weed Seeds Order, 2005 (SOR/2005-220)." Justice Laws Website. Modified June 7, 2013. http://laws-lois.justice.gc.ca/eng/regulations/SOR-2005-220/index.html.

Hanf, Martin. 1974. *Weeds and Their Seedlings*. BASF United Kingdom Limited, Agricultural Division, Ipswich.

Hawaiian Ecosystems at Risk Project (HEAR). 2008. "Invasive Species Information for Hawaii and the Pacific." Last updated June 3, 2008. http://www.hear.org.

Holm, LeRoy G., Donald L. Plucknett, Juan V. Pancho, and James P. Herberger. 1977. *The World's Worst Weeds: Distribution and Biology*. University Press of Hawaii, Honolulu.

James, Lynn F., et al. 1991. *Noxious Range Weeds*. Westview Press, Boulder, CO.

Kingsbury, J. M. 1964. *Poisonous Plants of the United States and Canada*. Prentice-Hall, Englewood Cliffs, NJ,.

Kummer, Anna P. 1951. *Weed Seedlings*. University of Chicago Press, Chicago.

Looman, J., and K. F. Best. 1987. *Budd's Flora of the Canadian Prairie Provinces*. Agriculture Canada, Ottawa.

Manitoba Agriculture, Food and Rural Initiatives. 1996. "Noxious Weeds Act." http://web2.gov.mb.ca/laws/regs/pdf/n110-035.96.pdf.

Marquez, M. H. R., and A. M. Castillo. 2001. "Listado florístico del Campo Experimental La Campana y usos de su flora." *Revista Mexicana de Ciencias Pecuarias* 39, no. 2: 105–25. http://www.tecnicapecuaria.org.mx/trabajos/200212173133.pdf.

Missouri Plants. 2007. "Photographs and Descriptions of the Flowering and Non-Flowering Plants of Missouri, USA." Last Modified June, 26, 2013. http://missouriplants.com/index.html.

Montgomery, F. H. 1964. *Weeds of Canada and the Northern United States*. Ryerson Press, Toronto.

Muenscher, Walter Conrad. 1980. *Weeds*. Comstock Publishing Associates, Ithaca, NY.

Mulligan, Gerald A. 1979. *The Biology of Canadian Weeds: Contributions 1–32*. Publication 1693. Communications Branch, Agriculture Canada, Ottawa.

———. 1984. *The Biology of Canadian Weeds: Contributions 33–61*. Publication

1765. Communications Branch, Agriculture Canada, Ottawa.

———. 1976. *Common Weeds of Canada.* McClelland and Stewart Limited, [Toronto].

Munro, Derek B. "Canadian Poisonous Plants Information System." Canadian Biodiversity Information Facility. Modified September 1, 2009. http://www.cbif.gc.ca/pls/pp/poison.

Nashua Conservation Commission. 2004. "New Hampshire Invasive Species Fact Sheets." http://www.nashuarpc.org /LMRLAC/documents/invasiveplants.pdf.

New Mexico State University. 2008. "NMSU Weed Information." Cite redesigned June 17, 2008. http://weeds.nmsu.edu.

Nielsen, C., H. P. Ravn, W. Nentwig, and M. Wade, eds. 2005. *The Giant Hogweed Best Practice Manual: Guidelines for the Management and Control of an Invasive Weed in Europe.* Hørsholm: Forest and Landscape Denmark. http://www.giant -alien.dk/pdf/Giant_alien_uk.pdf.

Ohio State University Extension. "Ohio Perennial and Biennial Weed Guide." Accessed February 15, 2012. http://www .oardc.ohio-state.edu/weedguide/listall.asp.

OMAFRA Staff. 2013. "Weeds in Ontario." Ontario Ministry of Agriculture and Food. Last Modified February 20, 2013. http:// www.omafra.gov.on.ca/english/crops /insects/weeds.html.

Ontario Ministry of Agriculture and Food. 2012. "Noxious Weeds in Ontario." Last modified July 4, 2012. http://www.omafra .gov.on.ca/english/crops/facts/noxious _weeds.htm.

Parker, Kittie F. *An Illustrated Guide to Arizona Weeds: A Complete Online Edition of the Printed Book.* University of Arizona Press, Tucson, 1972. http://www.uapress.arizona .edu/onlinebks/weeds/titlweed.htm.

Pennsylvania Department of Conservation and Natural Resources. 2006. "Invasive Plant Management Tutorial." Accessed February 15, 2012, http://www.dcnr.state .pa.us/forestry/plants/invasiveplants /invasiveplanttutorial/index.htm.

Saskatchewan Agriculture and Food. 2010.

"Designation of Prohibited, Noxious and Nuisance Weeds in Accordance with the Weed Control Act." http://www.agriculture .gov.sk.ca/default.aspx?dn=44fb5d3e-a7e0 -461c-84d5-876f651a3ed4.

Schlaphoff, Elmer, and Ellsworth Carlson. 1968. *Nebraska Weeds.* Bulletin No. 101-R. Nebraska Department of Agriculture, Lincoln.

Schmutz, Ervin M., Barry N. Freeman, and Raymond E. Reed. 1968. *Livestock-Poisoning Plants of Arizona.* University of Arizona, Tucson.

Schrader, Leonard. 1975. *South Dakota Weeds.* South Dakota State Weed Control Commission.

Stevens, P. F. 2001–. "Angiosperm Phylogeny Website." Version 12. Last updated June 21, 2013. http://www.mobot.org/MOBOT /research/APweb/.

Texas Invasive Plant and Pest Council. 2011. "Invasive Plant Database." http://www .texasinvasives.org/plant_database?index .php.

Thornberry, Halbert Houst. 1966. *Index of Plant Virus Diseases.* Agricultural Handbook No. 307. Agricultural Research Service, U.S. Department of Agriculture, Washington, DC.

U.S. Department of Agriculture. 1970. *Selected Weeds of the United States.* Agriculture Handbook No. 366. USDA Agricultural Research Service, Washington, DC.

Walters, Dirk R., and David J. Keil. 1996. *Vascular Plant Taxonomy.* 4th ed. Kendall /Hunt Publishing, Dubuque.

Washington State Department of Ecology. 1994-2013. "Water Quality: Aquatic Plants, Algae, and Lakes." Accessed February 15, 2012, http://www.ecy.wa.gov/programs /wq/links/plants.html.

Whitson, T. D., et al. 1991. *Weeds of the West.* Western Weed Science Society, Jackson, WY.

———. 1987. *Weeds and Poisonous Plants of Wyoming and Utah.* University of Wyoming, Laramie.

Wilkinson, R. E., and H. E. Jacques. 1959. *How to Know the Weeds.* 2nd ed. Wm. C. Brown Company, Dubuque.

ONLINE SOURCES

Alberta Agriculture and Rural Development:
  http://www1.agric.gov.ab.ca.
Canadian Food Inspection Agency: www
  .inspection.gc.ca.
Center for Invasive Species and Ecosystem
  Health: www.bugwood.org.
Early Detection and Distribution Mapping
  System: www.eddmaps.org.
Integrated Taxonomic Information System:
  www.itis.gov.
Invasive Plant Atlas of the United States:
  www.invasiveplantatlas.org.
Invasive Species in Canada: www
  .invasivespecies.gc.ca.
Nature Conservancy: www.nature.org.
Nova Scotia Department of Agriculture:
  http://www.gov.ns.ca/agri/.
Online Floras: www.efloras.org.
U.S. Department of Agriculture: www.plants
  .usda.gov.

Index to Common and Scientific Names

Note: **Bold numbers** *denote pages for primary species descriptions.*

bindweed, field, 281, 282, 609
bindweed, great, 280
bindweed, hedge, 280
bindweed, ivy, 608
bindweed, knot, 608
bindweed, lesser, 282
birdfoot deervetch, 364
birdgrass, 614
birdrape, 214
birdseed, 124
bird's eggs, 246
bird's-eye, 642
bird's nest, 28
bird's-tongue, 614
bitter buttons, 134
bittersweet, 712
bittersweet, European, 712
bittervine, 150
bitterweed, 56, 58, 98
blackberry, 671
blackberry, Himalayan, 671
black jacks, 536
black locust, 386, 701
black medic, 366, 509
blackseed, 366
black torch, 693
blackweed, 56
bladder ketmia, 468
blanket leaf, 692
blessed thistle, 77
blister plant, 650
blood stanch, 98
blood vine, 494
bloomfell, 364
bloom weed, 472
blowball, 136
bluebell, creeping, 220
bluebur, 172
blue buttons, 324
blue caps, 324
blue daisy, 92
blue devil, 170
bluegrass, 576
bluegrass, annual, 586
bluegrass, Canada, 587
bluegrass, Colorado, 568
bluejoint, 568
blue sailors, 92

bluet, 79
blueweed, 170
blueweed, Texas, 149
boatlily, 277
Bonaparte's crown, 348
borage family, 164–65
Boraginaceae, 164–65
bottlebrush, 336
bottle grass, 590
bouncing bet, 238, 253
boxthorn, African, 720
Brassica arvensis, 208
Brassica campestris, 214
Brassicaceae, 176–77
Brassica erucastrum, 192
Brassica integrifolia, 184
Brassica japonica, 184
Brassica juncea, **184–85**
Brassica kaber, 208
Brassica napus, 209
Brassica rapa, 209
Brassica rapa var. *rapa*, 214
Brassica sinapis, 208
Brassica sinapistrum, 208
Brassica willdenowii, 184
Brazilian pepper, 14
bristlegrass, giant, 588
bristlegrass, green, 590
bristlegrass, Japanese, 588
bristlegrass, tall green, 588
bristle grass, yellow, 580
brome, Austrian, 550
brome, awnless, 550
brome, cheatgrass, 552
brome, downy, 552
brome, Hungarian, 550
brome, nodding, 551
brome, Russian, 550
brome, slender, 552
brome, smooth, 550
Bromus anomalus, 551
Bromus inermis, **550–51**
Bromus japonicus, 553
Bromus pumpellianus, 550
Bromus tectorum, **552–53**
broom, 360
broom, English, 360
broom, French, 361, 383

clothbur, 62, 146
clover, alsike, 376
clover, black, 366
clover, cat's, 364
clover, Dutch, 376
clover, honey, 368
clover, hop, 366
clover, landino, 376
clover, pin, 390
clover, red, 377
clover, sour, 508
clover, tree, 368
clover, white, 376, 509
clover, wild white, 376
cluckenweed, 250
Clusiaceae, 414–15
coakrum, 524
coatbuttons, 140
cobbler's pegs, 68
Coccinia grandis, 302
cockle, China, 252
cockle, clammy, 244
cockle, corn, 232
cockle, cow, 239, 252
cockle, night-flowering, 244
cockle, purple, 232
cockle, spring, 252
cockle, sticky, 244
cockle, white, 242, 245
cocklebur, 146
cocklebur, broadleaved, 146
cocklebur, spiny, 147
cockle button, 62
cocksfoot, 558
cocksfoot panicum, 564
cockspur, 558
cockspur grass, 564
coco, 306
coco grass, 310
cocoyam, 50
codlins and cream, 494
coffee bush, 472
coffeeweed, 92, 374
cogon grass, 599
Colocasia esculenta, 50
Colorado bur, 716
coltsfoot, 144, 145
colt's tail, 98

Colubrina asiatica, 661
Columbus grass, 602
comfrey, prickly, 175
Commelina benghalensis, 276
Commelina caroliniana, 275
Commelinaceae, 270–71
Commelina communis, **272–73**
Commelina diffusa, **274–75**
Commelina gigas, 274
Commelina longicaulis, 274
compass plant, 114
confetti bush, 736
Conium maculatum, **26–27**
Conringia orientalis, 215
Convolvulaceae, 278–79
Convolvulus ambigens, 282
Convolvulus arvensis, 281, **282–83**, 609
Convolvulus incanus, 282
Convolvulus repens, 288
Convolvulus sepium, 280
Conyza canadensis, **98–99**
Conyza floribunda, 99
copal tree, 698
copperleaf, 342
copperleaf, rhombic, 342
copperleaf, Virginia, 342, 343
coralvine, 284
Coriandrum maculatum, 26
cornbind, 608
corn bind, 282
cornbind, dullseed, 608
cornbine, 282
cornflower, 77, 79
cornflower, low, 78
cornflower, mountain, 80
corn lily, 282
corn rose, 232
corn spurrey, 248, 267
Coronilla varia, 372
Coronopus squamatus, 215
cotton, wild, 38
cottonweed, 38, 466
couch grass, 556, 568
couchgrass, African, 598
coughwort, 144
cowbell, 246
cowgrass, 614
cow herb, 252

cow parsnip, 31
cow parsnip, giant, 30
cowquake, 248
crabgrass, large, 562
crabgrass, purple, 562
crabgrass, silver, 566
crabgrass, small, 563
crab's eye, 358
craneshill, 390
Crataegus monogyna, 670
crazyweed, 370
creeping charlie, 438
creeping Jenny, 282, 296, 644
Crepis capillaris, 101
Crepis tectorum, **100–101**
cress, Austrian yellow, 217
cress, bastard, 212
cress, bitter, 180
cress, creeping yellow, 217
cress, rocket, 180
cress, St. Barbara's, 180
cress, water, 208
cress, winter, 180
Critesion jubatum, 570
crofton weed, 148
crossflower, 214
Crotalaria pallida, 382
crowfoot, 652
crowfoot, cursed, 655
crowfoot, tall, 650
crowfoot family, 646–47
crowfoot grass, 562, 566
crowfoot grass, Durban, 560
crown-of-the-field, 232
crownweed, 58
crows foot grass, 566
crowsfoot grass, 561
crowtoes, 364
Cruciferae, 176–77
Crucifera thlaspi, 212
crunchweed, 208
crupina, common, 148
Crupina vulgaris, 148
Cuba grass, 592
cuckoo button, 62
cucumber, four-seeded bur, 296
cucumber, mock, 296
cucumber, squirting, 298

cucumber, star, 300
cucumber, wild, 296
Cucurbitaceae, 294–95
Cupaniopsis anacardioides, 688
Cuscuta approximata, 286
Cuscuta epilinum, 286, 287
Cuscuta epithymum, 286, 287
Cuscuta europaea, 286
Cuscuta gronovii, 286, 287
Cuscuta indecora, 286, 287
Cuscuta japonica, 286
Cuscuta pentagona, 286
Cuscuta reflexa, 286
Cuscuta ssp., **284–85**
Cuscuta umbellata, 286
Cylindropyrum cylindricum, 546
Cynanchum rossicum, 40
Cynodon dactylon, **556–57**
Cynodon plectostachyus, 557
Cynodon transvaalensis, 557
Cynoglossum hybridum, 168
Cynoglossum officinale, **168–69**
Cynosurus aegyptius, 560
Cynosurus indicus, 566
Cyperaceae, 304–5
Cyperus difformis, 312
Cyperus esculentus, **306–7**
Cyperus involucratus, 313
Cyperus iria, **308–9**
Cyperus purpuro-variegatus, 310
Cyperus repens, 306
Cyperus rotundus, **310–11**
Cyperus stoloniferum-pallidus, 310
Cyperus tetrastachyos, 310
Cyperus tuberosus, 306, 310
cypress, broom, 264
cypress, mock, 264
cypress, summer, 264
cyrillo, 250
Cytisus scoparius, **360–61**

Dactylis glomerata, **558–59**
Dactylis hispanica, 558
Dactyloctenium aegyptium, **560–61**
daisy, dog, 116
daisy, false, 102
daisy, field, 116
daisy, Irish, 136

daisy, Mexican, 140
daisy, midsummer, 116
daisy, oxeye, 116, 143
daisy, red, 108
daisy, shasta, 117
daisy, stinking, 60
daisy, swamp, 102
daisy, white, 116
daisy, wild, 140
damascisa, 484
dandelion, 136
dandelion, coast, 110
dandelion, common, 136
dandelion, false, 110
dandelion, red-seeded, 137
Danthonia sp., 575
darnel, 572
darnel, Persian, 572
darnel, poison, 573
Darwinia exaltata, 374
dasheen, 50
datura, 705
Datura inoxia, 705
Datura stramonium, **704–5**
Datura tatula, 704
Daucus carota var. *carota*, **28–29**
Daucus carota var. *sativa*, 29
dayflower, Asiatic, 272
dayflower, Benghal, 276
dayflower, Carolina, 275
dayflower, climbing, 274
dayflower, common, 272
dayflower, spreading, 274
daylily, orange, 457
daylily, tawny, 457
daza, 603
death camas, 454
death camas, meadow, 454
Decodon verticillatus, 461
delphinium, plains, 648
Delphinium bicolor, **648–49**
Delphinium glaucum, 649
dent de lion, 136
Deptford pink, 254
Descurainia incana ssp. *incana*, 191
Descurainia sophia, **190–91**
devil grass, 556
devil's grass, 90, 568

devil's grip, 482
devil's guts, 248, 280, 282
devil's horsewhip, 10
devil's-needles, 68
devil's paintbrush, 108
devil's plague, 28
devil's rattlebox, 246
devil's thorn, 620
devil's-tobacco, 692
devil's trumpet, 704
devil's vine, 280
dewtry, 704
Dianthus armeria, 254
Digitalis lanata, 540
Digitaria abyssinica, 598
Digitaria ischaemum, 563
Digitaria sanguinalis, **562–63**
dillweed, 60
Dioscorea alata, 318
Dioscorea batatas, 316
Dioscorea bulbifera, 319
Dioscoreaceae, 314–15
Dioscorea polystachya, **316–17**
Diplotaxis muralis, 215
Dipsacaceae, 320–21
Dipsacus fullonum ssp. *sylvestris*, **322–23**
Dipsacus laciniatus, 323, 326
Dipsacus sativus, 327
Dipsacus sylvestris, 322
ditchbur, 146
ditch moss, 404, 406
dock, cuckold, 62
dock, curled, 259, 618
dock, curly, 618
dock, Mexican, 619
dock, narrowleaf, 618
dock, sour, 618
dock, velvet, 150, 692
dock, yellow, 618
dodder, 284
dodder, bigseed alfalfa, 287
dodder, clover, 287
dodder, flax, 287
dodder, large alfalfa, 284
dodder, onion, 284
dogbane, flytrap, 36
dogbane, hemp, 42
dogbane, spreading, 36

flossflower, garden, 55
flossweed, 54
flower-of-an-hour, 468
fool hay, 576
forget-me-not, 175
forget-me-not, bur, 172
fox and cubs, 108
foxglove, Grecian, 540
foxtail, 570
foxtail, Chinese, 588
foxtail, giant, 588
foxtail, glaucous bristly, 580
foxtail, golden, 580
foxtail, green, 581, 589, 590
foxtail, nodding, 588
foxtail, yellow, 580, 591
foxtail grass, 570
foxtail rush, 336
Fragaria ssp., 665
Frangula alnus, 660
Fraxinus americana, 699
French berries, 658
frenchweed, 106, 212
frog-bit, American, 411, 412
frog-bit, European, 410
frog-bit, floating, 410
frostblite, 262
Fuller's herb, 238
furze, 378

Galarheous cyparissias, 348
Galega officinalis, **362–63**
Galeobdolon amplexicaule, 440
Galeopsis tetrahit, **436–37**
galingale, edible, 306
galinsoga, ciliate, 106
galinsoga, hairy, 107
galinsoga, small-flowered, 107
Galinsoga aristulata, 106
Galinsoga bicolorata, 106
Galinsoga caracasana, 106
Galinsoga ciliata, 106
Galinsoga parviflora, 107
Galinsoga quadriradiata, **106–7**
Galium agreste var. *echinospermum*, 674
Galium aparine, **674–75**
Galium spurium, 675
Galium spurium var. *vaillantii*, 674

Galium vaillantii, 674
garden huckleberry, 714
garden orache, 258
garget, 524
garlic, crow, 450
garlic, field, 450
garlic, stag's, 450
garlic, wild, 450
garlic root, 178
garlicwort, 178
Genista monspessulana, 361, 383
Geraniaceae, 388–89
geranium, Bicknell's, 392
geranium, Robert, 393
Geranium bicknellii, 392
geranium family, 388–89
Geranium robertianum, 393
gillale, 438
gill-over-the-ground, 438
ginger plant, 134
gingerroot, 144
giraffe head, 440
Githago segetum, 232
glasswort, 266, 268
glasswort, prickly, 266
Glaucium flavum, 520
Glechoma hederacea, **438–39**
Glechoma hirsuta, 438
Glinus lotoides, 484
glorybind, European, 282
glovewort, 168
Glycine abrus, 358
Glycyrrhiza lepidota, 363
goatgrass, 546
goatgrass, jointed, 546
goatgrass, low, 547
goatgrass, stout, 547
goathead, 748
goat's beard, 138
goat's rue, 362
goatweed, 416
gold cup, 650
golden buttons, 134
goldencreeper, 303
gold of pleasure, 187
goldthread vine, 284
goosefoot, lamb's-quarters, 262
goosefoot, white, 262

goosefoot family, 256–57
goosegrass, 566, 614, 674
goosegrass, Indian, 566
gorse, 378
gosmore, 110
gourd, bitter, 298
gourd, carilla, 298
gourd, ivy, 302
gourd family, 294–95
goutweed, bishop's, 32
Gramineae, 544–45
grand cousin, 474
grass family, 544–45
graveyard weed, 348
greasewood, 269
greasewood, black, 269
Greek pistra, 48
green ginger, 66
grimsel, 124
grip grass, 674
gromwell, 175
gros mil, 603
ground burnut, 748
ground ivy, 438
groundsel, 124
groundsel, red, 104
guinea-fowl grass, 602
gum succory, 90
gutweed, 128
gypsophila, perennial, 236
Gypsophila paniculata, **236–37**
gypsophill, tall, 236
gypsy-combs, 322
gypsy curtains, 22
gypsy flower, 168
gypsy rose, 324

halberleaf orach, 258
halogeton, 268
Halogeton glomeratus, 268
Haloragaceae, 394–95
Hancock's curse, 610
hardheads, 82, 122
hardock, 62
harebell, garden, 220
harebell family, 218–19
harmal peganum, 750
Hart's thorn, 658

haver-corn, 548
hawkbit, fall, 111, 137, 150
hawk's beard, annual, 100
hawk's beard, narrow-leaved, 100
hawk's beard, yellow, 100
hawkweed, mouse-ear, 149
hawkweed, orange, 108
hawkweed, smooth, 101
hawthorn, English, 670
hawthorn, single-seed, 670
hay-fever weed, 56
healall, 442
heart-of-the-earth, 442
heart pea, 686
heartsease, 742
heartseed, 686
heavenly bamboo, 163
heckhow, 26
hedgebell, 280
hedge bells, 282
hedge garlic, 178
hedgehog grass, 554
hedge lily, 280
hedgenettle, 444
hedgenettle, marsh, 444
hedgeparsley, spreading, 33
hedges pink, 238
hedge taper, 692
Helianthus ciliaris, 149
Helianthus tuberosus, 149
heliotrope, Indian, 174
Heliotropium indicum, 174
Hemerocallis fulva, 457
hemlock, deadly, 26
hemlock, poison, 24, 26
hemlock, water, 24
hemp, Colorado River, 374
hemp, Indian, 466
hemp, Queensland, 472
hemp, wild, 58
hemp nettle, 436
hemp nettle, brittle-stem, 436
henbane, 706
henbane, black, 706
henbit, 440
hen-plant, 536
Heracleum giganteum, 30
Heracleum mantegazzianum, **30–31**

mulberry, Russian, 488
mulberry, silkworm, 488
mulberry, white, 488
mulberry family, 486–87
mullein, black, 693
mullein, common, 692
mullein, corn, 232
mullein, moth, 694
mullein, orange, 693, 695
mullein pink, 232
murain grass, ribbed, 599
Murdannia keisak, 276
muskratweed, 24
musquash root, 24
mustard, ball, 204
mustard, blue, 214
mustard, brown, 184
mustard, Chinese, 184
mustard, corn, 208
mustard, dish, 212
mustard, dog, 192
mustard, field, 214
mustard, garlic, 178
mustard, hare's ear, 215
mustard, Indian, 184
mustard, Jim Hill, 210
mustard, leaf, 184
mustard, Marlahan, 198
mustard, mithridate, 212
mustard, poor man's, 178
mustard, purple, 214
mustard, tall hedge, 210, 211
mustard, treacle, 194, 215
mustard, tumbleweed, 210
mustard, tumbling, 210
mustard, wall, 215
mustard, wallflower, 194
mustard, water, 180
mustard, white, 185
mustard, wild, 207, 208
mustard, wormseed, 194
mustard, yellow, 208
mustard, yellow ball, 204
mustard family, 176–77
mustard root, 178
muster John Henry, 151
Myagrum paniculatum, 204
Myosotis scorpioides, 175

Myriophyllum aquaticum, **396–97**
Myriophyllum brasiliense, 396
Myriophyllum heterophyllum, 400
Myriophyllum sibiricum, 399
Myriophyllum spicatum, **398–99**
Myrophyllum proserpinacoides, 396

naked weed, 90
Nandina domestica, 163
nap-at-noon, 457
nassella polgrass, 574
Nassella trichotoma, **574–75**
Nasturtium officinale, 216
needle burr, 8
needlegrass, 576
Nepeta cataria, 446
Nepeta glechoma, 438
Nepeta hederacea, 438
neslia, 204
Neslia paniculata, **204–5**
nettle, bee, 436, 440
nettle, big sting, 730
nettle, blind, 440
nettle, bull, 710
nettle, California, 730
nettle, Carolina horse, 710
nettle, dead, 440
nettle, dog, 436, 733
nettle, flowering, 436
nettle, horse, 710, 717
nettle, slender, 730
nettle, small, 733
nettle, stinging, 730
nettle, tall, 730
nettle, wood, 732
Nicandra physalodes, **708–9**
nightshade, beaked, 716
nightshade, bitter, 712
nightshade, bittersweet, 712
nightshade, black, 714
nightshade, blue, 712
nightshade, climbing, 712
nightshade, deadly, 712, 714
nightshade, fetid, 706
nightshade, garden, 714
nightshade, hairy, 715
nightshade, horned, 716
nightshade, scrambling, 721

shepherd's clock, 642
shepherd's pouch, 188
shepherd's purse, 188, 213
shoofly, 468
shoofly plant, 708
shovel-plant, 188
Siberian oilseed, 186
Sicyos angulatus, **300–301**
Sicyos lobata, 296
sida, arrowleaf, 472
sida, heart-leaved, 478
sida, prickly, 473
sida, rhombic-leaved, 472
sida, spiny, 473, 479
Sida abutilon, 466
Sida cordifolia, 478
Sida retusa, 472
Sida rhombifolia, **472–73**
Sida spinosa, 473, 479
silene, night-flowering, 244
Silene alba, 242
Silene cucubalus, 246
Silene inflata, 246
Silene latifolia, 246
Silene latifolia ssp. *alba*, **242–43**, 245
Silene noctiflora, 243, **244–45**
Silene pratensis, 242
Silene vulgaris, **246–47**
silkweed, 38
silverberry, 331
silverweed, 665
Silybum marianum, **126–27**
Simaroubaceae, 696–97
simon's weed, 436
simson, 124
Sinapis alba, 185
Sinapis arvensis, 207, **208–9**
Sinapis juncea, 184
sisymbrium, tall, 210
Sisymbrium alliaria, 178
Sisymbrium altissimum, **210–11**
Sisymbrium loeselii, 211
Sisymbrium sophia, 190
Sium conium, 26
Sium suave, 25
sixweeks grass, 586
skeletonweed, 90
skeletonweed, rush, 91

skirt buttons, 250
skunktail grass, 570
skunk vine, 676
sleeping grass, 385
sleepy dick, 457
smartweed, dock leaf, 612
smartweed, nodding, 612
smartweed, pale, 612
smartweed, pink, 608
smooth rattlebox, 382
snake cuckoo, 242
snake flower, 170
snake grass, 336
snakeroot, 24
snakeroot, sticky, 148
snakeweed, 26, 346
snapdragon, dwarf, 530
snapdragon, perennial, 534
snapdragon, small, 530
snapdragon, wild, 534
snappery, 246
soapberry family, 684–85
soap plant, 454
soapwort, 238, 253
soapwort, cow, 252
Sodom apple, 718
Solanaceae, 702–3
Solanum carolinense, **710–11**, 717
Solanum cornutum, 716
Solanum dulcamara, **712–13**
Solanum elaeagnifolium, 720
Solanum khasianum var. *chatterjeeanum*, 718
Solanum nigrum, **714–15**
Solanum physalifolium, 715
Solanum rostratum, **716–17**
Solanum rubrum, 714
Solanum tampicense, 721
Solanum torvum, 721
Solanum viarum, **718–19**
soldierweed, 8
Sonchus arvensis ssp. *arvensis*, **128–29**
Sonchus arvensis ssp. *uliginosus*, 129
Sonchus asper, **130–31**, 133
Sonchus ciliatus, 132
Sonchus fallax, 130
Sonchus hispidus, 128
Sonchus laevis, 132
Sonchus oleraceus, 131, **132–33**

Verbenaceae, 734–35
Verbesina alba, 102
Verbesina prostrata, 102
vermouth, 64
Veronica officinalis, 543
vervain, Brazilian, 739
vervain family, 734–35
vetch, bird, 380
vetch, cow, 380
vetch, crown, 372
vetch, hairy, 381
vetch, narrow-leaved, 381
vetch, purple crown, 372
vetch, trailing crown, 372
vetch, tufted, 380
vetch, wild, 380
vetchling, tuberous, 383
Vicia cracca, **380–81**
Vicia graccha, 380
Vicia heteropus, 380
Vicia sativa ssp. *nigra*, 381
Vicia scheuchzeri, 380
Vicia semicincta, 380
Vicia tenuifolia, 380
Vicia versicolor, 380
Vicia villosa, 381
Vinca minor, 42
Vincetoxicum nigrum, 43
Vincetoxicum rossicum, **40–41**
Viola agrestis, 742
Viola arvatica, 742
Viola arvensis, **742–43**
Violaceae, 740–41
Viola deseglisei, 742
Viola lloydii, 742
Viola odorata, 744
Viola pentela, 742
Viola ruralis, 742
Viola timbali, 742
Viola tricolor, 745
Viola tricolor ssp. *arvensis*, 742
violet, dame's, 196
violet, English, 744
violet, field, 742
violet, sweet, 744
violet family, 740–41
viper's grass, 170
Virginia poke, 524

Virginia silk, 38
virgin's bower, Oriental, 654

walkgrass, 586
wallrocket, stinking, 215
walnut, black, 700
wandering Jew, 274
wandering Jew, green, 277
wandering Jew, white-flowered, 277
warmot, 64
wartremoving herb, 276
water celery, 413
watercress, 216
water feather, 396
water flag, 422
water grass, 274, 564
waterhemp, rough-fruited, 11
waterhemp, tall, 11
water hyacinth, 624
water hyacinth, rooted, 626
water hyacinth family, 622–23
water milfoil, 396
water milfoil, Brazilian, 396
water milfoil, Eurasian, 398
water milfoil, northern, 399
water milfoil, spiked, 398
water milfoil, two-leaf, 400
water milfoil family, 394–95
watermoss, eared, 682
water-orchid, 624
water pineapple, 413
water primrose, large-flowered, 498
water primrose, yellow, 498
water soldiers, 413
water spangles, 683
water thyme, 404, 406, 408
watertrough, 48
waterweed, Brazilian, 404
waterweed, broad, 406
waterweed, Canada, 406
waterweed, common, 404
waterweed, curly, 412
waterweed, Nuttall's, 407
waterweed, South American, 404
waterweed family, 402–3
waxballs, 342
waybread, 538
waygrass, 614